Set Theory, Logic and their

Set Theory, Logic and their Limitations

Moshé Machover
King's College London

Published by the Press Syndicate of the University of Cambridge
The Pitt Building, Trumpington Street, Cambridge CB2 1RP
40 West 20th Street, New York, NY 10011-4211, USA
10 Stamford Road, Oakleigh, Melbourne 3166, Australia

First published 1996

Printed in Great Britain at the University Press, Cambridge

A catalogue record for this book is available from the British Library

Library of Congress cataloguing in publication data available

ISBN 0 521 47493 0 hardback
ISBN 0 521 47998 3 paperback

KT

Contents

Preface

This is an edited version of lecture notes distributed to students in two of my courses, one on set theory, the other on quantification theory and limitative results of mathematical logic. These courses are designed primarily for philosophy undergraduates at the University of London who bravely choose the Symbolic Logic paper as one of their Finals options. They are also offered to mathematics undergraduates at King's College, London.

This then is a discourse addressed by a mathematician to an audience with a keen interest in philosophy. The style of technical presentation is mathematical. In particular, in logical notation and terminology I generally conform to the usage of mathematicians. (It seems that in this matter philosophers in any case tend follow suit – after some delay.) But philosophical and methodological issues are often highlighted instead of being glossed over, as is quite common in texts addressed primarily to students of mathematics.

A naive presentation of set theory may be in order if the main aim is instrumental: to acquaint would-be practitioners of mathematics with the basic tools of their chosen trade and to inculcate in them methods whereby nowadays the entire science is apparently reduced to set theory. In a course of that kind, the student is understandably not encouraged to scratch where it does not itch. But in the present course such an attitude would be out of place. To be sure, here as well set-theoretic concepts and results are needed as tools for formulating and proving results in mathematical logic. But it would be perverse not to alert would-be philosophers to the problematic aspects of set-theoretic reductionism.

These considerations have largely dictated the presentation of set theory: axiomatic, albeit unformalized. Critical notes about set theoretic reductionism are sounded from time to time as a leitmotiv, rounded off in a coda on Skolem's Paradox. Also, the technical exposition of set theory is accompanied by historical remarks, mainly because a historical perspective is needed in order to appreciate the emergence of reductionism and the anti-reductionist critique.

In the exposition of mathematical logic, I have drawn heavily on Chs. 1, 2, 3 and 7 of B&M (see Note below), which I had used for many years as a main text for a postgraduate logic course. However, considerable portions of the material presented in B&M had to be omitted, either because they are too hard or specialized, or simply for lack of time.

My greatest regret is that there is not enough time to include both linear and rule-based logical calculi (my own favourite is the tableau method). For certain technical reasons I had to sacrifice the latter. However, as partial compensation, the linear calculi are developed in a way that makes it clear that the logical axioms are mere stepping-stones towards rules of deduction: once these rules are established, the axioms can be shelved. Thus in practice the presentation comes quite close to being rule-based. The axiom schemes have been designed so as to make their connection with deduction rules quite direct and transparent.

(The connoisseur will note that the propositional axiom schemes have been chosen so that omitting one, two or three of them results in complete systems for intuitionistic implication and negation, classical implication, and intuitionistic implication. In particular, the only axiom scheme that is not intuitionistically valid is a purely implicational one.)

Propositional logic is studied with reference to a purely propositional language, rather than a first-order language as in B&M. This is done for didactic reasons: although propositional languages in themselves are of little interest, students are less intimidated by this approach.

For some tedious proofs that have been omitted, the reader is referred to B&M. These omissions are more than balanced by the addition of extensive methodological and explanatory comments.

A case in point is Lemma 10.10.12 (see Note below), which is the main technical result needed for the present version of the Gödel–Rosser First Incompleteness Theorem. I have omitted its proof, but added a detailed analysis of the meaning of the lemma and the reason why its proof works. When this is understood, the proof itself becomes a mere technicality, almost a foregone conclusion. The analysis is resumed after the proof of the Gödel–Rosser Theorem, to explain the meaning of the Gödel–Rosser sentence and the reason for its remarkable behaviour.

One major respect in which this course is not self-contained is its heavy borrowing from recursion theory. For further details, see Preview at the beginning of Ch. 9.

The Problems are an essential part of the text; the results contained in many of them are used later on.

Moshé Machover

Note

- Throughout 'B&M' refers to

  J. L. Bell and M. Machover, *A course in mathematical logic*, North-Holland, 1977 (second printing 1986).

- The system of cross-references used here is quite common in mathematical texts. It is illustrated by the following example. 'Def. 2.3.4' refers to the fourth numbered article (which in this case is a definition) in § 3 of Ch. 2. Within Ch. 2, this definition is referred to, more briefly, as 'Def. 3.4'.
- I would like to express my gratitude to Roger Astley, Michael Behrend and Tony Tomlinson of Cambridge University Press for their expert help in preparing the manuscript.

Warning

In the last three chapters of this book there is a systematic interplay between parallel sets of symbols; one set consisting of symbols in ordinary (feint) typeface:

'$=$', '$\neg$', '$\vee$', '$\wedge$', '$\rightarrow$', '$\exists$', '$\forall$', '$\times$', '$+$'

and the other of their bold-face counterparts:

'$\boldsymbol{=}$', '$\boldsymbol{\neg}$', '$\boldsymbol{\vee}$', '$\boldsymbol{\wedge}$', '$\boldsymbol{\rightarrow}$', '$\boldsymbol{\exists}$', '$\boldsymbol{\forall}$', '$\boldsymbol{\times}$', '$\boldsymbol{+}$'.

For explanations of the purpose of this system of notation, and warnings against confusing a feint symbol with its bold-face counterpart, see Warnings 8.1.2, 9.1.4 and 10.1.11 and Rem. 10.1.10.

Unfortunately the bold-face characters could not always be made as distinct from their feint counterparts as would be desirable. The reader is therefore urged to exercise special vigilance to discern which typeface is being used in each instance.

0

Mathematical induction

§ 1. Intuitive illustration; preliminaries

A familiar trick: dominoes standing on end are arranged in a row; then

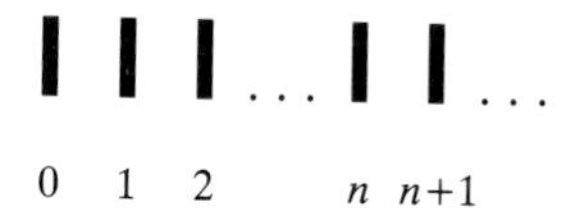

the initial domino (here labelled '0') is given a gentle push – and the whole row comes cascading down.

If you want to perform this trick, how can you make sure that all the dominoes standing in a row will fall? Clearly, the following two conditions are jointly sufficient.

1. The initial domino (domino 0) is made to fall to the right (for example, by giving it a push).
2. The dominoes are arranged in such a way that whenever any one of them (say domino n) falls to the right, it brings down the next domino after it (domino $n + 1$) and causes it also to fall to the right.

A moment's reflection shows that these two conditions are sufficient whether the row of dominoes is finite or proceeds *ad infinitum*. (In the former case, Condition 2 does not apply to the last domino.)

The reasoning that allows us to infer from Conditions 1 and 2 that all the dominoes will fall is based on the *Principle of Mathematical* (or *Complete*) *Induction*. This is a fundamental – arguably the *most* fundamental – fact about the so-called *natural numbers* (0, 1, 2, etc.). It has several equivalent forms, three of which will be presented here.

WARNING

The term 'induction' used here has nothing to do with inductive reasoning in the empirical sense.

We shall make use of the following terminology and notation.

By *number* we shall mean natural number. The class $\{0, 1, 2, \ldots\}$ of all numbers will be denoted by 'N'. We shall use lower-case italic letters as variables ranging over N.

If P is a property of numbers and n is any number, we write 'Pn' to mean that n has the property P. The *extension* of P is the class of all numbers n such that Pn. This class is denoted by '$\{n : Pn\}$'.

From an extensional point of view, P is identified with its extension: $P = \{n : Pn\}$; and hence Pn is equivalent to $n \in P$. (Here '$\in$' is short for 'is a member of'.)

We write '$\Rightarrow$' as short for 'implies that', 'iff' or '$\Leftrightarrow$' as short for 'if and only if', '$\forall$' as short for 'for all', and '$m \leqslant n$' as short for '$m < n$ or $m = n$'.

We state here as 'facts' the following elementary properties of the ordered system of numbers.

1.1. Fact

The relation $<$ *between numbers is* transitive: *whenever* $k < m$ *and* $m < n$, *then also* $k < n$.

1.2. Fact

The relation $<$ *obeys the* trichotomy: *for any numbers m and n, exactly one of the following three holds:*

$$m < n \text{ or } m = n \text{ or } n < m.$$

1.3. Fact

Every number n has an immediate successor $n + 1$, *such that, for any* m, $n < m$ *iff* $n + 1 \leqslant m$.

1.4. Fact

Zero is the least number: $0 \leqslant n$ *for all* n.

1.5. Fact

For any number $m \neq 0$, there is an n such that $m = n + 1$.

§2. Weak induction

Perhaps the most commonly used form of the Principle of Mathematical Induction is the so-called 'Weak' Principle of Induction. This asserts, for any property P of numbers, that in order to prove $\forall n Pn$ (i.e., that all numbers have the property P), it is sufficient to prove two things: first, $P0$ (i.e., that the number zero has P) and second, $\forall n[Pn \Rightarrow P(n+1)]$ (i.e., that whenever n is a number having the property P then its successor $n + 1$ also has P). In schematic form:

$$\frac{P0,\ \forall n[Pn \Rightarrow P(n+1)]}{\forall n Pn} \tag{2.1}$$

A proof of a statement $\forall n Pn$ by weak induction thus falls into two sections. One section, called the *basis* of the inductive proof, is a proof that $P0$ holds. The other section, called the *induction step*, is a proof that $\forall n[Pn \Rightarrow P(n+1)]$. When these two sections are completed (not necessarily in the above order), the proof that $\forall n Pn$ is complete.

In the induction step, in order to prove that $\forall n[Pn \Rightarrow P(n+1)]$, you have to show that if n is any number such that Pn holds, then $P(n+1)$ holds as well. In other words, you have to *deduce* $P(n+1)$ from the assumption that Pn holds. The latter assumption is called the *induction hypothesis*.

The induction step is therefore performed as follows. You consider an arbitrary number, say n, about which you make just one assumption: that Pn holds (the induction hypothesis). Using this assumption, you try to deduce that $P(n+1)$. When this is achieved, the induction step is complete.

In using the induction hypothesis Pn to deduce $P(n+1)$, you are merely considering an arbitrary hypothetical n for which Pn holds, without however committing yourself to the assumption that such a number exists; in other words, you are adopting Pn as a *provisional* hypothesis. If you succeed in deducing $P(n+1)$ from this provisional hypothesis, then you have established the *conditional* statement $Pn \Rightarrow P(n+1)$; and as you have established this for *arbitrary* number n, you are entitled to infer that $\forall n[Pn \Rightarrow P(n+1)]$.

Note that if you have completed the induction step only (without the basis – that is, you have not proved that $P0$) then you are *not* entitled to conclude that Pn holds for all numbers n; indeed you are not even entitled to conclude that there exist any numbers n for which Pn holds. For example, let P be the property of being a number that is greater than itself; so Pn means that $n > n$. Now, from the hypothesis $n > n$ it is easy to deduce $n + 1 > n + 1$ (for example, by adding 1 to both sides of the hypothesis); so we have shown that $\forall n[Pn \Rightarrow P(n + 1)]$. But it doesn't follow that there is any number greater than itself.

2.2. Remark

The Weak Principle of Induction was first invoked in 1653 by Pascal in the proof of one of the results (Corollary 12) in his *Traité du triangle arithmétique* (published in 1665). Pascal does not give an explicit formulation of the principle in general, for arbitrary P; but from his presentation of the method of proof it is clear that the general principle is being invoked. We shall not reproduce Pascal's proof here. Instead, we shall illustrate the use of weak induction in proving a simpler result.

2.3. Example

We shall prove that, for all n,

$$(*) \qquad 0 + 1 + 2 + \cdots + n = n(n + 1)/2.$$

PROOF

Define the property P by stipulating that Pn iff $(*)$ holds for n. We show by weak induction that $\forall nPn$.

Basis. For $n = 0$ the sum on the left-hand side reduces to 0, and the value of the right-hand side is 0. Thus $P0$.

Induction step. Let n be any number such that Pn; thus our induction hypothesis is that $(*)$ holds for this n. Then

$$\begin{aligned} 0 + 1 + 2 + \cdots + n + (n + 1) &= n(n + 1)/2 + (n + 1) \quad \text{by ind. hyp.,} \\ &= (n + 1)(n/2 + 1) \\ &= (n + 1)(n + 2)/2. \end{aligned}$$

(The last two steps consist of simple algebraic manipulation.) Thus from the induction hypothesis we have deduced that

$$0 + 1 + 2 + \cdots + (n + 1) = (n + 1)(n + 2)/2.$$

This equation says that $P(n + 1)$ – it is the same as (∗), but with $n + 1$ in place of n. So we have shown that $Pn \Rightarrow P(n + 1)$. ∎

§3. Strong induction

The so-called 'Strong' Principle of Induction can be stated schematically as follows:

(3.1)
$$\frac{\forall n[\forall m < nPm \Rightarrow Pn]}{\forall nPn}$$

Here, as before, P is any property of numbers. We have written '$\forall m < nPm$' as short for 'all numbers m smaller than n have the property P'.

Thus, to prove that all numbers have a given property P, it is enough to prove that $\forall n[\forall m < nPm \Rightarrow Pn]$. To do this, you have to show that if n is any number such that $\forall m < nPm$ holds, then Pn holds as well; in other words, you have to *deduce* Pn from the assumption that $\forall m < nPm$. This assumption is called the *induction hypothesis*.

Note that a proof by strong induction does not have a separate 'basis' section.

As in the case of weak induction, here too the induction hypothesis $\forall m < nPm$ is adopted provisionally, without presupposing it to be actually true.

However, unlike the case of weak induction, here there is one particular value of n for which the hypothesis $\forall m < nPm$ is in fact always automatically true. To see this, observe that there does not exist any m such that $m < 0$; this follows at once from Facts 1.2 and 1.4. Therefore any statement of the form 'for all $m < 0, \ldots$' (that is, '$\forall m < 0 \ldots$') is considered by convention to be *vacuously* true. In particular, $\forall m < 0Pm$ is always true.

3.2. Theorem

The Strong Principle of Induction follows from the Weak Principle of Induction.

PROOF

Assume that P is a property of numbers such that $\forall n[\forall m < nPm \Rightarrow Pn]$ holds. We shall show, using weak induction, that $\forall nPn$ holds as well. To this end, we define a new property Q by stipulating that, for any number n,

$$(*) \qquad Qn \Leftrightarrow_{\text{df}} \forall m < nPm.$$

(The subscript 'df' is short for 'definition'.) Note that our assumption regarding P can now be rewritten as

$$(**) \qquad \forall n[Qn \Rightarrow Pn].$$

We shall apply weak induction to Q, to prove that $\forall nQn$ holds.

First, observe that by $(*)$ $Q0$ is the same as $\forall m < 0Pm$, which – as we have noted – is vacuously true.

Next, let n be a number and suppose (as induction hypothesis) that Qn holds. From this hypothesis we shall deduce that $Q(n + 1)$ holds as well.

Using our induction hypothesis we infer from $(**)$ that Pn holds. We therefore have both Qn and Pn. But by $(*)$ Qn means $\forall m < nPm$. Therefore what we have shown is that

$$(***) \qquad Pm \text{ holds for all } m \leqslant n.$$

From Facts 1.2 and 1.3 it is easy to see that $m \leqslant n$ is equivalent to $m < n + 1$, hence $(***)$ can be rephrased as

$$Pm \text{ holds for all } m < n + 1,$$

which, by the definition $(*)$ of Q, means that $Q(n + 1)$ holds. This completes the proof of $\forall nQn$ by weak induction.

From $\forall nQn$, which we have just proved, together with $(**)$ it follows at once that Pn holds for all n. ■

§4. The Least Number Principle

Let M be any class of numbers; that is, $M \subseteq N$ (M is a subclass of N). By a *least* member of M we mean a number $a \in M$ such that $a \leqslant m$ for all $m \in M$.

Using Fact 1.2, it is easy to see that M cannot have more than one least member; so if M has a least member we can refer to the latter as *the* least member of M.

The Least Number Principle (LNP) states:

If $M \subseteq N$ and M is non-empty then M has a least member.

4.1. Theorem

The LNP follows from the Strong Principle of Induction.

PROOF

Let $M \subseteq N$ and suppose that M does *not* have a least member. We must show M is empty. To this end, let P be the property of *not* belonging to M. Thus, for any n,

$$Pn \Leftrightarrow_{\text{df}} n \notin M.$$

To show that M is empty is tantamount to showing that $\forall n Pn$ holds. We shall do so by applying strong induction to P.

So let n be any number, and assume (as induction hypothesis) that $\forall m < n Pm$ holds. By the definition of P, our induction hypothesis means that for all $m < n$ we have $m \notin M$. This is equivalent to saying that $m < n$ is *not* the case for any $m \in M$. But by Fact 1.2 this means that $n \leqslant m$ for all $m \in M$. Therefore n cannot belong to M, otherwise it would be the *least* member of M, contrary to our assumption that M has no such member. Hence Pn holds, and our induction is complete.

■

We shall now complete the cycle by proving:

4.2. Theorem

The Weak Principle of Induction follows from the LNP.

PROOF

Let P be a property of numbers such that $P0$ and $\forall n[Pn \Rightarrow P(n+1)]$ hold. We must prove that $\forall n Pn$ holds. This amounts to showing that the class

$$M =_{\text{df}} \{n : Pn \text{ does not hold}\}$$

is empty. By the LNP, it is enough to show that M has no least member.

Suppose that M *does* have a least member, m. Since $P0$ holds, 0 is not in M; hence $m \neq 0$. Therefore by Fact 1.5 there is a number n such that $m = n + 1$.

From Fact 1.3 it follows at once that $n < m$. If n were in M, then we would have $m \leqslant n$, because m is the least member of M; but $m \leqslant n$ is excluded by Fact 1.2, since we already have $n < m$. Therefore n cannot be in M, which means that Pn must hold.

From our assumption that $\forall n[Pn \Rightarrow P(n+1)]$ it now follows that $P(n+1)$ holds; in other words, Pm holds. But then m cannot be a member of M, let alone the *least* member. Thus our assumption that M has a least member leads to contradiction. ■

We have thus shown that the Weak Principle of Induction, the Strong Principle of Induction and the LNP are equivalent to one another.

4.3. Remark

While there is no evidence that the ancient Greek mathematicians knew the Principles of Weak and Strong Induction, they did use mathematical induction in the form of the LNP. We shall quote here from a proof of Proposition 31 in Euclid's *Elements*, Book VIII.

First we need a few definitions. By *arithmós* (plural: *arithmoi*) the Greeks meant what we call *natural number greater than* 1. An arithmos b is said to *measure* an arithmos a if $b < a$ and b goes into a (in modern terminology: b is a *proper divisor* of a). An arithmos a is said to be *composite* if there is an arithmos that measures it; otherwise, a is said to be *prime*.

In Proposition 31 of Book VII, Euclid claims that every composite arithmos is measured by some prime arithmos. He writes:

> 'Let a be a composite arithmos. I say that it is measured by some prime arithmos. For since a is composite, it will be measured by an arithmos, and let b be the least of the arithmoi measuring it.'

Here the LNP is clearly invoked. The proof is now easily concluded: b must be prime; otherwise, it would be measured by some smaller arithmos c, which must then also measure a – contrary to the choice of b as the *least* of the arithmoi measuring a.

Euclid also gives another proof of the same proposition, in which he uses yet another form of the Principle of Induction: *There does not exist an infinite decreasing sequence of natural numbers.*[1]

[1] On these matters see David Fowler, 'Could the Greeks have used Mathematical Induction? Did they use it?', *Physis*, vol. 31 1994 pp. 252–265.

1

Sets and classes

§ 1. Introduction

1.1. Preview

Set theory occupies a fundamental position in the edifice of modern mathematics. Its concepts and results are used nowadays in virtually all standard mathematical discourse – not only in pure mathematics, but also in applied mathematics and hence in all the mathematics-based deductive sciences. In particular, set theory is used extensively in technical discussions of logic and analytical philosophy.

The purpose of Chs. 1–6 is to present a minimal core of set theory, adequate for the kind of application just mentioned. In particular, we shall provide the set-theoretical vocabulary, notation and results needed in later chapters, devoted to Symbolic Logic.

We shall not venture into the higher reaches of the theory, which are of interest to specialist set-theorists. Nor shall we attempt a systematic logical-axiomatic investigation of set theory itself.

1.2. Further reading

There are hundreds of books on set theory, many of them very good. Among those pitched at a level similar to this course, there are two classics:

> Abraham A Fraenkel, *Abstract set theory*,
> Paul R Halmos, *Naive set theory*.

Both contain more material than our course. Fraenkel's book is suitable for readers with relatively little previous mathematical knowledge. If you are mathematically more experienced, you may find it too slow or verbose. Halmos is then likely to be more suitable.

For a more advanced, logical-axiomatic study of set theory, the two original masterpieces are:

> Kurt Gödel, *The consistency of the continuum hypothesis* (1940),
> Paul J Cohen, *Set theory and the continuum hypothesis* (1966).

An alternative exposition of Gödel's results and some additional related material is in Chapter 10 of B&M. An alternative exposition of Cohen's results and much additional related material is in John L Bell, *Boolean-valued models and independence proofs in set theory*.

1.3. Intuitive explanation

Intuitively speaking, a *set* is a definite collection, a plurality of objects of any kind, which is itself apprehended as a single object.

For example, think of a lot of sheep grazing in a field. They are a collection of sheep, a plurality of individual objects. However, we may (and often do) think of them – it – as a single object: a *herd* of sheep.[1]

Note that in order to qualify as a set, the collection in question must be *definite*. By this we mean that, if *a* is any object whatsoever, then *a* either definitely belongs to the collection or definitely does not. For this reason there is no such thing as the set of all blue cars, if 'blue' and 'car' are understood in their everyday fuzzy sense: my car is sort of blu*ish*, and a friend of mine has a vehicle that is half-way between a car and a sad joke. (Most collections and concepts that are used in everyday thinking and discourse are fuzzy; some philosophers have therefore attempted to construct a theory of so-called *fuzzy sets* – which are clearly not sets at all in the present sense of the term. This difficult subject lies outside the scope of our course.)

From now on, whenever we speak of a *collection* (or *plurality*) we shall tacitly take it to be definite, in the sense just explained. We shall also use the word *class* as synonymous with *collection*.

The objects belonging to a class may be of any kind whatsoever – physical or mental, real or ideal. In fact, being an object (in the sense in which we shall use this term) is tantamount to being capable of belonging to a collection.

In particular, since a set is a class *regarded as a single object*, it can itself belong to a class. So we can have a class some, or even all, of

[1] Cf. Eric Partridge, *Usage and abusage*: 'COLLECTIVE NOUNS; . . . Such collective nouns as can be used either in the singular or in the plural (*family*, *clergy*, *committee*, *Parliament*), are singular when unity (a unit) is intended; plural, when the idea of plurality is predominant.'

whose members are sets. If such a class, in turn, is regarded as a single object, we get a set having sets as (some of its) members. Thus, there are sets of sets (sets all of whose members are sets), sets of sets of sets, and so on.

The objects dealt with by set theory are therefore of two sorts: sets, and objects that are not sets. An object of the latter sort is called an *individual*; the German term *Urelement* (plural: *Urelemente*) is often used as well for such an object. Somewhat surprisingly, it has turned out that, as far as applications to pure mathematics are concerned, individuals are in principle dispensable, so that set theory can confine itself to sets only. We shall not make any ruling on this matter. Unless otherwise stated, what we shall say will apply regardless of whether, or how many, individuals are present.

1.4. Definition

We write '$a \in A$' as short for '[the object] a belongs to [the class] A'. The same proposition is also expressed by saying that a is a *member* of A, or an *element* of A, or that A *contains* a. We write '$a \notin A$' to negate the proposition that $a \in A$.

A class is specified by means of a definite property, say P, for which it is stipulated that the condition Px is necessary and sufficient for any object x's membership in the class.

1.5. Definition

If P is any definite property, such that the condition Px is meaningful for an arbitrary object x, then the *extension of* P, denoted by

$$`\{x : Px\}',$$

is the class of all objects x such that Px. Thus $a \in \{x : Px\}$ iff Pa.

Classes having exactly the same members are regarded as identical. Let us state this more formally:

1.6. Principle of Extensionality (PX)

If A and B are any classes such that, for every object x,

$$x \in A \Leftrightarrow x \in B,$$

then $A = B$.

For example, the two classes

$$\{x : x \text{ is an integer such that } x^2 = x\},$$

$$\{y : y \text{ is an integer such that } -1 < y < 2\}$$

are equal: although the two defining conditions differ in meaning, they are satisfied by the same objects – the integers 0 and 1.

1.7. Remark

Set theory (along with other parts of present-day mathematics) is dominated by a *structuralist* ideology, which entails an *extensionalist* view of properties. This means that properties having equal extensions are considered to be equal; thus a property and its extension uniquely determine each other.

§ 2. The antinomies; limitation of size

Since ancient times, mathematicians have dealt with infinite pluralities as a matter of course – an obvious example is the class of positive integers. However, until well into the 19th century there was great reluctance to regard such pluralities as single objects, as *sets* in the sense explained in 1.3. The infinitude of a class meant that more and more of its members could be constructed or conceived of, without limit. But to apprehend such a plurality as a *single object* seems to imply that all its members have 'already' been constructed or conceived of, or at least that they are somehow all 'out there'. This idea of a *completed* or *actual* – rather than *potential* – infinity was (rightly!) regarded with utmost suspicion.

However, the needs of mathematics as it developed in the 19th century drove Georg Cantor (1845–1918) to create his *Mengenlehre*, set theory, which admits infinite classes as objects. Despite early hostility, set theory was soon accepted by the majority of mathematicians as a powerful and indispensable tool; indeed, many regard it as a framework and foundation for the whole of mathematics.

The success of set theory first lured its adherents into assuming that *every* class can be regarded as a set. This assumption, known as the *Comprehension Principle*, is however untenable: it leads to certain logical contradictions or *antinomies*. The first such antinomy to be discovered is called the *Burali-Forti Paradox*, after the person who first published it, in 1897; but Cantor himself had been aware of it at least two years earlier. The antinomy results directly from the assumption that the class W of all ordinals is a set. (The theory of ordinals is an important but quite technical part of set theory. In Ch. 4, when we study the ordinals, we shall prove that W cannot be a set.) Similar antinomies were later discovered by Cantor himself and by others.

Cantor was not too disturbed by these discoveries. He noticed that the antinomies arose from applying the Comprehension Principle to classes that were not just infinite but extremely vast. (An early result of his set theory was that not all infinite classes have the same 'size'.) He concluded that some classes are not merely infinite but *absolutely* infinite, hence simply too large to be comprehended as a single object. Set theory would be on safe ground if the Comprehension Principle were restricted to classes of moderate size.[1] However, he did not specify precisely how to draw the line between moderately large infinite classes, which can be regarded as sets with impunity, and vast ones, which cannot be so regarded.

Matters came to a head in 1903, when Bertrand Russell published a new antinomy, *Russell's Paradox*, which he had discovered two years earlier. Whereas previous antinomies arose in rather technical reaches of set theory and therefore required lengthy expositions, Russell's Paradox checkmated the Comprehension Principle in two simple moves, as follows. Let

$$S =_{\text{df}} \{x : x \text{ is a set such that } x \notin x\}.$$

Assuming that S is a set, it follows that $S \in S$ iff S satisfies the defining condition of S – that is, iff $S \notin S$. This is absurd.

The fact that an antinomy follows so easily from apparently sound assumptions plunged set theory and logic (which cannot be sharply demarcated from set theory) into a crisis.

In 1908, two solutions were proposed to this crisis. Both amounted to imposing restrictions on the Comprehension Principle – but in two very different ways. The first, proposed by Russell himself and embodied in his *type theory*, refused to accept $\{x : Px\}$ as an object if the condition Px is *impredicative* (that is, refers to a totality to which the object, if it did exist, would belong).[2] Russell's type theory, elaborated

[1] See Michael Hallett, *Cantorian set theory and limitation of size*.

[2] Russell's paper, 'Mathematical logic as based on the theory of types', is reprinted in van Heijenoort, *From Frege to Gödel*.

by Whitehead and him in their three-volume *Principia Mathematica* (1910, 1912, 1913) as a total system for logic and mathematics, turned out to be quite complicated and cumbersome; and, at least in part because of this, has won very few adherents.

The other solution, proposed by Ernst Zermelo, embodied an idea similar to that entertained by Cantor: limitation of size.[1] Zermelo proceeded to develop set theory *axiomatically*: he laid down postulates, or [extralogical] axioms, from which the theorems of set theory were to be deduced by elementary logical means. Besides an Axiom of Extensionality (for sets), Zermelo's axioms include certain particular cases of the Comprehension Principle, which are regarded as safe because – as far as one can tell – they do not allow the formation of over-large sets and do not give rise to antinomies. In addition, Zermelo postulated a special axiom, the Axiom of Choice, which is not a restricted form of the Comprehension Principle, but is needed for proving certain important results in set theory itself and in other branches of mathematics.[2]

In 1921–2, Abraham Fraenkel, Thoralf Skolem and Nels Lennes (independently of one another) proposed one further postulate, the Axiom of Replacement, which is vital for the internal needs of set theory rather than for applications to other branches of mathematics. This postulate is another apparently safe special case of the Comprehension Principle.[3]

The resulting theory – known as *Zermelo–Fraenkel set theory (ZF)* – has proved to be very convenient and has been adopted almost universally by users of set theory.

While Zermelo's axiomatic approach is, as far as we can tell, sufficient for blocking the *logical* antinomies, such as the Burali-Forti and Russell Paradoxes, it does not ward against another sort of antinomy, which may be called *linguistic* or *semantic*.

Here is a modified version of a linguistic antinomy published in 1906 by Russell, who attributed it to G. G. Berry. Some English expressions define natural numbers; for example, 'zero', 'the square of eighty-seven', 'the least prime number greater than eighty-seven million'.

1 Russell too had briefly toyed with the same idea in 1905.

2 A translation of Zermelo's paper, 'Investigations in the foundations of set theory I', is printed in van Heijenoort, *From Frege to Gödel*.

3 This postulate, as well as Zermelo's Axiom of Separation and Axiom of Union Set, had in fact been foreshadowed in 1899 by Cantor, in a letter to Dedekind, a translation of which is printed in van Heijenoort, *From Frege to Gödel*.

Only finitely many numbers can be defined by English expressions that use fewer than 87 letters, since clearly there are only finitely many such expressions. Hence the class M of natural numbers not so definable must be non-empty. By the Least Number Principle (see §4 of Ch. 0), M has a unique least member: *the least natural number not definable by an English expression using fewer than eighty-seven letters*. But observe: the italicized part of the previous sentence is an English expression using just 86 letters, which (presumably) defines a number that cannot be defined by an English expression using less than 87 letters!

On the face of it, this antinomy affects arithmetic rather than set theory. However, as we shall see in §3 of Ch. 4 and §1 of Ch. 6, the arithmetic of natural numbers can be simulated within set theory, so that Berry's antinomy threatens set theory as well.

We cannot go here into a detailed discussion of the linguistic antinomies. Suffice it to say that the source of the trouble is that the notion of *definite property*, and hence also that of *class* (as the extension of such a property) has been left too loose and vague. Thus, for example, the property of being *definable by an English expression using fewer than eighty-seven letters* does not have a rigorously defined meaning.

These antinomies can be blocked by laying down precise conditions as to what may count as a definite property (or a class).[1] This may be done by specifying a formal language with precise structure and rules, and allowing as *definite properties* only such as can be expressed formally in this language. For a formalized presentation of ZF see, for example, Chapter 10 of B&M.

We shall present a fairly rigorous but unformalized version of ZF. However, if desired it would be easy in principle (though tedious in practice) to formalize our treatment.

§3. Zermelo's axioms

Here we present (with minor modifications) Zermelo's axioms except for the Axiom of Choice, which we shall discuss in Ch. 5.

First, we shall assume that our universe of discourse – the class of all

[1] The first to formulate such precise conditions was Hermann Weyl in *Das Kontinuum* (1918). A similar (and somewhat more formal) characterization was given independently by Skolem in a 1922 paper whose translation, 'Some remarks on axiomatized set theory', is printed in van Heijenoort, *From Frege to Gödel*.

objects with which set theory deals – is non-empty. We do not announce this assumption officially as a special postulate, because it is conventional to consider it as a *logical* presupposition.

The objects in the universe of discourse are of two distinct sorts: *sets* and *individuals*. Classes are admitted as extensions of properties: if P is a definite property of objects, then we admit the class $A = \{x : Px\}$. Note that, by Def. 1.5, to say that $a \in A$ is just another way of saying that Pa (the object a has the property P).

In order to block the semantic antinomies we must however insist that P be defined in purely set-theoretic terms, without using extraneous concepts.

The universe of discourse itself can be presented as a class according to this format: it is $\{x : x = x\}$.

Although we refer to a class in the singular, this is merely a manner of speaking and does not imply that the class is necessarily a single object. From the axioms it will follow, however, that certain classes *are* sets, and hence objects of set theory. Each set is identified with the class of all its members.

The universe may also contain other objects, called *individuals*. An individual is not a set and has no members. As we shall see shortly, there is also a set that has no members – the *empty* set.

A class that is not a set is called a *proper* class; a proper class is not an object, and therefore cannot be a member of any class.

As our first postulate we adopt the Principle of Extensionality 1.6. We shall refer to it briefly as 'PX'.

Zermelo postulated PX for sets only, as he did not consider classes (except the universe of discourse) and used properties instead.

Before stating our next postulate, we introduce a useful piece of notation.

3.1. Definition

If n is any natural number and $a_1, a_2, \ldots, a_n$ are any objects, not necessarily distinct, we put

$$\{a_1, a_2, \ldots, a_n\} =_{\text{df}} \{x : x \neq x \text{ or } x = a_1 \text{ or } x = a_2 \text{ or } \ldots \text{ or } x = a_n\}.$$

In particular, for $n = 0$ we get the *empty class* $\{\ \} = \{x : x \neq x\}$, which we denote by '$\varnothing$'. (No object can differ from itself!)

3.2. Axiom of Pairing (A2)

For all objects a and b the class $\{a, b\}$ *is a set.*

3.3. Remarks

(i) This set is called *the pair of a and b*. By PX we have $\{a, b\} = \{b, a\}$.

(ii) For any object a we clearly have $\{a\} = \{a, a\}$, which is a set by A2. This set is called *the singleton of a*.

(iii) From our assumption that there exists at least one object a, it now follows that there exists at least one set, namely $\{a\}$. Note however that we cannot prove the existence of an individual: our postulates are neutral on this matter.

3.4. Definition

Let A and B be classes. If every member of B is also a member of A, we say that B is a *subclass* of A (also, B is *included* in A, or A *includes* B), briefly: $B \subseteq A$.

If $B \subseteq A$ but $A \neq B$, we say that B is a *proper* subclass of A (also, B is *properly* included in A, or A *properly* includes B), briefly: $B \subset A$.

3.5. Warnings

(i) Beware of confusing 'contains' and 'includes'; the former refers to the relation of membership $\in$ while the latter refers to the relation $\subseteq$ just defined.

(ii) However, this terminological distinction is not observed by all authors, so watch out for other usages.

(iii) Also, the notation introduced in Def. 3.4 is not universally accepted. Some authors use '$\subset$' instead of '$\subseteq$' for not-necessarily-proper inclusion; and '$\subsetneq$' instead of '$\subset$' for proper inclusion.

The following postulate was one of Zermelo's central ideas.

3.6. Axiom of Subsets (AS)

If $B \subseteq A$ *and* A *is a set then so is* B.

3.7. Definition

If A is a class and P is a definite property such that the condition Px is meaningful for any object x, we put

$$\{x \in A : Px\} =_{\mathrm{df}} \{x : x \in A \text{ and } Px\}.$$

3.8. Remarks

(i) Zermelo's formulation of AS, clearly equivalent to the one used here, said (in effect) that if A is a set then the class $\{x \in A : Px\}$ is always a set. Since this class *separates* or *singles out* those members of A that have the property P, he called AS the *Axiom of Separation* (*Aussonderung*). This name is still in current use.

(ii) The intuitive idea behind AS is clear: if $B \subseteq A$ and A is not too vast, then B cannot be too vast either.

3.9. Theorem

$\varnothing$ is a set.

PROOF

Clearly $\varnothing$ is included in any class, and in particular in any set. By Rem. 3.3(iii) there exists a set. Hence $\varnothing$ is included in some set, and by AS is itself a set. ■

3.10. Theorem

The class of all objects (the universe of discourse) and the class of all sets are proper classes.

PROOF

We saw in § 2 that Russell's class,

$$\{x : x \text{ is a set such that } x \notin x\}$$

cannot be a set. Since Russell's class is included in the class of all sets, the latter cannot be a set by AS. The same applies to the universe of discourse. ■

3.11. Definition

If A is any class, we put

$$\bigcup A =_{\text{df}} \{x : x \in y \text{ for some } y \in A\}.$$

$\bigcup A$ is called *the union class of* A.

3.12. Axiom of Union set (AU)

If A is a set then so is $\bigcup A$.

3.13. Remarks

(i) The members of $\bigcup A$ are the members of the members of A.

(ii) Intuitively, the idea behind AU is that if A is a set then it does not have 'too many' members; and each of these, being an object (an individual or a set), in turn does not have 'too many' members. Therefore $\bigcup A$ – obtained by pooling together not-too-many collections, none of which is too vast – cannot itself be too vast.

3.14. Definition

For any classes A and B, we put

$$A \cup B =_{\text{df}} \{x : x \in A \text{ or } x \in B\}.$$

$A \cup B$ is called *the union* (or *join*) *of A and B*.

3.15. Theorem

$A \cup B$ is a set iff both A and B are sets.

PROOF

If A and B are sets, then $A \cup B = \bigcup\{A, B\}$, which is a set by A2 and AU. The converse follows easily from AS. ■

3.16. Theorem

If n is any natural number and $a_1, a_2, \ldots, a_n$ are any objects, the class $\{a_1, a_2, \ldots, a_n\}$ is a set.

PROOF

By (weak) induction on n.

Basis. For $n = 0$ the assertion of our theorem is Thm. 3.9.

Induction step. By Def. 3.14,

$$\{a_1, a_2, \ldots, a_n, a_{n+1}\} = \{a_1, a_2, \ldots, a_n\} \cup \{a_{n+1}\},$$

which is a set by the induction hypothesis, Rem. 3.3(ii) and Thm. 3.15. ■

3.17. Definition

If A is any class, we put

$$\mathsf{P}A =_{\text{df}} \{x : x \text{ is a set such that } x \subseteq A\}.$$

$\mathsf{P}A$ is called *the power class of* A.

3.18. Axiom of Power set (AP)

If A is a set then so is $\mathsf{P}A$.

3.19. Remark

Intuitively, the idea behind AP is that although $\mathsf{P}A$ can be very large – in fact, much larger than A – its size is nevertheless bounded provided A itself is not too vast.

3.20. Problem

Prove that if A is a class of sets (that is, a class all of whose members are sets) such that $\bigcup A$ is a set, then A is a set as well.

The last axiom we shall postulate here is

3.21. Axiom of Infinity (AI)

There exists a set Z such that $\varnothing \in Z$ and such that for every set $x \in Z$ also $x \cup \{x\} \in Z$.

3.22. Remarks

(i) Without AI it is impossible to prove that there are infinite sets. On the other hand, it is easy to see intuitively that any set Z satisfying the conditions imposed by AI must be infinite. We shall be able to *prove* this rigorously when we have a rigorous definition of *infiniteness*.

(ii) A2, AS, AU and AP are clearly particular cases of the Principle of Comprehension: they say that certain classes are sets. Although AI as it stands is not of this form, we shall see later that it is *equivalent* to the proposition that a certain class, ω, is a set.

§ 4. Intersections and differences

The following definitions will be needed later on.

4.1. Definition

If A is any class,

$$\bigcap A =_{\mathrm{df}} \{x : x \in y \text{ for every } y \in A\}.$$

$\bigcap A$ is called *the intersection class of A*.

4.2. Definition

If A and B are classes,

$$A \cap B =_{\mathrm{df}} \{x : x \in A \text{ and } x \in B\}.$$

$A \cap B$ is called *the intersection* (or *meet*) *of A and B*.

4.3. Definition

If A is any class,

$$A^{\mathrm{c}} =_{\mathrm{df}} \{x : x \notin A\}.$$

A^{c} is called *the complement of A*.

4.4. Definition

If A and B are any classes,

$$A - B =_{\mathrm{df}} A \cap B^{\mathrm{c}}.$$

$A - B$ is called *the difference between A and B*.

4.5. Problem

(i) Prove that if A is a non-empty class then $\bigcap A$ is a set. What is $\bigcap \varnothing$?

(ii) Prove that if A or B is a set then so is $A \cap B$.

(iii) Prove that A and A^c cannot both be sets.

2

Relations and functions

§ 1. Ordered n-tuples, cartesian products and relations

1.1. Preview

By Def. 1.1.5, the extension of a property P of objects is the class $\{x : Px\}$. Recall (Rem. 1.1.7) that from an extensionalist point of view a property and its extension determine each other uniquely; so that—wielding Occam's razor, the structuralist mathematician's favourite instrument—one can *identify* the two and pretend that a property simply *is* its extension. As set theory developed, it transpired that a similar procedure could be applied to other fundamental mathematical notions such as *relation* (among objects) and *function*: instead of taking these as independent primitive notions, as had been done in the early days of set theory, they could be reduced to classes and the membership relation. In this and the next section we shall see how this is done.

For any two objects a and b, not necessarily distinct, we need a unique object $\langle a, b \rangle$ called *the ordered pair of a and b* [in this order]. It is not really important how the ordered pair is defined, so long as the following condition is satisfied:

(1.2) $$\langle a, b \rangle = \langle c, d \rangle \Leftrightarrow a = c \text{ and } b = d.$$

1.3. Warning

The ordered pair $\langle a, b \rangle$ must not be confused with the set $\{a, b\}$, sometimes known as an *unordered* pair, whose members are just a and b. For example, the sets $\{a, b\}$ and $\{b, a\}$ are always equal (see Rem. 1.3.3(i)), but by (1.2) the ordered pairs $\langle a, b \rangle$ and $\langle b, a \rangle$ are equal only if $a = b$. However, when there is no risk of confusion we shall often omit the adjective 'ordered' and say 'pair' when we mean *ordered* pair.

As part of the reductionist programme aiming to reduce all mathematical concepts to the notion of class and the membership relation, the following rather artificial definition, first proposed by Kazimierz Kuratowski in 1921, has been widely accepted.

1.4. Definition

For any objects a and b,

$$\langle a, b \rangle =_{\text{df}} \{\{a\}, \{a, b\}\}.$$

1.5. Problem

Prove that (1.2) follows from Def. 1.4.

More generally, for any number n and any n objects $a_1, a_2, \ldots, a_n$ —not necessarily distinct—we need a unique object $\langle a_1, a_2, \ldots, a_n \rangle$ called the *ordered n-tuple of* $a_1, a_2, \ldots, a_n$ [in this order]. Again, it is not really important how ordered n-tuples are defined, so long as the following condition—of which (1.2) is a special case—is satisfied:

$$\text{(1.6)} \qquad \langle a_1, a_2, \ldots, a_n \rangle = \langle b_1, b_2, \ldots, b_n \rangle$$
$$\Leftrightarrow a_i = b_i \text{ for } i = 1, 2, \ldots, n.$$

Again, we shall often say 'n-tuple' as short for 'ordered n-tuple'.

The following definitions deliver the goods. Proceeding inductively, we supplement Def. 1.4 by:

1.7. Definition

For any $n \geqslant 2$ and objects $a_1, a_2, \ldots, a_n, a_{n+1}$,

$$\langle a_1, a_2, \ldots, a_n, a_{n+1} \rangle =_{\text{df}} \langle \langle a_1, a_2, \ldots, a_n \rangle, a_{n+1} \rangle.$$

1.8. Problem

Prove (1.6) for all $n \geqslant 2$. (Use weak induction on n, taking $n = 2$ as basis.)

There remain the cases $n = 1$ and $n = 0$. For $n = 1$, condition (1.6) reduces to:

$$\langle a \rangle = \langle b \rangle \Leftrightarrow a = b.$$

The simplest way to satisfy this is to adopt the following.

1.9. *Definition*

$\langle a \rangle =_{\text{df}} a.$

As for $n = 0$, condition (1.6) reduces to the unconditional equality $\langle\rangle = \langle\rangle$, which will hold trivially, no matter how we define $\langle\rangle$. Since $\varnothing$ is the simplest object, the simplest convention to adopt is

1.10. *Definition*

$\langle\rangle =_{\text{df}} \varnothing.$

1.11. *Remark*

The equality which was decreed by Def. 1.7 for $n \geqslant 2$, now holds also for $n = 1$ by virtue of Def. 1.9. However, it does *not* hold for $n = 0$, because by Def. 1.9 $\langle a \rangle = a$, whereas by Def. 1.10 $\langle\langle\rangle, a\rangle = \langle\varnothing, a\rangle$.

We proceed to define the notions of *cartesian product* and *cartesian power*.

1.12. *Definition*

(i) For any classes $A_1, A_2, \ldots, A_n$, not necessarily distinct, their *cartesian product* [in this order] is the class

$A_1 \times A_2 \times \cdots \times A_n =_{\text{df}}$

$$\{\langle x_1, x_2, \ldots, x_n\rangle : x_1 \in A_1, x_2 \in A_2, \ldots, x_n \in A_n\},$$

that is, the class of all n-tuples whose i-th component belongs to A_i for $i = 1, 2, \ldots, n$.

(ii) The *n-th cartesian power* of a class A is the cartesian product of A with itself n times:

$$A^n =_{\text{df}} \underbrace{A \times A \times \cdots \times A}_{n \text{ times}},$$

that is, the class of all n-tuples of members of A. In particular, $A^1 = A$ and $A^0 = \{\langle\rangle\} = \{\varnothing\}$.

1.13. Remarks

(i) In Def. 1.12(i) we have used a convenient generalization of the class notation introduced in Def. 1.1.5. Although it is almost self-explanatory, let us spell it out.

Suppose $F(x_1, x_2, \ldots, x_n)$ is an object whenever $x_1, x_2, \ldots, x_n$ are objects; and suppose $P(x_1, x_2, \ldots, x_n)$ is a condition involving $x_1, x_2, \ldots, x_n$. Then

$$\{F(x_1, x_2, \ldots, x_n) : P(x_1, x_2, \ldots, x_n)\}$$

is defined to be the class

$$\{y : \text{there exist } x_1, x_2, \ldots, x_n \text{ such that } F(x_1, x_2, \ldots, x_n) = y \text{ and } P(x_1, x_2, \ldots, x_n)\}.$$

(ii) It is easy to see that, for any $n \geqslant 1$, $A_1 \times A_2 \times \cdots \times A_n = \varnothing$ iff $A_i = \varnothing$ for at least one i.

Intuitively, if $n \geqslant 1$ and R is an n-ary relation on a class A, then for any n-tuple of members of A it is meaningful to say that R *holds* or *does not hold* for it. The class of all those n-tuples for which R does hold is known as the *extension* of R. From an extensionalist point of view, two relations are identical iff they have the same extension. Thus, a relation and its extension uniquely determine each other. In the spirit of the reductionist programme mentioned above, a relation is simply identified with its extension. Hence the following

1.14. Definition

(i) For any $n \geqslant 1$ and any class A, an *n-ary relation on A* is a class of n-tuples of members of A—that is, a subclass of A^n.

(ii) In particular, a *property on A* is a unary relation on A—that is, a subclass of A.

1.15. Remarks

(i) If R is an n-ary relation we shall often write '$R(a_1, a_2, \ldots, a_n)$' as short for '$\langle a_1, a_2, \ldots, a_n\rangle \in R$'. In the special case where R is a binary relation we shall often write 'aRb' for '$\langle a, b\rangle \in R$'.

(ii) We could extend Def. 1.14(i) to the case $n = 0$, but the resulting notion of 0-*ary relation* is found to be of little use.

§2. Functions; the axiom of replacement

Intuitively, if f is a function (or map, or mapping) then f assigns to any object x at most one object fx as *value*. The class of all objects x to which a value fx *is* assigned by f is called the *domain [of definition]* of f and denoted by 'dom f'.

The *graph* of f is then the class $\{\langle x, fx \rangle : x \in \operatorname{dom} f\}$. Note that the graph of a function is a class of pairs. But not every class of pairs can be the graph of a function: a class G of pairs is the graph of a function iff for any object x there is at most one object y such that $\langle x, y \rangle \in G$.

From an extensionalist point of view, two functions are identical if they have the same graphs. In the spirit of reductionism, we can therefore identify a function with its graph:

2.1. Definition

A *function* (a.k.a. *map* or *mapping*) is a class f of ordered pairs satisfying the *functionality condition*: whenever both $\langle x, y \rangle \in f$ and $\langle x, z \rangle \in f$ then $y = z$.

2.2. Definition

Let f be a function.

(i) The *domain* of f is the class

$$\operatorname{dom} f =_{\text{df}} \{x : \langle x, y \rangle \in f \text{ for some } y\}.$$

(ii) If $x \in \operatorname{dom} f$, then the *value of f at x* – usually denoted by 'fx' – is the [necessarily unique] y such that $\langle x, y \rangle \in f$.

(iii) The *range* of f is the class

$$\operatorname{ran} f =_{\text{df}} \{fx : x \in \operatorname{dom} f\}.$$

2.3. Problem

Verify that from Defs. 2.1 and 2.2 it follows that a function f is equal to its own graph; that is,

$$f = \{\langle x, fx \rangle : x \in \operatorname{dom} f\}.$$

Hence prove that functions f and g are equal iff $\operatorname{dom} f = \operatorname{dom} g$ and $fx = gx$ for every x in their common domain.

2.4. *Definition*

Let f be a function.

(i) We say that f is a map *from A to B* (or that f maps A *into* B) if $\operatorname{dom} f = A$ and $\operatorname{ran} f \subseteq B$.

(ii) We say that f is a *surjection* from A to B (or that f maps A *onto* B) if $\operatorname{dom} f = A$ and $\operatorname{ran} f = B$.

(iii) We say that f is an *injection* (or a *one-to-one* map) if whenever x and y are distinct members of $\operatorname{dom} f$ then fx and fy are also distinct.

(iv) We say that f is a *bijection* from A to B if it is an injection as well as a surjection from A to B (that is, a one-to-one map from A onto B).

We shall now enquire when a relation or a function is a set.

2.5. *Lemma*

Let A and B be non-empty classes. Then $A \times B$ is a set iff both A and B are sets.

PROOF

Let a and b be any members of A and B respectively. Then by Defs. 1.4 and 1.12 we have

$$\{a, b\} \in \{\{a\}, \{a, b\}\} = \langle a, b \rangle \in A \times B.$$

Therefore by Def. 1.3.11

$$\{a, b\} \in \bigcup (A \times B).$$

Since both a and b belong to $\{a, b\}$, it follows, again by Def. 1.3.11, that both are members of $\bigcup\bigcup(A \times B)$. Thus we have shown that $A \subseteq \bigcup\bigcup(A \times B)$ and $B \subseteq \bigcup\bigcup(A \times B)$, hence $A \cup B \subseteq \bigcup\bigcup(A \times B)$.

Also, it is easy to see that $\bigcup\bigcup(A \times B) \subseteq A \cup B$. Therefore by PX we have

$$\bigcup\bigcup(A \times B) = A \cup B.$$

If $A \times B$ is a set, it follows from AU and Thm. 1.3.15 that A and B are sets as well.

Conversely, if A and B are sets, then by Thm. 1.3.15 and Prob. 1.3.20 it follows that $A \times B$ is a set as well. ■

2.6. Theorem

Let $n \geqslant 1$, and let $A_1, A_2, \ldots, A_n$ *be non-empty classes. Then* $A_1 \times A_2 \times \cdots \times A_n$ *is a set iff* A_i *is a set for each* $i = 1, 2, \ldots, n$.

PROOF

By weak induction on n.

Basis. For $n = 1$ the assertion of our theorem is trivial, since in this case $A_1 \times A_2 \times \cdots \times A_n$ is simply A_1 (see Defs. 1.12(i) and 1.9).

Induction step. It is easy to see that

$$A_1 \times A_2 \times \cdots \times A_n \times A_{n+1} = (A_1 \times A_2 \times \cdots \times A_n) \times A_{n+1}$$

(use Defs. 1.12(i) and 1.7 and Rem. 1.11). Hence, by Lemma 2.5 and the induction hypothesis, $A_1 \times A_2 \times \cdots \times A_n \times A_{n+1}$ is a set iff A_i is a set for each $i = 1, 2, \ldots, n, n + 1$. ■

2.7. Corollary

If A is a set and R is an n-ary relation on A (for some $n \geqslant 1$) then R is a set as well.

PROOF

By Def. 1.14 we have $R \subseteq A^n$. If $A = \varnothing$ then $A^n = \varnothing$ by Def. 1.12(ii) and Rem. 1.13(ii); hence $R = \varnothing$. If A is a non-empty set then A^n is a set by Thm. 2.6, hence R is a set by AS. ■

2.8. Theorem

Let f be a function. Then f is a set iff both dom f *and* ran f *are sets.*

PROOF

It is easy to verify that

$$\bigcup\bigcup f = \operatorname{dom} f \cup \operatorname{ran} f.$$

From this the required result follows, using the same argument as in the proof of Lemma 2.5. ■

At this point we introduce

2.9. Axiom of Replacement (AR)

If f is a function and dom f *is a set then* ran f *is a set as well.*

2.10. Remarks

(i) AR is clearly a particular case of the Comprehension Principle.

(ii) In view of Thm. 2.8, AR is equivalent to the proposition that if f is a function such that dom f is a set then f itself is a set. The intuitive idea behind AR is that f has exactly 'as many' members as does dom f : for each $a \in \operatorname{dom} f$, f contains the corresponding pair $\langle a, fa \rangle$. Therefore if dom f is not too vast, neither is f itself.

(iii) In mathematical applications, a function f is almost always defined as a mapping from A to B, where both A and B are known in advance to be sets. It then follows from AS and Thm. 2.8 that ran f and f itself are sets. AR is not needed for this. But as we shall see AR plays an important role within set theory itself.

§ 3. Equivalence and order relations

3.1. Preview

In this section we discuss two kinds of relation that are of particular importance, not only in set theory but in mathematics as a whole.

Throughout the section, A is an arbitrary class.

3.2. Definition

R is an *equivalence* relation on A if R is a binary relation on A such that, for any members x, y and z of A, the following three conditions are satisfied:

xRx (reflexivity),
if xRy then also yRx (symmetry),
if xRy and yRz then also xRz (transitivity).

3.3. Example

The paradigmatic example of an equivalence relation on A is the binary relation $\{\langle x, x\rangle : x \in A\}$, called the *identity* (or *diagonal*) relation on A, and denoted by 'id_A'. By the way, id_A is clearly a function; indeed, it is a bijection from A to itself.

3.4. Definition

Let R be an equivalence relation on A. For each $a \in A$ we put

$$[a]_R =_{\text{df}} \{x : xRa\}.$$

We call $[a]_R$ the *R-class* of a, or the *equivalence class of a modulo R*. Where there is no risk of confusion we omit the subscript 'R' and write simply '$[a]$'.

3.5. Theorem

Let R be an equivalence relation on A and let a and b be any members of A. Then $[a] = [b]$ iff aRb.

PROOF

($\Rightarrow$). By reflexivity, aRa, so $a \in [a]$. If $[a] = [b]$ then by PX also $a \in [b]$, so that aRb.

($\Leftarrow$). Suppose aRb. If $x \in [a]$, then xRa, hence by transitivity xRb, so that $x \in [b]$. Thus we have shown that $[a] \subseteq [b]$.

Also, from aRb it follows by symmetry that bRa, so the argument we have just used shows that $[b] \subseteq [a]$. Hence by PX $[a] = [b]$. ■

3.6. Corollary

Let R be an equivalence relation on A and let a be any member of A. Then a belongs to exactly one R-class, namely $[a]$.

PROOF

We have seen that $a \in [a]$. If also $a \in [b]$ then by Def. 3.4 aRb, so by Thm. 3.5 it follows that $[a] = [b]$. ■

3.7. Definition

(i) S is a *sharp partial order* on A if S is a binary relation on A such that, for any members x, y and z of A, the following two conditions are satisfied:

if xSy, then ySx does not hold (anti-symmetry),
if xSy and ySz then also xSz (transitivity).

(ii) B is a *blunt partial order* on A if B is a binary relation on A such that, for any members x, y and z of A, the following three conditions are satisfied:

xBx (reflexivity),
if xBy and yBx then $x = y$ (weak anti-symmetry),
if xBy and yBz then also xBz (transitivity).

3.8. Example

Let A be a class of sets (that is, all the members of A are sets rather than individuals). Let S and B be the restrictions to A of $\subset$ and $\subseteq$ respectively; that is,

$$S =_{\mathrm{df}} \{\langle x, y \rangle \in A^2 : x \subset y\} \quad \text{and} \quad B =_{\mathrm{df}} \{\langle x, y \rangle \in A^2 : x \subseteq y\}.$$

Then it is easy to see that S and B are a sharp and a blunt partial order, respectively, on A.

3.9. Problem

Let S and B be a sharp and a blunt partial order, respectively, on A. Put

$$S^{\flat} =_{\mathrm{df}} S \cup \mathrm{id}_A \quad \text{and} \quad B^{\sharp} =_{\mathrm{df}} B - \mathrm{id}_A.$$

(For the definitions of id_A and $-$ see Ex. 3.3 and Def. 1.4.4.)

(i) Prove that $S^{\flat}$ and $B^{\sharp}$ are a blunt and a sharp order on A, respectively.
(ii) Verify that $S^{\flat\sharp} = S$ and $B^{\sharp\flat} = B$.

3.10. Remarks

(i) The qualifications 'sharp' and 'blunt' are often omitted and a partial order of either kind is referred to simply as a 'partial order'. There is no real harm in this, for two reasons. First, because it is usually clear from the context which kind of partial order is meant. Second, as shown in Prob. 3.9, there is a natural mutual association between a sharp partial order and a blunt partial order, whereby the latter is obtained from the former by applying $\flat$ and the former from the latter by applying $\sharp$.

(ii) Sharp partial orders are often denoted by symbols such as '$<$' or '$\prec$'; the corresponding blunt partial orders are then denoted by symbols such as '$\leqslant$' or '$\preccurlyeq$' respectively.

3.11. *Definition*

(i) S is a *sharp total order* on A if S is a binary relation on A such that, for any members x, y and z of A, the following two conditions are satisfied:

exactly one of the following three disjuncts holds
xSy or $x = y$ or ySx (trichotomy),
whenever xSy and ySz then also xSz (transitivity).

(ii) B is a *blunt total order* on A if B is a binary relation on A such that, for any members x, y and z of A, the following three conditions are satisfied:

xBy or yBx (connectedness),
if xBy and yBx then $x = y$ (weak anti-symmetry),
if xBy and yBz then also xBz (transitivity).

3.12. *Problem*

Let S and B be a sharp and a blunt total order, respectively, on A. Prove that

(i) S is a sharp partial order, (ii) $S^\flat$ is a blunt total order,
(iii) B is a blunt partial order, (iv) $B^\sharp$ is a sharp total order,
on A.

§4. Operations on functions

The following definitions will be needed later on.

4.1. *Definition*

If f and g are functions such that $\operatorname{ran} f \subseteq \operatorname{dom} g$, we put

$$g \circ f =_{\text{df}} \{\langle x, gy\rangle : \langle x, y\rangle \in f\}.$$

$g \circ f$—often denoted briefly 'gf'— is called the *composition of f and g*. (Note reading from right to left!)

4.2. Problem

Show: $g \circ f$ is a function, $\operatorname{dom}(g \circ f) = \operatorname{dom} f$ and $\operatorname{ran}(g \circ f) \subseteq \operatorname{ran} g$. Moreover, for any x in $\operatorname{dom}(g \circ f)$—which is also $\operatorname{dom} f$—check that

$$(g \circ f)x = g(fx).$$

4.3. Definition

If f is an injective (that is, one-to-one) function we put

$$f^{-1} =_{\text{df}} \{\langle y, x\rangle : \langle x, y\rangle \in f\}.$$

f^{-1} is called the *inverse* of f.

4.4. Problem

Verify that f^{-1} itself is an injective function and, moreover,

$$\operatorname{dom}(f^{-1}) = \operatorname{ran} f, \qquad \operatorname{ran}(f^{-1}) = \operatorname{dom} f,$$

$$f^{-1} \circ f = \operatorname{id}_{\operatorname{dom} f}, \qquad f \circ f^{-1} = \operatorname{id}_{\operatorname{ran} f}.$$

(For the definition of id see Ex. 3.3.)

4.5. Problem

Prove that if f is a function from a proper class to a set, then f is not injective.

4.6. Definition

If f is a function and $C \subseteq \operatorname{dom} f$, we put
 (i) $f \restriction C =_{\text{df}} \{\langle x, fx\rangle : x \in C\}$,
 (ii) $f[C] =_{\text{df}} \{fx : x \in C\}$.

$f \restriction C$ is called the *restriction of f to C* and $f[C]$ is called the *image of C under f*.

4.7. Problem

Verify that $f \restriction C$ is a function, $\operatorname{dom}(f \restriction C) = C$ and $\operatorname{ran}(f \restriction C) = f[C]$. Moreover, $(f \restriction C)x = fx$ for every $x \in C$.

4.8. Problem

Let F be a class whose members are functions. Show that $\bigcup F$ is a function iff the following *coherence condition* is fulfilled: $fx = gx$ for all f and g in F and all $x \in \operatorname{dom} f \cap \operatorname{dom} g$. Assuming this condition holds, what are $\operatorname{dom} F$ and $\operatorname{ran} F$?

3

Cardinals

§ 1. Equipollence and cardinality

We start by defining a binary relation $\approx$ on the class of all sets:

1.1. Definition

Let A and B be sets. We say that *A and B are equipollent*, briefly: $A \approx B$, if there exists a bijection from A to B (that is, a one-to-one map from A onto B).

1.2. Theorem

Equipollence is an equivalence relation on the class of sets.

PROOF

For any set A, id_A is a bijection from A to itself; so $\approx$ is reflexive.

If f is a bijection from A to B then clearly f^{-1} is a bijection from B to A; so $\approx$ is symmetric.

Finally, if f is a bijection from A to B and g is a bijection from B to C, then $g \circ f$ is a bijection from A to C; so $\approx$ is transitive. ■

It is convenient to introduce the following

1.3. Definition (incomplete)

To each set A we assign an object $|A|$, called the *cardinality* of A, such that for any two sets A and B, $|A| = |B|$ iff $A \approx B$.

An object of the form $|A|$ for some set A is called a *cardinal*.

1.4. Remarks

(i) Def. 1.3 is incomplete, because we have not specified what the object $|A|$ is or how it is to be chosen.

Cantor regarded cardinals as special abstract entities of a new kind. In effect, this amounted to introducing the notion of cardinal as a separate primitive notion.

However, it would obviously be more convenient – and conform to the reductionist programme – if cardinals were among the hitherto posited objects of set theory. In this spirit, Frege proposed in 1884 the elegant idea of defining $|A|$ as $[A]_{\approx}$, the equivalence class of A modulo $\approx$ (see Def. 2.3.4). The condition required by Def. 1.3 – $|A| = |B| \Leftrightarrow A \approx B$ – would then follow at once by Thm. 2.3.5.

This procedure, novel at the time, was to become standard practice, used with respect to various equivalence relations that arise in numerous mathematical situations.

Ironically, Frege's procedure does not work at all well in the present case, where the equivalence relation is $\approx$. Unaware that the Comprehension Principle had to be restricted, he assumed as a matter of course that $[A]_{\approx}$ is always a set, hence an object. Unfortunately, this is in general false. For example, if A is a singleton, then $[A]_{\approx}$ is the class of all singletons, and hence $\bigcup[A]_{\approx}$ is the class of all objects, the entire universe of discourse, which is a proper class by Thm. 1.3.10. Hence by AU $[A]_{\approx}$ must be a proper class as well. This is very inconvenient, because we would like to be able to form classes of cardinals, which is impossible if cardinals are proper classes.

Fortunately there are other ways of defining cardinals, satisfying the requirement of Def. 1.3, while ensuring that the cardinals are sets. Later on, in Ch. 6, we shall follow one such procedure. In each $\approx$-class we shall be able to select a unique 'distinguished' member. Then, for any set A, we can take $|A|$ to be the distinguished member of $[A]_{\approx}$ rather than that class itself. Then Thm. 2.3.5 ensures that the requirement of Def. 1.3 is satisfied.

(ii) For the time being, let us take it on trust that Def. 1.3 can be completed in a satisfactory way. This is not asking too much, since our reference to cardinals may be regarded as a mere convenience: everything that we shall say in this chapter in terms of cardinals can easily be rephrased (at the cost of some circumlocution) in terms of sets and mapping between sets.

(iii) The cardinality $|A|$ of a set A is a measure of its size. Cardinals can be regarded intuitively as generalized natural numbers. Indeed, if A is a finite set of the form $\{a_1, a_2, \ldots, a_n\}$, where the a_i are distinct, then we could take $|A|$ to be n, the number of members of A. Thus, each natural number may be regarded intuitively as the cardinality of a finite set.

(iv) However, we shall not assume formally that the natural numbers are in fact cardinals. Rather, in §3 we shall posit for each n a *corresponding* cardinal $\boldsymbol{n}$, without necessarily identifying the two.

§2. Ordering the cardinals; the Schröder–Bernstein Theorem

We define a binary relation $\leqslant$ on the class of cardinals, which, as we shall soon see, is a [blunt] partial order on that class:

2.1. Definition

Let λ and μ be cardinals. Let A and B be sets such that $|A| = \lambda$ and $|B| = \mu$. We say that λ is *smaller-than-or-equal-to* μ – briefly: $\lambda \leqslant \mu$ – if there is an injection from A to B.

2.2. Remark

This definition is in need of legitimation: we must make sure that the criterion it provides for asserting that $\lambda \leqslant \mu$ depends only on these cardinals themselves rather than on the choice of particular sets A and B such that $|A| = \lambda$ and $|B| = \mu$. This is done as follows. Let A, A', B, B' be sets such that $|A| = |A'|$ and $|B| = |B'|$. Given an injection from A to B, it is easy to show – DIY! – that there is also an injection from A' to B'.

2.3. Theorem

Let λ and μ be cardinals and let B be a set such that $|B| = \mu$. Then $\lambda \leqslant \mu$ iff B has a subset whose cardinality is λ.

PROOF

Let A be a set such that $|A| = \lambda$. By Def. 2.2.4, an injection from A to B is the same thing as a bijection from A to a subset of B. ∎

2.4. Theorem

The relation $\leqslant$ on the class of cardinals is reflexive and transitive.

PROOF

DIY. ■

To show that $\leqslant$ is a partial order, it remains to establish that it is weakly anti-symmetric (see Def. 2.3.7). This fact was conjectured by Cantor and proved independently by F. Bernstein and E. Schröder. The proof we shall present here, due to Zermelo, uses a lemma that is of some interest in its own right.

2.5. Definition

A map $\mathcal{g}$ from a class of sets to a class of sets is *monotone* if whenever X and Y are sets in dom $\mathcal{g}$ such that $X \subseteq Y$ then $\mathcal{g}X \subseteq \mathcal{g}Y$.

2.6. Lemma

Let A be a set and let $\mathcal{g}$ be a monotone map from $\mathsf{P}A$ to itself. Then A has a subset G such that $\mathcal{g}G = G$.

PROOF

For any subset X of A, the value $\mathcal{g}X$ is also a subset of A. Let us say X is a *good* set if it is a subset of A such that $\mathcal{g}X \subseteq X$. (For example, A itself is clearly good.)

Note that if X is good then $\mathcal{g}X$ is good as well. Indeed, if $\mathcal{g}X \subseteq X$ then by the monotonicity of $\mathcal{g}$ we get $\mathcal{g}(\mathcal{g}X) \subseteq \mathcal{g}X$, which means that $\mathcal{g}X$ is good.

Let G be the intersection of all good subsets of A, that is:

$$G = \bigcap\{X \in \mathsf{P}A : \mathcal{g}X \subseteq X\}.$$

(See Def. 1.4.1.) We claim that G itself is good. To show this, let X be any good set. Then $G \subseteq X$ because G is the intersection of *all* good sets. Therefore by the monotonicity of $\mathcal{g}$ we have $\mathcal{g}G \subseteq \mathcal{g}X$. Also, since X is good, we have $\mathcal{g}X \subseteq X$; hence $\mathcal{g}G \subseteq X$. Thus we see $\mathcal{g}G$ is included in every good set. Hence $\mathcal{g}G$ must also be included in the *intersection* of all good sets. But this intersection is G itself; this means that $\mathcal{g}G \subseteq G$, so G is good, as claimed.

It now follows that $\mathcal{g}G$ is good as well. But G, the intersection of *all* good sets, is included in each of them and in particular in the good set $\mathcal{g}G$. So we have shown both $\mathcal{g}G \subseteq G$ and $G \subseteq \mathcal{g}G$. Thus $\mathcal{g}G = G$. ■

2.7. Theorem (Schröder–Bernstein)

If λ and μ are cardinals such that $\lambda \leqslant \mu$ and $\mu \leqslant \lambda$ then $\lambda = \mu$.

PROOF

Let A be a set such that $|A| = \mu$. Since $\lambda \leqslant \mu$, according to Thm. 2.3 A has a subset, say B, such that $|B| = \lambda$. Since also $\mu \leqslant \lambda$, according to Def. 2.1 there is an injection, say, f, from A to B.

The claim that $\lambda = \mu$ will be proved if we show that there is a bijection from A to B.

Define a map $\mathcal{g}$ from $\mathsf{P}A$ into itself by putting, for any $X \subseteq A$,

$$\mathcal{g}X = (A - B) \cup f[X].$$

(For the definitions of $A - B$ and $f[X]$, cf. Def. 1.4.4 and Def. 2.4.6.) It is easy to see that $\mathcal{g}$ is monotone. By Lemma 2.6, there exists some $G \subseteq A$ such that $G = \mathcal{g}G$. Thus

$$G = (A - B) \cup f[G].$$

Note that $f[G] \subseteq B$ because f maps the whole of A into B. (See Fig. 1. The large rectangle represents A; like Gaul, it is divided into three parts.)

Now, $f \restriction G$ is an injection from G to B and a bijection from G to $f[G]$ (see Prob. 2.4.7). Let us put

$$h = (f \restriction G) \cup \mathrm{id}_{A-G}.$$

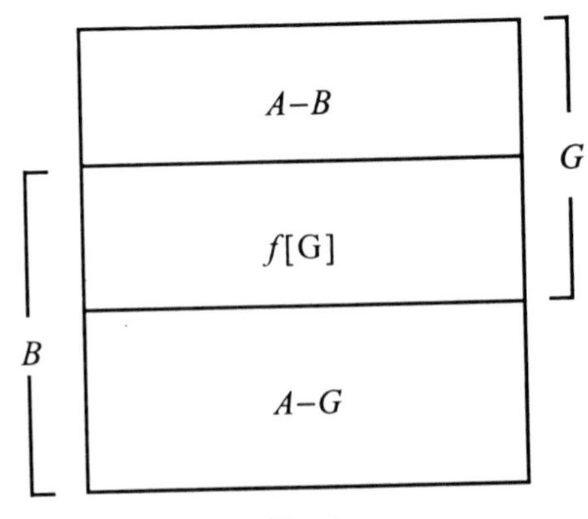

Fig. 1

Thus h is a map whose domain in the whole of A, such that

$$hx = \begin{cases} fx & \text{if } x \in G, \\ x & \text{if } x \in A - G. \end{cases}$$

It is obvious that h is a bijection from A to B. ■

2.8. *Remarks*

(i) In view of Thms. 2.4 and 2.7, $\leqslant$ is a [blunt] partial order on the class of cardinals.

(ii) As usual in such cases, we denote by '$<$' the sharp partial order associated with $\leqslant$. (Thus $<$ is $\leqslant^{\#}$; see Prob. 2.3.9.) If λ and μ are cardinals such that $\lambda < \mu$ we say that λ is *smaller than* μ.

(iii) Later on we shall prove (using the Axiom of Choice) that $\leqslant$ is a *total* order on the class of cardinals.

§3. Cardinals for natural numbers

3.1. *Definition*

If n is a natural number and $a_1, a_2, \ldots, a_n$ are distinct objects, we put

$$\boldsymbol{n} =_{\text{df}} |\{a_1, a_2, \ldots, a_n\}|.$$

In particular, $\mathbf{0} = |\varnothing|$ and $\mathbf{1} = |\{a\}|$, where a is any object. We call $\boldsymbol{n}$ *the cardinal for* (or *corresponding to*) n.

3.2. *Remarks*

(i) To legitimize Def. 3.1 we must verify that if $a_1, a_2, \ldots, a_n$ are distinct objects and $b_1, b_2, \ldots, b_n$ are likewise distinct objects then

$$\{a_1, a_2, \ldots, a_n\} \approx \{b_1, b_2, \ldots, b_n\}.$$

This is easy: $\{\langle a_1, b_1\rangle, \langle a_2, b_2\rangle, \ldots, \langle a_n, b_n\rangle\}$ is clearly a bijection from $\{a_1, a_2, \ldots, a_n\}$ to $\{b_1, b_2, \ldots, b_n\}$.

(ii) By Thm. 2.3, $\mathbf{0} \leqslant \mu$ for every cardinal μ.

3.3. *Problem*

Define c_n by induction on n as follows:

$$c_0 = \varnothing \text{ and } c_{n+1} = \{c_n\} \text{ for each } n.$$

Prove that, for each n, the objects $c_0, c_1, \ldots, c_n$ are distinct. (Use induction on n.)

Thus for any natural number n there exist n distinct objects, and hence the corresponding cardinal $\boldsymbol{n}$ exists.

3.4. Theorem

Let $a_1, a_2, \ldots, a_n$ be any objects. Then there does not exist an injection from the set $\{a_1, a_2, \ldots, a_n\}$ to any proper subset of itself.

PROOF

By induction on n. For $n = 0$ our theorem is trivial, since $\varnothing$ has no proper subset.

For the induction step, consider a set $A = \{a_1, a_2, \ldots, a_n, a_{n+1}\}$. We may assume that the objects $a_1, a_2, \ldots, a_n, a_{n+1}$ are all distinct; otherwise, by eliminating one duplication we can write A in the form '$\{b_1, b_2, \ldots, b_n\}$' and the required result follows at once by the induction hypothesis.

Suppose f is an injection from A to some $B \subseteq A$. If $B \subset A$ then at least one member of A must be outside B; and (by relabelling the a's if necessary) we may assume that $a_{n+1} \notin B$.

Since fa_{n+1} must be in B, it cannot be a_{n+1} itself; and (again, by relabelling if necessary) we may assume that $fa_{n+1} = a_1$. Therefore $a_1 \in B$. Also, since f is injective, a_{n+1} is the *only* $x \in A$ such that $fx = a_1$.

It would then follow that $f \upharpoonright \{a_1, a_2, \ldots, a_n\}$ is an injection from the set $\{a_1, a_2, \ldots, a_n\}$ to its proper subset $B - \{a_1\}$ – contrary to the induction hypothesis. Thus B cannot be a proper subset of A. ■

3.5. Theorem

For any natural numbers n and m:

(i) *if* $m \leqslant n$ *then* $\boldsymbol{m} \leqslant \boldsymbol{n}$; (ii) *if* $m \neq n$ *then* $\boldsymbol{m} \neq \boldsymbol{n}$.

(WARNING. The two '$\leqslant$' here mean different things: the first denotes the usual order among natural numbers, while the second denotes the partial order on the cardinals.)

PROOF

(i) Assume $m \leqslant n$. Take n distinct objects $a_1, a_2, \ldots, a_n$ (which exist by Prob. 3.3). Since $\{a_1, a_2, \ldots, a_m\}$ is clearly a subset of $\{a_1, a_2, \ldots, a_n\}$, we have $\boldsymbol{m} \leqslant \boldsymbol{n}$ by Thm. 2.3.

(ii) Let $m \neq n$. Without loss of generality we may assume $m < n$. Take n distinct objects $a_1, a_2, \ldots, a_n$. By Thm. 3.4 there is no bijection from $\{a_1, a_2, \ldots, a_n\}$ to its proper subset $\{a_1, a_2, \ldots, a_m\}$. Therefore $\boldsymbol{m} \neq \boldsymbol{n}$. ■

3.6. Remark

A subtle matter: we have not shown that *being a natural number* is a notion of set theory. Rather, we have taken this notion to be understood in advance, prior to the development of set theory. Therefore Def. 3.1 cannot be regarded as a single definition within this theory. Rather, it is a *definition scheme*, a sequence of definitions whereby each of the cardinals **0**, **1**, **2**, **3**, etc., in turn may be defined separately. Similar caveats apply to the whole of this section as well as to definitions like 1.3.1 and 2.1.7 and theorems like 1.3.16.

§4. Addition

In this section we shall see how cardinals may be added. But first we introduce a useful bit of terminology.

4.1. Definition

If $A \cap B = \varnothing$, we say that A and B are *disjoint*.

4.2. Lemma

For any sets A and B, there are disjoint sets A' and B', such that $|A| = |A'|$ and $|B| = |B'|$.

PROOF

Take any two distinct objects a and b (for example, $\varnothing$ and $\{\varnothing\}$; see Prob. 3.3). Then let

$$A' = \{a\} \times A = \{\langle a, x\rangle : x \in A\}, \quad B' = \{b\} \times B = \{\langle b, x\rangle : x \in B\}.$$

Using (2.1.2) it is easy to see that $A' \cap B' = \varnothing$. Also, a bijection f from A' to A is obtained by putting $f\langle a, x\rangle = x$ for every $x \in A$; so $|A| = |A'|$. Similarly, $|B| = |B'|$. ■

4.3. Lemma

Let A, B, A', B' be sets such that $A \cap B = A' \cap B' = \varnothing$, $|A| = |A'|$ and $|B| = |B'|$. Then $|A \cup B| = |A' \cup B'|$.

PROOF

Let f and g be bijections from A to A' and from B to B' respectively. Then it is clear that $f \cup g$ is a bijection from $A \cup B$ to $A' \cup B'$. ■

4.4. Definition

For any cardinals λ and μ, we define the *sum of λ and μ*:

$$\lambda + \mu =_{\text{df}} |A \cup B|,$$

where A and B are disjoint sets such that $|A| = \lambda$ and $|B| = \mu$.

4.5. Remarks

(i) Def. 4.4 is legitimized by Lemma 4.3.

(ii) In the proof of Thm. 2.7 we made use of a special case of Lemma 4.3. We had there $A = G \cup (A - G)$ and $B = f[G] \cup (A - G)$, where the unions in both cases are between disjoint sets. Also, $|G| = |f[G]|$ because f is injective. Hence we concluded that $|A| = |B|$.

4.6. Theorem

If k, m and n are natural numbers and $k + m = n$, then $\mathbf{k} + \mathbf{m} = \mathbf{n}$.

PROOF

DIY. (WARNING. The two '+' here mean different things. The first denotes the operation of addition of numbers. The second denotes addition of cardinals.) ■

4.7. Problem

Verify, for all cardinals $\varkappa$, λ and μ:

(i) $\varkappa+(\lambda+\mu)=(\varkappa+\lambda)+\mu$ (associativity of addition),
(ii) $\lambda + \mu = \mu + \lambda$ (commutativity of addition),
(iii) $\lambda + \mathbf{0} = \lambda$ (neutrality of $\mathbf{0}$ w.r.t. addition),
(iv) $\lambda \leqslant \mu \Rightarrow \varkappa + \lambda \leqslant \varkappa + \mu$ (weak monotonicity of addition).

4.8. Warning

Although cardinal addition behaves in many ways like ordinary addition of natural numbers, not *all* rules of ordinary arithmetic apply here. For example, as we shall see later, from $\varkappa + \lambda = \varkappa$ it does not always follow that $\lambda = \mathbf{0}$. Hence the *cancellation law* does not apply in general (from $\varkappa + \lambda = \varkappa + \mu$ it does not always follow that $\lambda = \mu$); nor is addition of cardinals *strongly* monotone (from $\lambda < \mu$ it does not always follow that $\varkappa + \lambda < \varkappa + \mu$).

Instead of adding just a pair of cardinals at a time, it is possible to define the sum of many – even infinitely many – cardinals simultaneously. However, the legitimation of this definition requires the Axiom of Choice (AC, see Ch. 5). We shall explain the definition here, leaving its legitimation for later. First, we need some new notation:

4.9. Definition

If B is a function whose domain is a set X, we sometimes denote the value of B at $x \in X$ by 'B_x' rather than by 'Bx' and denote B itself by

$$\text{`}\{B_x \mid x \in X\}\text{'}.$$

In this connection we refer to X as the *index set* and to B as the *family of the* B_x, *indexed by* X.

4.10. Remark

Many authors use the vertical stroke '|' instead of the colon for class abstraction (as in Def. 1.1.5) and so use some other notation for indexed families.

4.11. Definition

Let $\{B_x \mid x \in X\}$ be an indexed family of sets (that is, all the B_x are sets). Let $\mu_x = |B_x|$ for each $x \in X$. We put:

$$\sum\{\mu_x \mid x \in X\} =_{\text{df}} |\bigcup\{\{x\} \times B_x : x \in X\}|.$$

This is called the *sum of the [family of the] μ_x, indexed by X.*

4.12. Remarks

(i) Thus, to add up all the μ_x simultaneously, we form the cartesian product $\{x\} \times B_x$ for each $x \in X$. (Note that these products are pairwise disjoint: if $x \neq y$ then $\{x\} \times B_x$ and $\{y\} \times B_y$ are disjoint, although B_x and B_y need not be disjoint and may even be equal.) Then we take the union of all these products. Using AR and AU it is easy to verify that this union is a set. The cardinality of this set is the required sum.

(ii) To legitimize this definition one must show that if A is another indexed family of sets with the same index set X such that $|A_x| = |B_x|$ for all $x \in X$, then

$$\bigcup\{\{x\} \times A_x : x \in X\} \approx \bigcup\{\{x\} \times B_x : x \in X\}.$$

This can easily be done, using AC (see Rem. 5.1.3(iii) below).

(iii) We need to define the sum of a *family*, rather than a set, of cardinals because in a set of cardinals each cardinal can occur at most once: a given cardinal either does or does not belong to a given set. However, we must not forbid multiple occurrence of a cardinal in a sum. This is taken care of by our definition, since in the family $\{\mu_x \mid x \in X\}$ we can have $\mu_x = \mu_y$ for $x \neq y$.

(iv) Def. 4.4 is obtained as a special case of Def. 4.11 by taking the index set X to have just two members.

(v) The set $\bigcup\{\{x\} \times B_x : x \in X\}$ is called the *direct sum* of the indexed family $\{B_x \mid x \in X\}$.

§5. Multiplication

5.1. Definition

For any cardinals λ and μ, we define the *product of λ and μ*:

$$\lambda \cdot \mu =_{\text{df}} |A \times B|,$$

where A and B are any sets such that $|A| = \lambda$ and $|B| = \mu$. We often abbreviate '$\lambda \cdot \mu$' as '$\lambda\mu$'.

5.2. Remarks

(i) $A \times B$ is a set by Rem. 2.1.13(ii) and Lemma 2.2.5.

(ii) Def. 5.1 is legitimized by the easily proved fact that if $A' \approx A$ and $B' \approx B$, then also $A' \times B' \approx A \times B$.

For natural numbers m and n, the product mn equals the sum obtained when n is added to itself m times (this is why the product is read as 'm times n'). A similar result also holds in cardinal arithmetic, in the following sense:

5.3. Theorem

Let λ and $\varkappa$ be any cardinals and let $\{\mu_a \mid a \in A\}$ be an indexed family of cardinals such that $\mu_a = \varkappa$ for every $a \in A$ and such that $|A| = \lambda$. Then

$$\sum\{\mu_a \mid a \in A\} = \lambda\varkappa.$$

PROOF

Let D be a set such that $|D| = \varkappa$. Applying Def. 4.11 to the indexed family of sets $\{B_a \mid a \in A\}$ such that $B_a = D$ for every $a \in A$, we obtain

$$\sum\{\mu_a \mid a \in A\} = \left|\bigcup\{\{a\} \times D : a \in A\}\right|.$$

However, it is not difficult to verify (DIY!) that

$$\bigcup\{\{a\} \times D : a \in A\} = A \times D.$$

Hence $\sum\{\mu_a \mid a \in A\} = |A \times D| = \lambda\varkappa$. ■

5.4. Theorem

If k, m and n are natural numbers and $km = n$, then $\mathbf{km} = \mathbf{n}$.

PROOF

DIY. ■

5.5. Problem

Verify, for all cardinals $\varkappa$, λ and μ:

(i) $\varkappa(\lambda\mu) = (\varkappa\lambda)\mu$ (associativity of multiplication),
(ii) $\lambda\mu = \mu\lambda$ (commutativity of multiplication),
(iii) $\lambda\mathbf{1} = \lambda$ (neutrality of **1** w.r.t. multiplication),
(iv) $\lambda \leqslant \mu \Rightarrow \varkappa\lambda \leqslant \varkappa\mu$ (weak monotonicity of multiplication),
(v) $(\varkappa + \lambda)\mu = \varkappa\mu + \lambda\mu$
(distributivity of multiplication over addition),
(vi) $\lambda\mu = \mathbf{0} \Leftrightarrow \lambda = \mathbf{0}$ or $\mu = \mathbf{0}$ (absorptive property of **0**).

5.6. Problem

Prove the following generalization of Prob. 5.5(v): if $\{\lambda_x \mid x \in X\}$ is any indexed family of cardinals and μ is any cardinal then

$$\left(\sum\{\lambda_x \mid x \in X\}\right) \cdot \mu = \sum\{\lambda_x \cdot \mu \mid x \in X\}.$$

5.7. Warning

The same as 4.8, *mutatis mutandis*.

As in the case of addition, multiplication can be defined for a whole family of cardinals rather than just a pair of cardinals. (Legitimation again requires AC.) We start from a simple observation:

5.8. Lemma

Let C and D be any sets and let u and v be distinct objects. Let P be the class

$\{f : f$ is a function such that $\operatorname{dom} f = \{u, v\}$ and $fu \in C$ and $fv \in D\}$.

Then P is a set equipollent to $C \times D$.

PROOF

It is quite easy to show, without using AR, that P is a set. However, we shall not bother to do so. Instead, we shall define a bijection F from the set $C \times D$ to P. Thus by AR the latter is also a set. We put, for each $c \in C$ and $d \in D$,

$$F\langle c, d\rangle = \{\langle u, c\rangle, \langle v, d\rangle\}.$$

It is easy to verify that F is indeed a bijection from $C \times D$ to P. ■

The following definition generalizes the construction of Lemma 5.8 to an arbitrary family of sets.

5.9. Definition

If $\{B_x \mid x \in X\}$ is an indexed family of sets, the class

$$\{f : f \text{ is a function such that } \operatorname{dom} f = X \text{ and } fx \in B_x \text{ for all } x \in X\}$$

is denoted by

$$`\times\{B_x \mid x \in X\}\text{'}$$

and called the *direct product* of the family $\{B_x \mid x \in \mathrm{X}\}$.

5.10. Lemma

If $\{B_x \mid x \in X\}$ is any indexed family of sets, then $\times\{B_a \mid x \in X\}$ is a set.

PROOF

Recall (Def. 4.9) that $\{B_x \mid x \in X\}$ is the function having the index set X as its domain, whose value at each $x \in X$ is B_x. Therefore the range of this function is

$$\{B_x : x \in X\}$$

and this range is a set by AR. Now let us put

$$U = \bigcup\{B_x : x \in X\}.$$

U is a set by AU. Next, observe that by Def. 5.9, if f is any member of $\times\{B_x \mid x \in X\}$ then f is a map from X to U. Hence $f \subseteq X \times U$, which means that $f \in \mathsf{P}(X \times U)$. Thus we have shown that

$$\times\{B_x \mid x \in X\} \subseteq \mathsf{P}(X \times U).$$

Since $X \times U$ is a set (cf. Rem. 5.2(i)), it follows that $\mathsf{P}(X \times U)$ is a set by AP. Hence $\times\{B_x \mid x \in X\}$ is a set by AS. ■

5.11. Definition

Let $\{B_x \mid x \in X\}$ be a family of sets and let $\mu_x = |B_x|$ for each $x \in X$. We put

$$\prod\{\mu_x \mid x \in X\} =_{\mathrm{df}} |\times\{B_x | x \in X\}|.$$

This is called the *product of the [family of]* μ_x*, indexed by* X.

5.12. Remarks

(i) Using AC it is easy to legitimize this definition by showing that if A is another indexed family of sets with the same index set X such that $|A_x| = |B_x|$ for all $x \in X$, then

$$\times\{A_x \mid x \in X\} \approx \times\{B_x \mid x \in X\}.$$

(ii) Def. 5.1 can be regarded as a special case of Def. 5.11. Indeed, if C and D are any sets, whose cardinalities are $\varkappa$ and λ respectively, take $X = \{u, v\}$, where u and v are distinct objects, and let $\{B_x \mid x \in X\}$ be the family such that $B_u = C$ and $B_v = D$. Then Lemma 5.8, rewritten in the notation of Def. 5.9, says that

$$\times\{B_x \mid x \in X\} \approx C \times D.$$

So in this case we have

$$|\times\{B_x \mid x \in X\}| = |C \times D|,$$

which is what Def. 5.1 says $\varkappa\lambda$ should be.

§6. Exponentiation; Cantor's Theorem

6.1. Definition

Let A and B be any sets. Then

$$\operatorname{map}(A, B) =_{\mathrm{df}} \{f : f \text{ is a map from } A \text{ to } B\}.$$

6.2. Remarks

(i) If f is any member of $\operatorname{map}(A, B)$ then $f \subseteq A \times B$, hence f is a member of $\mathsf{P}(A \times B)$. Thus $\operatorname{map}(A, B) \subseteq \mathsf{P}(A \times B)$, and $\operatorname{map}(A, B)$ is a set.

(ii) Perhaps more instructively, the same result can be derived from Lemma 5.10, as follows. Consider the indexed family

$\{D_a \mid a \in A\}$ such that $D_a = B$ for every $a \in A$. Then $\times\{D_a \mid a \in A\}$ – which is a set by Lemma 5.10 – is, by Def. 5.9 equal to

$$\{f : f \text{ is a function such that } \operatorname{dom} f = A \text{ and } fa \in B \text{ for all } a \in A\}.$$

By Def. 6.1 this is exactly map(A, B).

6.3. *Definition*

For any cardinals λ and μ, we define *μ to the [power of] λ*:

$$\mu^\lambda = |\operatorname{map}(A, B)|,$$

where A and B are sets such that $|A| = \lambda$ and $|B| = \mu$.

6.4. Remarks

(i) This definition is legitimized by the easily verified fact that if $A \approx A'$ and $B \approx B'$ then $\operatorname{map}(A, B) \approx \operatorname{map}(A', B')$.

(ii) From Rem. 6.2(ii) it follows that exponentiation (raising to a power) can be achieved by repeated multiplication, in the following sense: if $\{\varkappa_a \mid a \in A\}$ is an indexed family of cardinals such that $\varkappa_a = \mu$ for all $a \in A$, and if $|A| = \lambda$, then

$$\prod\{\varkappa_a \mid a \in A\} = \mu^\lambda.$$

6.5. Problem

Let k, m be natural numbers, and let $n = m^k$. Verify that $\boldsymbol{n} = \boldsymbol{m}^k$.

6.6. Problem

Verify that for any cardinals $\varkappa$, λ and μ:

(i) $\mu^{\mathbf{0}} = \mathbf{1}$,
(ii) $\mu^{\mathbf{1}} = \mu$,
(iii) $\mu^\varkappa \mu^\lambda = \mu^{\varkappa+\lambda}$,
(iv) $(\mu^\lambda)^\varkappa = \mu^{\varkappa\lambda}$,
(v) $(\lambda\mu)^\varkappa = \lambda^\varkappa \mu^\varkappa$.

6.7. *Theorem*

For any set A, $|\mathsf{P}A| = \mathbf{2}^{|A|}$.

PROOF

By Def. 6.3, what we have to show is that $\mathsf{P}A$ is equipollent to $\operatorname{map}(A, B)$, where B is a set having exactly two members. Let us take $B = \{\varnothing, \{\varnothing\}\}$. Define a map F from $\operatorname{map}(A, B)$ to $\mathsf{P}A$, by putting, for every $f \in \operatorname{map}(A, B)$,

$$Ff = \{a \in A : fa = \varnothing\}.$$

It is easy to verify that F is a bijection from $\operatorname{map}(A, B)$ to $\mathsf{P}A$. ■

6.8. Cantor's Theorem

For any set A, $|A| < |\mathsf{P}A|$.

PROOF

First, we show that $|A| \leq |\mathsf{P}A|$. We define a map f from A into $\mathsf{P}A$ by putting $fa = \{a\}$ for each $a \in A$. Clearly, f is an injection from A to $\mathsf{P}A$.

We show that $|A| \neq |\mathsf{P}A|$ by *reductio*. Let g be any map from A to $\mathsf{P}A$. For each $x \in A$, then, gx is a member of $\mathsf{P}A$—that is, a subset of A. Put

$$D = \{x \in A : x \notin gx\}.$$

Then D is a subset of A—that is, a member of $\mathsf{P}A$. If g were to map A *onto* $\mathsf{P}A$, there would be some $d \in A$ for which $gd = D$. Then $d \in gd \Leftrightarrow d \in D$.

But from the definition of D we see that $d \in D \Leftrightarrow d \notin gd$.

Thus, d belongs to gd iff it doesn't. This contradiction shows that g cannot map A onto $\mathsf{P}A$, and hence cannot be a bijection from A to $\mathsf{P}A$. ■

6.9. Remark

The idea of Russell's Paradox derives from this proof. Indeed, if A is the class of all sets, then it is easy to see that $\mathsf{P}A \subseteq A$. Thus id_A is in fact a bijection from A to a class—A itself—that includes $\mathsf{P}A$. Taking id_A as the g in Cantor's proof, the D of that proof becomes Russell's paradoxical class of all sets that do not belong to themselves.

4
Ordinals

§ 1. Intuitive discussion and preview

The introduction of the set-theoretical cardinals was motivated by the wish to generalize the natural numbers in their capacity as *cardinal* numbers, answering the question 'how many?'. But the natural numbers are also used, in arithmetic as well as in ordinary life, in other capacities. In my local bank branch there is a number dispenser: on entering the branch, each customer collects from the dispenser a piece of paper showing a number. This number is not (at least, not directly) an answer to a 'how many?' question, but an *ordinal* number, fixing the place of the customer in the queue.

A finite set can always be arranged as a queue – and if we ignore the identity of the elements being ordered, this can done in just one way. For example, the first three customers in the bank, arranged according to the numbers assigned to them by the dispenser, always form the following pattern:

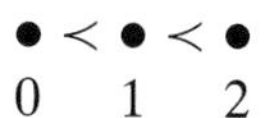

We can use the number three as an *ordinal* number, to describe this general abstract pattern, the *order type* of three objects arranged in a queue. Note that three is also the number to be assigned to the *next* customer, who is about to join the queue. This is quite general: the ordinal number assigned to each customer is the order-type (the queue pattern) of the queue of all *preceding* customers.

Cantor wished to extend this idea of finite queues and finite ordinal numbers into the transfinite. Imagine that all the old (finite) ordinal numbers have been dispensed. We have now got an infinite queue forming the pattern

(*)
$$\bullet < \bullet < \bullet < \bullet < \ldots$$
$$0 \quad 1 \quad 2 \quad 3$$

We need a new ordinal to describe the order type of this infinite queue. Cantor denoted this new ordinal by 'ω'. We can assign this ordinal to the next 'customer' and extend the queue by placing that customer behind all the finite-numbered ones:

$$\bullet < \bullet < \bullet < \bullet < \ldots < \bullet$$
$$0 \quad 1 \quad 2 \quad 3 \qquad\qquad \omega$$

The new order type just formed is described by the next ordinal, which Cantor denoted by '$\omega + 1$'. We can continue in this way, getting not only $\omega + n$ for every natural n but also $\omega + \omega$, then $\omega + \omega + 1$ and so on and on and on.

Examining the 'queues' formed in this way, Cantor saw that they are not merely totally ordered, but have a special property not shared by all totally ordered sets: every non-empty subset of the queue has a least (first) member. Cantor called such queues *well-ordered*.

An example of a total order that is not a well-ordering is provided by the integers, ordered according to magnitude:

$$\ldots < -3 < -2 < -1 < 0 < 1 < 2 < 3 < \ldots.$$

Note that the fact that the pattern (*), described by the ordinal ω, is well-ordered is just the Least Number Principle, a form of the Principle of Mathematical Induction (see § 4 of Ch. 0).

Cantor introduced the ordinals as a new and separate sort of abstract entity, just as he did with cardinals. However, in 1923 John von Neumann pointed out that among all well-ordered sets having a given Cantorian ordinal as their order-type there is a particular one with some very special properties. In the spirit of reductionism, this particular set can then be taken to *be* the ordinal of that order type.

We shall present von Neumann's theory of ordinals as streamlined by Raphael M Robinson and others.

§ 2. Definition and basic properties

2.1. Definition

Let $<$ be a [sharp] partial order on a class A and let $B \subseteq A$. If $b \in B$ and $b < x$ for every other $x \in B$, we say that b is *least in B with respect to* $<$.

2.2. Remarks

(i) Instead of demanding that $b < x$ for every *other* $x \in B$, we may equivalently demand that $b \leqslant x$ for *every* $x \in B$. Here $\leqslant$ is of course $<^{\mathrm{b}}$, the blunt partial order associated with $<$ (see Prob. 2.3.9 and Rem. 2.3.10).

(ii) When there is no risk of confusion, we omit the phrase 'with respect to $<$'.

(iii) Since $<$ is anti-symmetric, if B does have a least member it is unique and we may therefore refer to it as *the* least member of B.

2.3. Definition

A *well-ordering* on a class A is a partial order on A such that every non-empty set included in A has a least member.

2.4. Lemma

If $<$ is a well-ordering on a class A then $<$ is a [sharp] total order on A.

PROOF

According to Def. 2.3.11, we must show that $<$ fulfils the trichotomy and transitivity conditions. The latter condition is fulfilled because by Def. 2.3 $<$ is a partial order; so it only remains to verify the trichotomy.

Let x and y be any members of A. We must show that exactly one of the three disjuncts

$$x < y \text{ or } x = y \text{ or } y < x$$

holds. That no two of these disjuncts can hold simultaneously follows at once from the anti-symmetry of $<$. On the other hand, the set $\{x, y\}$ is included in A and so must have a least member; hence at least one of the three disjuncts must hold. ∎

2.5. Definition

If A is any class, we define the binary relation $\in_A$ on A, called *the restriction of* $\in$ to A, by putting

$$\in_A =_{\mathrm{df}} \{\langle x, y\rangle \in A^2 : x \in y\}.$$

2.6. Remark

The relation $\in_A$ can also be characterized by the fact that, for all x and y,

$$x \in_A y \Leftrightarrow x \in A \text{ and } y \in A \text{ and } x \in y.$$

2.7. Definition

We say that a class A is *$\in$-well-ordered* if the relation $\in_A$ is a well-ordering on A.

2.8. Problem

(i) Let A be a class such that $\in_A$ is a sharp total order on A; let $B \subseteq A$ and $b \in B$. Prove that b is least in B w.r.t. $\in_A$ iff b is either an individual or a set such that $b \cap B = \varnothing$.

(ii) Hence verify that a class A is $\in$-well-ordered iff the following two conditions are satisfied:

(1) $\in_A$ is a sharp total order on A.

(2) Every non-empty set u included in A has a member v such that v is either an individual or a set such that $v \cap u = \varnothing$.

(iii) Prove that in (ii) we may replace (1) by the weaker condition:

(1′) For any members x and y of A, at least one of the following three disjuncts holds:

$$x \in y \text{ or } x = y \text{ or } y \in x.$$

(Show that if two of these disjuncts hold simultaneously then the set $u = \{x, y\}$ violates (2). To verify that $\in_A$ is transitive, let x, y and z be members of A such that $x \in y \in z$ and apply (2) to the set $u = \{x, y, z\}$.)

(iv) Hence (or directly from Def. 2.7) prove that if $B \subseteq A$ and A is $\in$-well-ordered, then so is B.

2.9. Theorem

If A is an $\in$-well-ordered class and B is a non-empty subclass of A, then B has a least member w.r.t. $\in_A$.

PROOF

Take any $z \in B$. If z is the least member of B, we need look no further. So let us suppose z is *not* the least member of B. Therefore by Prob. 2.8(i) z is a set rather than an individual and $z \cap B \neq \varnothing$.

By Prob. 1.4.5(ii) $z \cap B$ is a set; we have just seen that it is non-empty; and it is clearly included in B and hence also in A. So by Def. 2.7 $z \cap B$ must have a least member w.r.t. $\in_A$.

Let y be the least member of $z \cap B$. We claim that this y is also the least member of B. Indeed, if this were untrue, then (applying to y the argument we have just applied to z) we would find an x such that $x \in y \cap B$. Then $x \in y$ as well as $y \in z$ and by the transitivity of $\in_A$ it would follow that $x \in z$, hence $x \in z \cap B$. But this is impossible, because $x \in y$ and y is the least member of $z \cap B$. ■

2.10. *Definition*

A class A is *transitive* if, for all y,

$$y \in A \Rightarrow y \subseteq A.$$

2.11. *Remarks*

(i) Note that every member of a transitive class must be a set rather than an individual, because by Def. 1.3.4 $y \subseteq A$ holds only if y is a class. So a class A is transitive iff:
 (1) all its members are sets and
 (2) $\bigcup A \subseteq A$; that is, for all x and y, $x \in y \in A \Rightarrow x \in A$.

(ii) Unfortunately, 'transitivity' is used with two meanings: the present one and that applicable to binary relations (as, for example, in Def. 2.3.2). In practice no confusion shall arise, as the context will indicate which meaning is intended.

2.12. *Definition*

An *ordinal* is a transitive and $\in$-well-ordered set. The class of all ordinals is denoted by 'W'.

2.13. *Examples*

The empty set $\varnothing$ is, vacuously, an ordinal. It is also easy to verify that $\{\varnothing\}$ and $\{\varnothing, \{\varnothing\}\}$ are ordinals.

2.14. Convention

We shall use lower-case Greek letters – mainly 'α', 'β', 'γ', 'λ', 'ξ' and 'η' – as variables ranging over the ordinals.

2.15. Theorem

All members of an ordinal are ordinals; thus, if α is an ordinal, $\alpha = \{\xi : \xi \in \alpha\}$.

PROOF

Let $y \in \alpha$. Since α is transitive, we have $y \subseteq \alpha$. Since α is an $\in$-well-ordered set, it follows from Prob. 2.8(iv) that its subset y is also $\in$-well-ordered. It remains to show that y is transitive.

So let $u \in x \in y$. Using the fact that α is a transitive set, we have $x \in \alpha$ and then in turn also $u \in \alpha$. Hence u and x, as well as y, are members of α; so by the transitivity of the relation $\in_\alpha$ we infer from $u \in x \in y$ that $u \in y$. ∎

2.16. Lemma

If y is any transitive subset of an ordinal α then y itself is an ordinal; moreover, $y = \alpha$ or $y \in \alpha$.

PROOF

That y is an ordinal follows at once from Prob. 2.8(iv). Moreover, let $u = \alpha - y$. If $u = \varnothing$ then $y = \alpha$. If u is non-empty, then it has a (unique) least member x w.r.t. $\in_\alpha$. We shall show that $y = x$.

First, let $z \in x$. Since $x \in \alpha$ and α is transitive, it follows that $z \in \alpha$. But z cannot be in u, because $z \in x$, and x is the *least* member of u; thus z must be in y. This proves that $x \subseteq y$.

Conversely, let $z \in y$. Then $z = x$ is impossible because $x \notin y$. Also, $x \in z$ is impossible because, by the transitivity of y, it would imply $x \in y$. Hence by Lemma 2.4 we must have $z \in x$. This proves that $y \subseteq x$. Thus $y = x \in \alpha$. ∎

2.17. Theorem

The class W of all ordinals is transitive and $\in$-well-ordered.

PROOF

The transitivity of W follows at once from Thm. 2.15. To prove that W is $\in$-well-ordered, we shall make use of Prob. 2.8(iii).

To verify that condition (1′) of Prob. 2.8(iii) holds for W, let α and β be any ordinals. Since both α and β are transitive, it is easy to see that $\alpha \cap \beta$ is also transitive. Thus by Lemma 2.16 $\alpha \cap \beta$ is an ordinal, say γ; moreover, $\gamma = \alpha$ or $\gamma \in \alpha$. Likewise, $\gamma = \beta$ or $\gamma \in \beta$.

But we cannot have both $\gamma \in \alpha$ and $\gamma \in \beta$ because then $\gamma \in \alpha \cap \beta$ – that is, $\gamma \in \gamma$; and this would violate the anti-symmetry of the well-ordering relation $\in_\gamma$ on γ. Therefore $\gamma = \alpha$ or $\gamma = \beta$. Hence $\alpha = \beta$ or $\alpha \in \beta$ or $\beta \in \alpha$, which proves condition (1′) for W.

Now let u be any non-empty set of ordinals. We must prove that there exists an ordinal $\xi \in u$ such that $\xi \cap u = \varnothing$. Take any $\alpha \in u$. If $\alpha \cap u = \varnothing$, we are through.

On the other hand, suppose $\alpha \cap u \neq \varnothing$. Since α is $\in$-well-ordered, there must exist some member ξ of $\alpha \cap u$ such that $\xi \cap \alpha \cap u = \varnothing$. But $\xi \in \alpha$ and α is transitive; so $\xi \subseteq \alpha$. Hence $\xi \cap u = \xi \cap \alpha \cap u = \varnothing$.

■

2.18. Corollary

W is a proper class (that is, not a set).

PROOF

If W were a set, then by Def. 2.12 and Thm. 2.17 it would be an ordinal, hence $W \in W$, in violation of the anti-symmetry of the well-ordering relation $\in_W$. ■

2.19. Remarks

(i) The (naive) assumption that W is a set led to a contradiction. This was the Burali-Forti Paradox (see § 2 of Ch. 1). Cor. 2.18 is a 'tame' version, within ZF, of the paradox. Similarly, Thm. 1.3.10 is a 'tame' ZF version of Russell's Paradox.

(ii) In the proofs of Thm. 2.17 and Cor. 2.18 we used the argument that an ordinal γ cannot be a member of itself because this would violate the anti-symmetry of the well-ordering relation $\in_\gamma$ on γ. In mathematical practice it is often convenient to posit a further postulate – the *Axiom of Foundation* (or *Regularity*), first proposed by Dimitry Mirimanoff in 1917 – one of whose effects is to exclude *any* set that belongs to itself. On the other hand, in some special applications of set theory – notably in so-called situation semantics, developed by Jon Barwise and others, and in abstract computation theory – it is convenient to use an extension of ZF proposed by Peter Aczel, which negates the Axiom of Foundation and admits some sets that belong to themselves. In the present course we do not commit ourselves either way.

2.20. *Corollary*

Any class of ordinals is $\in$-well-ordered.

PROOF

Immediate from Thm. 2.17 and Prob. 2.8(iv). ∎

2.21. *Definition*

The $\in$-well-ordering on W shall be denoted by '$<$'. Thus for any ordinals α and β,

$$\alpha < \beta \Leftrightarrow \alpha \in \beta.$$

2.22. *Remarks*

(i) As usual, we denote by '$\leqslant$' the blunt version of $<$. Thus

$$\alpha \leqslant \beta \Leftrightarrow \alpha \in \beta \text{ or } \alpha = \beta.$$

(ii) Thm. 2.15 can now be read as saying that if α is any ordinal then $\alpha = \{\xi : \xi < \alpha\}$.

(iii) From now on, whenever we use order-related terminology in connection with ordinals, we shall take it for granted that the order relation referred to is the $\in$-well-ordering, unless otherwise stated.

2.23. *Definition*

Let $<$ be a partial order on a class A and let $B \subseteq A$.

(i) If $u \in A$ and $x \leqslant u$ for all $x \in B$, then u is said to be an *upper bound of* (or *for*) B *with respect to* $<$.

(ii) If u is the least member of the class of upper bounds for B w.r.t. $<$ – that is, if u is an upper bound for B w.r.t. $<$ and if $u < v$

whenever v is any other upper bound for B w.r.t. $<$ – then u is said to be the *least* upper bound (abbreviated '*lub*') for B w.r.t. $<$.

2.24. Remarks

(i) The phrase 'with respect to $<$' is omitted when there is no danger of confusion.

(ii) A subclass B of A need not in general have any upper bound, let alone a lub; but if it has a lub, it is unique.

2.25. Theorem

If A is a set of ordinals then its union-set $\bigcup A$ is an ordinal. Moreover, $\bigcup A$ is the lub of A.

PROOF

To show that $\bigcup A$ is transitive, assume that $x \in y \in \bigcup A$. Then for some ordinal α we have $x \in y \in \alpha \in A$. Since α is transitive, it follows that $x \in \alpha \in A$; hence $x \in \bigcup A$.

By Thm. 2.15, all the members of $\bigcup A$ are ordinals; so by Cor. 2.20 $\bigcup A$ is $\in$-well-ordered. Thus $\bigcup A$ is an ordinal.

If $\alpha \in A$ then $\alpha \subseteq \bigcup A$, since $\bigcup A$ is a transitive set. Therefore by Lemma 2.16 $\alpha \leqslant \bigcup A$. This means that $\bigcup A$ is an upper bound for A.

Finally, if β is any upper bound for A, then for each $\alpha \in A$ we have $\alpha \leqslant \beta$ – that is, $\alpha \in \beta$ or $\alpha = \beta$. By the transitivity of the set β it follows that in either case $\alpha \subseteq \beta$. Since this holds for *each* $\alpha \in A$, it follows that also $\bigcup A \subseteq \beta$. By Lemma 2.16 we now have $\bigcup A \leqslant \beta$ – which proves that $\bigcup A$ is the *least* upper bound for A. ■

2.26. Definition

For any ordinal α we put $\alpha' =_{\text{df}} \alpha \cup \{\alpha\}$. We call α' *the immediate successor of* α. (This terminology is justified by the following theorem.)

2.27. Theorem

For any α, α' is an ordinal. Moreover, for any β, $\beta \leqslant \alpha$ iff $\beta < \alpha'$ (equivalently: $\alpha < \beta$ iff $\alpha' \leqslant \beta$). Hence $\alpha < \beta$ iff $\alpha' < \beta'$.

PROOF

Easy – DIY. ■

2.28. Definition

(i) An ordinal of the form α' is called a *successor* ordinal.
(ii) An ordinal that is neither $\varnothing$ nor a successor ordinal is called a *limit* ordinal.

§3. The finite ordinals

3.1. Definition

An ordinal α is said to be *finite* if no ordinal $\xi \leqslant \alpha$ is a limit ordinal. Otherwise, α is said to be an *infinite* ordinal. We put

$$\omega =_{\mathrm{df}} \{\alpha : \alpha \text{ is a finite ordinal}\}.$$

3.2. Theorem

ω is transitive.

PROOF

Let α be a finite ordinal. We must show that every member of α is also a finite ordinal. This is easily done – DIY, using Rem. 2.22(ii). ■

3.3. Theorem

(i) *$\varnothing$ is a finite ordinal.*
(ii) *If α is a finite ordinal then so is α'.*

PROOF.

(i) We know that $\varnothing$ is an ordinal (Ex. 2.13). But by Def. 2.28(ii) $\varnothing$ is not a limit ordinal. Since $\varnothing$ has no members, the only ξ such that $\xi \leqslant \varnothing$ is $\varnothing$ itself. Hence $\varnothing$ is a finite ordinal.
(ii) Let α be a finite ordinal and let $\xi \leqslant \alpha'$. We must show that ξ is not a limit ordinal. Now, α' itself is a successor ordinal, hence not a limit ordinal. It remains to consider the case where $\xi < \alpha'$. By Thm. 2.27 this means that $\xi \leqslant \alpha$. Since α is a finite ordinal, ξ is not a limit ordinal. ■

3.4. *Theorem*

ω is a set.

PROOF

Using the Axiom of Infinity (Ax. 1.3.21), take a set Z such that $\varnothing \in Z$ and such that whenever $x \in Z$, then also $x \cup \{x\} \in Z$. Thus if an ordinal α belongs to Z then (by Def. 2.26) so does α'.

Consider the class $\omega - Z$, the class of all finite ordinals *not* belonging to Z. If this class is non-empty, then by Thm. 2.9 it must have a least member, say β. Now, β cannot be $\varnothing$, because $\varnothing$ does belong to Z. Also, β, being a finite ordinal, cannot be a limit ordinal. So it must be a successor ordinal, say $\beta = \alpha' = \alpha \cup \{\alpha\}$. But in this case α itself is a finite ordinal (by Thm. 3.2), such that $\alpha < \beta$. Since β was supposed to be the *least* finite ordinal not belonging to Z, it follows that $\alpha \in Z$. Therefore by the assumption on Z also $\alpha' \in Z$. But this is impossible, because $\alpha' = \beta$, which is the least finite ordinal *not* belonging to Z.

So $\omega - Z$ must be empty. Thus $\omega \subseteq Z$; hence ω is a set by AS. ■

3.5. *Corollary*

ω is the unique set X having the following three properties:

(i) $\varnothing \in X$;
(ii) whenever $\alpha \in X$ then also $\alpha' \in X$;
(iii) $X \subseteq Z$ for any set Z such that $\varnothing \in Z$ and such that whenever $\alpha \in Z$ then also $\alpha' \in Z$.

PROOF

Thm. 3.3 says that ω has properties (i) and (ii). The proof of Thm. 3.4 shows that ω has also property (iii). The uniqueness of ω follows by PX, because if X is any set having the three properties then both $\omega \subseteq X$ and $X \subseteq \omega$. ■

3.6. *Remarks*

(i) Our first use of AI was to prove that ω is a set. Conversely, if we *postulate* that ω is a set, then by Thm. 3.3 ω is a set satisfying the conditions that AI lays down for Z. This shows that (in the presence of the other postulates) AI is *equivalent* to the proposition that ω is a set, which is a special case of the Comprehension Principle.

(ii) In fact, it now transpires (Cor. 3.5) that ω is simply the *smallest* set satisfying the conditions of AI.

We restate the fact that ω satisfies condition (iii) of Cor. 3.5 as a principle in its own right:

3.7. Corollary (Weak Principle of Induction on Finite Ordinals)

Let Z be any set such that $\varnothing \in Z$ and such that whenever $\alpha \in Z$ then also $\alpha' \in Z$. Then $\omega \subseteq Z$. ■

3.8. Remarks

(i) We see that the set ω of finite ordinals, with its $\in$-well-ordering, simulates, within the confines of ZF set theory, the behaviour that characterizes the system of natural numbers. We can take $\varnothing$ as the counterpart of the number 0 and the $\in$-well-ordering on ω as the counterpart of the usual ordering of the natural numbers. Just as each natural number n has an immediate successor, $n + 1$, so every finite ordinal α has an immediate successor, α'. Moreover, the basic facts about the ordering of the natural numbers (Facts 0.1.1–0.1.5) are mimicked by theorems about the finite ordinals and their $\in$-well-ordering. And, most importantly, the Principle of Mathematical Induction is mimicked by the Principle of Induction on Finite Ordinals. Certainly, *within* ZF ω impersonates, plays the role of, 'the set of natural numbers'. In fact, Cor. 3.5 reproduces within ZF Richard Dedekind's famous characterization of the natural numbers.[1]

(ii) The obvious reductionist step at this point is to *identify* the ZF-set ω of finite ordinals as *the* 'true' (hitherto intuitive) set N of natural numbers. This would be a grand reduction indeed, because work done during the 19th century by several mathematicians (including Hamilton, Bolzano, Weierstrass, Dedekind and Cantor) showed that all the concepts of mathematical analysis could be reduced to those of *natural number, set* and *membership* (plus concepts such as *relation* and *function* that we have by

[1] *Was sind und was sollen die Zahlen?*, 1888. (English translation in *Essays on the theory of numbers* edited by W. W. Beman, 1901.)

now reduced to set-theoretic concepts). Thus a huge part, if not the whole, of mathematics would be reduced to set theory.

Many (perhaps most) mathematicians, under the influence of the dominant structuralist ideology, do proceed in this way, and frame (or think of) their mathematical discourse as taking place within set theory.

3.9. *Warning*

This reduction, although extremely successful in a formal sense, is by no means unproblematic, as Skolem pointed out in 1922, when he published his famous paradox. (We shall discuss Skolem's Paradox in the Appendix.)

3.10. *Theorem*

ω is the least infinite ordinal and the least limit ordinal.

PROOF

That ω is an ordinal follows at once from Cor. 2.20 and Thms. 3.2 and 3.4. Also, ω cannot be a *finite* ordinal, because that would mean that $\omega \in \omega$ – which is impossible for an ordinal. Thus ω must be an infinite ordinal. On the other hand, if $\xi < \omega$ – that is, $\xi \in \omega$ – then by Def. 3.1 ξ is a finite ordinal; hence ω must be the *least* infinite ordinal.

If $\xi \in \omega$ then, as we have just seen, ξ is a finite ordinal, hence *a fortiori*, not a limit ordinal. If ω itself were not a limit ordinal then by Def. 3.1 it would follow that ω is a finite ordinal, contrary to what we have proved. Thus ω must be a limit ordinal. As we have just observed, no ordinal smaller than ω can be a limit ordinal. Hence ω is the *least* limit ordinal. ∎

3.11. *Preview*

We have yet to justify the adjectives *finite* and *infinite* introduced in Def. 3.1 in connection with ordinals. Dedekind defined a set as *infinite* if there exists an injection from it to a proper subset of itself, and as *finite* if there is no such injection. We will not adopt Dedekind's definition, but we shall show that finite and infinite ordinals in the sense of Def. 3.1 are finite and infinite respectively in Dedekind's sense.

3.12. Theorem

There does not exist an injection from a finite ordinal to a proper subset of itself.

PROOF

We proceed by weak induction on finite ordinals (Cor. 3.7). The proof is a formal (or 'internalized') version of the proof of Thm. 3.3.4.

Let Z be the set of all finite ordinals α such that there is no injection from α to a subset of itself. In order to prove our theorem it is enough to show that $\varnothing \in Z$ and that if $\alpha \in Z$ then also $\alpha' \in Z$.

That $\varnothing \in Z$ is obvious, since $\varnothing$ has no proper subsets. Now assume, as induction hypothesis, that $\alpha \in Z$ and let f be an injection from α' – that is, from $\alpha \cup \{\alpha\}$ – to a subset B of itself. If B is a *proper* subset of α' then the set $\alpha' - B$ is non-empty.

Without loss of generality we may assume that α belongs to $\alpha' - B$ rather than to B. (In the contrary case, where $\alpha \in B$, take any member β of $\alpha' - B$ and let g be the bijection from α' to itself that interchanges β and α but leaves all other members of α' fixed: thus, $g\beta = \alpha$, $g\alpha = \beta$ and $g\xi = \xi$ for any $\xi \in \alpha'$ other than β and α. Then use $g \circ f$ instead of f itself: it is an injection from α' to its proper subset $g[B] = (B - \{\alpha\}) \cup \{\beta\}$.)

Our assumption that $\alpha \in \alpha' - B$ means that $B \subseteq \alpha$. Next, let $\gamma = f\alpha$; then γ must belong to B, since f is a map to B. It now follows that $f \restriction \alpha$ is an injection from α to its proper subset $B - \{\gamma\}$. This contradicts the induction hypothesis. So B cannot be a proper subset of α'. ∎

3.13. Theorem

If α is an infinite ordinal then there is an injection from α to a proper subset of itself.

PROOF

First, consider ω. Define a map f on ω (that is, with ω as its domain) by putting $f\xi = \xi'$ for every finite ordinal ξ. Then f is injective. Indeed, if ξ and η are distinct, say $\xi < \eta$, then by Thm. 2.27 $\xi' < \eta'$, hence ξ' and η' are also distinct. Also, f maps ω to (in fact, onto) its proper subset $\omega - \{\varnothing\}$.

Now let α be any infinite ordinal. By Thm. 3.10 we have $\omega \leqslant \alpha$, which means that $\omega \in \alpha$ or $\omega = \alpha$; and since α is a transitive set, it follows that $\omega \subseteq \alpha$. Then the map $f \cup \mathrm{id}_{\alpha-\omega}$ (with f as before) is clearly an injection from α to its proper subset $\alpha - \{\varnothing\}$. ■

3.14. Theorem

A finite ordinal is not equipollent to any other ordinal.

PROOF

Let α be a finite ordinal and let β be another ordinal. First, suppose β is finite as well. We have $\alpha < \beta$ or $\beta < \alpha$ – that is, $\beta \in \alpha$ or $\beta \in \alpha$ – and since ordinals are transitive sets it follows that $\alpha \subset \beta$ or $\beta \subset \alpha$; hence by Thm. 3.12 α and β cannot be equipollent.

Now suppose β is an infinite ordinal. By Thm. 3.13 there exists an injection, say g, from β to a proper subset of itself. If f were a bijection from α to β, then clearly $f^{-1} \circ g \circ f$ would be an injection from α to a proper subset of itself – which is impossible. ■

3.15. Definition

A set is *finite* if it is equipollent to a finite ordinal (in the sense of Def. 3.1). Otherwise, it is *infinite*.

3.16. Remarks

(i) By virtue of Thm. 3.14, an ordinal is finite (or infinite) in the sense of Def. 3.1 iff it is finite (or infinite, respectively) in the sense of Def. 3.15; so there in no conflict between the two definitions.

(ii) By Thm. 3.14, a finite set is equipollent to a *unique* finite ordinal.

3.17. Problem

(i) Prove that there does not exist an injection from a finite set to a proper subset of itself. (Use Thm. 3.12.)

(ii) Prove that if A is a non-empty finite set of ordinals, then A has a *greatest* member – that is, an ordinal $\alpha \in A$ such that $\xi \leqslant \alpha$ for each $\xi \in A$. (Otherwise, define a map f on A by taking, for each $\alpha \in A$, $f\alpha$ as the least $\xi \in A$ such that $\alpha < \xi$. Show that f would be an injection from A to a proper subset of itself.)

3.18. Problem

Let n be a natural number. Show that for any objects $a_1, a_2, \ldots, a_n$, the set $\{a_1, a_2, \ldots, a_n\}$ is finite. (Use weak mathematical induction on the number n.)

§4. Transfinite induction

Various forms of the Principle of Mathematical Induction have analogues that apply to ordinals. These analogues collectively are known as the Principle of Transfinite Induction. First, by virtue of the fact that W is well-ordered, we have immediately by Thm. 2.9:

4.1. Theorem (Least Ordinal Principle)

If X is a non-empty class of ordinals, then X has a least member. ■

Hence other forms of the Principle of Transfinite Induction can be deduced.

4.2. Theorem (Strong Principle of Transfinite Induction)

If X is a class of ordinals such that for every ordinal ξ

$$(*) \qquad \eta \in X \textit{ for every } \eta < \xi \Rightarrow \xi \in X,$$

then $X = W$.

PROOF

Let $Y = W - X$. If Y were non-empty, it would have a least member, say ξ. So for each $\eta < \xi$ we would have $\eta \in X$. But then by $(*)$ $\xi \in X$, which is impossible. Thus Y must be empty. ■

4.3. Remark

By Rem. 2.22(ii) the antecedent, $\eta \in X$ *for every* $\eta < \xi$, in condition $(*)$ of Thm. 4.2 is equivalent to the statement that $\xi \subseteq X$.

4.4. Theorem (Weak Principle of Transfinite Induction)

If X is a class of ordinals satisfying the following three conditions

(i) $\varnothing \in X$,

(ii) *for every ordinal* ξ, $\xi \in X \Rightarrow \xi' \in X$,
(iii) *for every limit ordinal* λ, $\lambda \subseteq X \Rightarrow \lambda \in X$,

then $X = W$.

PROOF

Assume X satisfies these three conditions. Then by (i) and (iii) X satisfies condition $(*)$ of Thm. 4.2 for $\varnothing$ and for limit ordinals.

Now suppose $\xi' \subseteq X$. By Def. 2.26 it follows that $\xi \in X$; hence by (ii) $\xi' \in X$. Thus X satisfies $(*)$ also for successor ordinals. ∎

4.5. Remarks

(i) These principles have *restricted* forms, in which X is assumed to be a subset of some (arbitrary) given ordinal α rather than a subclass of W. Thus, the form of Thm. 4.1 restricted to an arbitrary ordinal α says that a non-empty subset of α has a least member. The restricted form of Thm. 4.2 says that if X is a subset of α such that for all $\xi < \alpha$ we have $\xi \subseteq X \Rightarrow \xi \in X$, then $X = \alpha$.

(ii) The Principle of Transfinite Induction restricted to the particular ordinal ω is precisely the Principle of Induction on Finite Ordinals.

4.6. Problem

Prove the restricted form of Thm. 4.2. Formulate and prove a form of Thm. 4.4 restricted to an arbitrary ordinal.

§5. The Representation Theorem

5.1. Preview

In this section we shall show that every well-ordered set is similar in its ordering to a unique ordinal.

5.2. Definition

A *partially ordered set* (briefly, *poset*) is a pair $\langle A, < \rangle$, where A is a set and $<$ is a [sharp] partial order on A. A *totally ordered set* is a poset $\langle A, < \rangle$, in which $<$ is a total order on A. A *well-ordered set* is a poset $\langle A, < \rangle$, in which $<$ is a well-ordering on A.

5.3. Remarks

(i) This is just a convenient way of packaging a set A together with a particular partial order on A into a single object. It saves us having to keep saying 'such-and-such a set with such-and-such a partial order on it'.

(ii) However, we shall often refer, somewhat inaccurately, to A itself as the poset (or ordered set, or well-ordered set) when, strictly speaking, we have in mind the pair $\langle A, < \rangle$. We shall only commit this peccadillo when it is clear from the context which relation $<$ is involved. Thus, we refer to an ordinal α as a well-ordered set, when strictly speaking we mean the pair $\langle \alpha, < \rangle$, where $<$ is $\in_\alpha$, the $\in$-well-ordering on α.

5.4. Definition

A *similarity map* (a.k.a. *isomorphism*) from a poset $\langle A, < \rangle$ to a poset $\langle A', <' \rangle$ is a bijection f from A to A' such that, for all x and y in A,

$$x < y \Leftrightarrow fx <' fy.$$

If such a map exists, $\langle A, < \rangle$ is said to be *similar* (or *isomorphic*) to $\langle A', <' \rangle$.

5.5. Remark

It is easy to see that the identity map id_A is a similarity map from $\langle A, < \rangle$ to itself. Also if f is a similarity map from $\langle A, < \rangle$ to $\langle A', <' \rangle$ then its inverse f^{-1} is a similarity map from $\langle A', <' \rangle$ to $\langle A, < \rangle$. Finally, if f is a similarity map from $\langle A, < \rangle$ to $\langle A', <' \rangle$ and g is a similarity map from $\langle A', <' \rangle$ to $\langle A'', <'' \rangle$ then the composition $g \circ f$ is a similarity map from $\langle A, < \rangle$ to $\langle A'', <'' \rangle$.

It follows that similarity is an equivalence relation on the class of posets.

5.6. Theorem

If f is a similarity map from an ordinal α to an ordinal β then f is the identity map id_α, *hence* $\alpha = \beta$.

PROOF

First, we prove by strong transfinite induction (restricted to α) that $\xi \leqslant f\xi$ for every $\xi \in \alpha$.

Let $\xi \in \alpha$. By the induction hypothesis, if $\eta < \xi$ then $\eta \leqslant f\eta$. But if $\eta < \xi$ then also $f\eta < f\xi$, since f is a similarity map. Thus for every $\eta < \xi$ we have $\eta < f\xi$. In particular, $\eta \neq f\xi$ for every $\eta < \xi$; in other words, $f\xi < \xi$ is impossible. This proves that $\xi \leqslant f\xi$ and completes the induction.

Now, f^{-1} is a similarity map from β to α; therefore by the same token we have also $\zeta \leqslant f^{-1}\zeta$ for all $\zeta \in \beta$. Taking ζ to be $f\xi$, where $\xi \in \alpha$, we obtain $f\xi \leqslant f^{-1}f\xi = \xi$. Thus $f\xi \leqslant \xi$ as well as $\xi \leqslant f\xi$, which shows that f must be the identity id_α. ■

5.7. Corollary

For any poset $\langle A, < \rangle$, there exists at most one similarity map from $\langle A, < \rangle$ to an ordinal.

PROOF

If f and g are isomorphisms from $\langle A, < \rangle$ to α and β respectively, then the composition $g \circ f^{-1}$ is clearly an isomorphism from α to β. Therefore $\alpha = \beta$ and $g \circ f^{-1}$ is the identity mapping, which means that $f = g$. ■

5.8. Preliminaries

(i) For the rest of this section, we consider a fixed but otherwise arbitrary well-ordered set $\langle A, < \rangle$.

(ii) If $B \subseteq A$, then B is clearly well-ordered by the relation $< \cap B^2$, that is:

$$\{\langle x, y \rangle : x \in B, \text{ and } y \in B, \text{ and } x < y\},$$

which is called the *restriction of* $<$ to B. Whenever we refer to a subset B of A as well-ordered, we shall mean B with this well-ordering, inherited by B from A.

(iii) For each $a \in A$, the *segment of A determined by a* is the set

$$A_a =_{\mathrm{df}} \{x \in A : x < a\}.$$

(iv) We define a class F as follows:

$$F =_{\mathrm{df}} \{\langle x, \xi \rangle : x \in A, \text{ and } \xi \text{ is an ordinal, and } A_x \text{ is similar to } \xi\}.$$

By Cor. 5.7, F is a function (see Def. 2.2.1). We may therefore use functional notation in connection with F. Thus '$Fx = \xi$' means the same as '$\langle x, \xi \rangle \in F$'.

Clearly, dom F is a subset of A. By AS dom F is a set; hence by AR ran F is a set as well. Note that all the members of ran F are ordinals.

5.9. Lemma

Let $Fa = \alpha$. Then for any ordinal $\beta < \alpha$ there exists some $b < a$ such that $Fb = \beta$. Conversely, if $b < a$ then b belongs to dom F *and Fb is some ordinal $\beta < \alpha$.*

PROOF

Let f be the similarity map from A_a to α. Suppose $\beta < \alpha$. This means that $\beta \in \alpha$. Therefore $fb = \beta$ for some $b \in A_a$ – that is, $b < a$. Note that by the transitivity of α we have $\beta \subseteq \alpha$. It is easy to verify that $f \restriction A_b$, the restriction of f to A_b, is a similarity map from A_b to β. Hence $Fb = \beta$.

Conversely, suppose that $b < a$. This means that $b \in A_a$. Therefore $fb = \beta$ for some $\beta \in \alpha$ – that is, $\beta < \alpha$. As before, it follows that $Fb = \beta$. ■

5.10. Lemma

F is injective.

PROOF

Let a and b be two distinct members of dom F. We have to show that $Fa \neq Fb$. Without loss of generality, we may assume $b < a$. Let $Fa = \alpha$. Then by Lemma 5.9 it follows that Fb is some ordinal $\beta < \alpha$. ■

5.11. Lemma

The set ran F *is an ordinal.*

PROOF

As a set of ordinals, ran F is $\in$-well-ordered. It remains to prove that it is a transitive set. Let $\alpha \in$ ran F; thus $Fa = \alpha$ for some $a \in A$. Now let $\beta \in \alpha$ – that is, $\beta < \alpha$. Then by Lemma 5.9 β also belongs to ran F, showing that this set is transitive. ■

5.12. Theorem (Representation Theorem for well-ordered sets)

There exists a unique similarity map from the well-ordered set A to an ordinal.

PROOF

Uniqueness follows from Cor. 5.7. To prove existence, we shall show that F is a similarity map from A to the ordinal ran F. By Lemmas 5.9 and 5.10, F is a similarity map from dom F, which is a subset of A, to ran F; so it only remains to establish that dom F is the whole of A.

Suppose not. Then, since A is well-ordered, there would be a *least* $b \in A$ such that $b \notin \text{dom } F$. Thus, if $a \in A$ such that $a < b$ then a must belong to dom F. On the other hand, if $b < a$ then a cannot be in dom F because if it were then by the second half of Lemma 5.9 b would also be in that domain.

It would follow that dom F is exactly A_b. But then F is a similarity map from A_b to ran F. Thus A_b is similar to the ordinal ran F. By the definition of F it would then follow that $\langle b, \text{ran } F \rangle \in F$, hence $b \in \text{dom } F$, contradicting the choice of b. ■

5.13. Definition

A set is *denumerable* if it is equipollent to ω. A set is *countable* if it is finite or denumerable.

5.14. Problem

(i) Let D be a subset of an ordinal α. By Cor. 2.20, D is $\in$-well-ordered; and by Thm. 5.12, D is similar to an ordinal β. Prove that $\beta \leqslant \alpha$. (Let f be a similarity map from β to D. Show that $\xi \leqslant f\xi$ for every $\xi \in \beta$.)

(ii) Prove that a set is countable iff it is equipollent to a subset of ω. (Use (i) to show that every subset of ω is countable.)

§6. Transfinite recursion

6.1. Preview

In this section we validate a powerful method of defining functions on W (that is, having W as domain). Roughly speaking, $F\xi$, the value of the function F at ξ, is defined in terms of the 'behaviour' of F for all ordinals smaller than ξ.

6.2. Convention

Throughout this section we let C be a fixed but arbitrary function such that dom C is the class of all sets.

6.3. Definition

We shall write '$\mathcal{R}_C(F, \alpha)$' as short for the statement:

$$F \text{ is a function and } \alpha' \subseteq \operatorname{dom} F \text{ and } F\xi = C(F \restriction \xi) \text{ for all } \xi \leqslant \alpha.$$

The equation '$F\xi = C(F \restriction \xi)$' is called an *ordinal recursion equation*.

6.4. Remarks

(i) Recall that $\alpha' = \{\xi : \xi \leqslant \alpha\}$.

(ii) Note that $F \restriction \xi = \{\langle \eta, F\eta \rangle : \eta \in \xi\}$. Therefore the recursion equation determines $F\xi$ in terms of the 'previous behaviour' of F – the restriction of F to the set of all ordinals $\eta < \xi$. Note also that even if F is a proper class, $F \restriction \xi$ is always a set by AR and Thm. 2.2.8.

(iii) $\mathcal{R}_C(F, \alpha)$ means that F is defined and satisfies the recursion equation for all ordinals up to α inclusive. Hence

$$\mathcal{R}_C(F, \alpha) \Rightarrow \mathcal{R}_C(F, \beta) \text{ for all } \beta \leqslant \alpha.$$

6.5. Lemma

If both $\mathcal{R}_C(F, \alpha)$ and $\mathcal{R}_C(G, \alpha)$ then $F\xi = G\xi$ for all $\xi \leqslant \alpha$.

PROOF

By (strong) transfinite induction, restricted to α'. Let ξ be any ordinal $\leqslant \alpha$ (that is, $\xi < \alpha'$) and assume, as induction hypothesis, that $F\eta = G\eta$ for all $\eta < \xi$ – that is, for all $\eta \in \xi$. This means that $F \restriction \xi = G \restriction \xi$, hence $C(F \restriction \xi) = C(G \restriction \xi)$. It now follows from $\mathcal{R}_C(F, \alpha)$ and $\mathcal{R}_C(G, \alpha)$ that $F\xi = G\xi$. ■

6.6. Lemma

For any ordinal α there exists a unique function f_α such that dom $f_\alpha = \alpha' = \{\xi : \xi \leqslant \alpha\}$ *and such that $\mathcal{R}_C(f_\alpha, \alpha)$.*

PROOF

Uniqueness follows from Lemma 6.5. We prove existence by strong transfinite induction. Assume as induction hypothesis that for each $\beta < \alpha$ there exists a (necessarily unique) function f_β whose domain is $\beta' = \{\xi\colon \xi \leq \beta\}$ such that $\mathcal{R}_C(f_\beta, \beta)$.

If $\gamma \leq \beta < \alpha$ then by Rem 6.4(iii) we have $\mathcal{R}_C(f_\beta, \gamma)$ and hence by Lemma 6.5 $f_\beta(\xi) = f_\gamma(\xi)$ for all $\xi \leq \gamma$. This means that f_β and f_γ agree wherever both of them are defined; in fact, it is easy to see that $f_\gamma \subseteq f_\beta$. By Prob. 2.4.8, we can therefore glue all the f_β together to obtain a single function: we put

$$f = \bigcup\{f_\beta\colon \beta < \alpha\}.$$

Clearly, f is a function whose domain is $\{\beta\colon \beta < \alpha\}$ – that is α itself – and it satisfies the recursion equation $f\beta = C(f \upharpoonright \beta)$ for all $\beta < \alpha$. Finally, we extend f to a function defined for all $\beta \leq \alpha$:

$$f_\alpha = f \cup \{\langle \alpha, C(f)\rangle\}.$$

Then $\operatorname{dom} f_\alpha = \alpha'$. Also, $f = f_\alpha \upharpoonright \alpha$ and hence $f_\alpha(\alpha) = C(f) = C(f_\alpha \upharpoonright \alpha)$. Thus f_α satisfies the required recursion equation for all $\beta \leq \alpha$. ∎

6.7. Theorem (Definition by transfinite recursion)

We can define a (necessarily unique) function F such that $\operatorname{dom} F = W$ *and such that $F\xi = C(F \upharpoonright \xi)$ for all $\xi \in W$.*

PROOF

To define F, note that the f_α of Lemma 6.6 satisfy the recursion equation wherever they are defined, and any two of them agree with each other wherever both are defined. Therefore all we have to do is glue them together:

$$F =_{\text{df}} \bigcup\{f_\alpha\colon \alpha \in W\}.$$

It is easy to see that indeed $\operatorname{dom} F = W$ and $F\xi = C(F \upharpoonright \xi)$ for every $\xi \in W$. Moreover, these two conditions fulfilled by F imply that $\mathcal{R}_C(F, \alpha)$ for all α; hence F is unique by Lemma 6.5. ∎

6.8. Remarks

(i) Note the phrasing of Thm. 6.7: it does not claim that such-and-such an F *exists* but that *we can define* it. To say, in set theory, that F 'exists' would mean that it is an object of the theory – which is false, since F is a proper class. In fact, Thm. 6.7 is not a single theorem of set theory, but a *meta-theorem* or a *theorem scheme* which shows how, for any given class C fulfilling a certain condition (Convention 6.2), we can define a class F fulfilling certain other conditions. The same applies to any other theorem, postulate and definition in which general statements or stipulations are made concerning classes – for example Def. 1.3.4 and Ax. 1.3.6 (AS): they are not individual statements of set theory, but *schemes*. (Compare Rem. 3.3.6.)

(ii) From Thm. 6.7 (or directly from Lemma 6.6) it is easy to obtain a version of definition by transfinite recursion *restricted* to any given ordinal α, in which $\operatorname{dom} F$ is α instead of W and the recursion equation $F\xi = C(F \restriction \xi)$ is satisfied for all $\xi < \alpha$.

5

The axiom of choice

§ 1. From the axiom of choice to the well-ordering theorem

1.1. Definition

A *choice function* on a class $\mathcal{S}$ of sets is a function g with $\operatorname{dom} g = \mathcal{S}$, such that $gX \in X$ for every $X \in \mathcal{S}$.

1.2. The axiom of choice (AC) states:

If $\mathcal{S}$ is a set of non-empty sets then there exists a choice function on $\mathcal{S}$.

1.3. Remarks

(i) AC was the first postulate of set theory (apart from PX) to be stated as such. Its first known explicit formulation is due to Giuseppe Peano (1890), who however rejected it as untenable. It was first proposed as a new valid mathematical principle by Beppo Levi in 1902, although it had been used inadvertently by Cantor and others long before that. Zermelo, who was told about AC by Erhard Schmidt, used it almost at once in his first (1904) proof of the Well-Ordering Theorem (WOT, Cor. 1.6 below), a result that had been conjectured by Cantor. Our formulation of AC is essentially that used by Zermelo in his 1904 paper.

(ii) In his 1908 paper on the foundations of set theory, in which the theory is given its first fully fledged axiomatic presentation, Zermelo does not state AC in this form but in a more restricted version. He assumes that $\mathcal{S}$ is a set of non-empty sets that are pairwise disjoint—that is, $X \cap Y = \varnothing$ for any two distinct members of $\mathcal{S}$ (see Def. 3.4.1). He then postulates the existence of a set A such that, for any $X \in \mathcal{S}$, the intersection $A \cap X$ has exactly one member.

This restricted version follows at once from AC. Indeed, if $\mathcal{S}$ is a set of non-empty pairwise disjoint sets, then by AC there exists a choice function g on $\mathcal{S}$. It is then easy to see that, for any $X \in \mathcal{S}$, $\operatorname{ran} g \cap X = \{gX\}$.

Conversely, AC in the form we have stated it follows from the restricted version. To show this, let $\mathcal{S}$ be any set of non-empty sets. Put

$$\mathcal{T} = \{\{X\} \times X : X \in \mathcal{S}\}.$$

It is easy to verify that $\mathcal{T}$ is a set of non-empty and pairwise disjoint sets. According to the restricted version, there exists a set A whose intersection with each member of $\mathcal{T}$ is a singleton. We now define a function g on $\mathcal{S}$ as follows. For any $X \in \mathcal{S}$, the set $\{X\} \times X$ belongs to $\mathcal{T}$ and hence its intersection with A has exactly one member. This member must be of the form $\langle X, x_0 \rangle$, where x_0 is some member of X. We put $gX = x_0$. Then g is a choice function on $\mathcal{S}$.

(iii) Using AC, Def. 3.4.11 is easily legitimized. If $|A_x| \approx |B_x|$ for each $x \in X$, then by AC there exists a family $f = \{f_x \mid x \in X\}$ such that, for each x, f_x is a bijection from $\{x\} \times A_x$ to $\{x\} \times B_x$. Then it is easy to see that $\bigcup \operatorname{ran} f$ is a bijection from $\bigcup\{\{x\} \times A_x : x \in X\}$ to $\bigcup\{\{x\} \times B_x : x \in X\}$. A similar argument applies to Def. 3.5.11.

(iv) AC has been regarded with suspicion because it is a *purely* existential postulate. It asserts the existence of a set – a choice function – without characterizing it as the extension of some previously specified property. In other words, AC is not a special case of the Principle of Comprehension. In this respect AC is markedly different from all other existential postulates of set theory. For example, the Power-set Axiom asserts that, for each set A, there exists the power-set $\mathrm{P}A$, which is characterized as the extension of the property *being a subset of A*.

(v) In 1938 Gödel proved that AC is consistent relative to the other, commonly accepted, postulates of set theory, in the sense that if they are consistent, then the addition of AC does not result in inconsistency. In 1963 P. J. Cohen proved that the same holds also for the negation of AC.

(vi) AC has some weird (counter-intuitive) consequences. However, its negation has even weirder ones: for example, the direct product of a family of non-empty sets may well be empty. Note also that the *finite* version of AC – in which the set $\mathcal{S}$ is assumed to be finite – can be deduced from the remaining postulates of ZF. Thus AC is only needed as an additional postulate for the case where $\mathcal{S}$ is infinite. It therefore appears as a natural extension to the infinite case of a principle that must in any case be accepted in the finite case.

(vii) Most mathematicians regard AC as indispensable: without it, many results in modern mathematics as well as in set theory itself would be unprovable. However, in view of its somewhat controversial status, when the AC is needed for proving a mathematical result, it is customary to point this out.

1.4. Preview

Starting from AC, we shall prove a chain of other major principles, all of which turn out to be equivalent to each other and to AC. The first of these principles, which is also the most important, is a corollary of the following theorem.

1.5. Theorem

Every set is equipollent to an ordinal.

PROOF

Let A be a set, and let $\mathcal{S}$ be the set $\mathsf{P}A - \{\varnothing\}$ of all non-empty subsets of A. By AC there exists a choice function g on $\mathcal{S}$. Since A is a set, it cannot be the universal class (Thm. 1.3.10); so there exists an object b that does *not* belong to A.

We now define a function C whose domain is the class of all sets, as follows: for any set x we put

$$(*) \qquad Cx = \begin{cases} g(A - \operatorname{ran} x) & \text{if } x \text{ is a map such that } \operatorname{ran} x \subset A, \\ b & \text{otherwise.} \end{cases}$$

Using transfinite recursion (Thm. 4.6.7), we get a function F with W as domain, satisfying the recursion equation $F\xi = C(F \upharpoonright \xi)$ for all $\xi \in W$. Combining this equation with $(*)$, we obtain for all ξ:

$$F\xi = \begin{cases} g(A - \operatorname{ran}(F \upharpoonright \xi)) & \text{if } \operatorname{ran}(F \upharpoonright \xi) \subset A, \\ b & \text{otherwise.} \end{cases}$$

Let ξ be any ordinal such that $F\xi \neq b$. This means that $F \restriction \xi$ must be a map from ξ to A, and

$$F\xi = g(A - \text{ran}(F \restriction \xi)) \in A - \text{ran}(F \restriction \xi).$$

Thus $F\xi$ is a 'fresh' member of A, different from $F\eta$ for all $\eta < \xi$. (What happens is that so long as A is not exhausted by previous values of F, the new value $F\xi$ is chosen, using the choice function g, as a fresh member of A.)

If $F\xi \neq b$ for *all* ordinals ξ, it would follow that F is an injection from the proper class W (Cor. 4.2.18) to the set A. This is impossible by Prob. 2.4.5. So there must exist some ordinal ξ for which $F\xi = b$.

Let α be the *least* ordinal such that $F\alpha = b$. Such an α exists by the Least Ordinal Principle (Thm. 4.4.1). Then it is easy to see that $F \restriction \alpha$ is an injection from α – that is, from the set $\{\xi : \xi < \alpha\}$ – to A. Also, $\text{ran}(F \restriction \alpha)$ cannot be a *proper* subset of A. Thus $F \restriction \alpha$ is in fact a bijection from α to A. ∎

1.6. Corollary (Well-Ordering Theorem)

For every set A there exists a well-ordering on A.

PROOF

By Thm. 1.5, there exists a bijection F from an ordinal α to A. Now put

$$< =_{\text{df}} \{\langle F\xi, F\eta \rangle : \xi < \eta < \alpha\}.$$

This means that for any members x and y of A, $x < y$ iff $\xi < \eta$, where ξ and η are the (necessarily unique) ordinals $< \alpha$ such that $x = F\xi$ and $y = F\eta$. Clearly, $<$ is a well-ordering on A. ∎

1.7. Remarks

(i) With F, α and $<$ as above, $F \restriction \alpha$ is a similarity map from α to the well-ordered set $\langle A, < \rangle$.

(ii) Thms. 1.5 and Cor. 1.6 are equivalent to each other. Indeed, the former can easily be deduced from the latter using the Representation Theorem 4.5.12. We shall therefore refer also to both Thm. 1.5 and Cor. 1.6 as *the WOT*.

Another important consequence of Thm. 1.5 is that the class of cardinals is *totally* ordered (see Def. 2.3.11(ii)):

1.8. Corollary

For any sets A and B, $|A| \leqslant |B|$ *or* $|B| \leqslant |A|$.

PROOF

By Thm. 1.5, A and B are equipollent to ordinals, say α and β respectively. Since the class of ordinals is $\in$-well-ordered, it follows (see Lemma 4.2.4) that $\alpha \in \beta$ or $\alpha = \beta$ or $\beta \in \alpha$. But ordinals are transitive sets, hence $\alpha \subseteq \beta$ or $\beta \subseteq \alpha$. ■

§2. From the WOT via Zorn's Lemma back to AC

We start by proving two simple lemmas about finite sets, which do not depend on AC.

2.1. Lemma

If $B \subset A$ *and* A *is equipollent to a finite ordinal* α, *then* B *is equipollent to an ordinal* $\beta < \alpha$. *Hence every subset of a finite set is finite.*

PROOF

Let $B \subset A$, where A is equipollent to a finite ordinal α. Then B is clearly equipollent to some $D \subset \alpha$. By Prob. 4.5.14(i), D is similar – and hence equipollent – to some ordinal $\beta \leqslant \alpha$. However, since here α is finite, Thm. 4.3.12 excludes the possibility that $\beta = \alpha$. Therefore $\beta < \alpha$. ■

2.2. Lemma

If f is a map such that dom f *is finite then* ran f *is finite as well.*

PROOF

By Def. 4.3.15, dom f is equipollent to a finite ordinal α. Without loss of generality we may therefore assume that dom f is α itself. (Otherwise, replace f by $f \circ h$, where h is a bijection from α to dom f.) Define a map g from ran f to α by putting, for each $x \in$ ran f,

$$gx =_{\text{df}} \text{the least } \xi \in \alpha \text{ such that } f\xi = x.$$

It is easy to see that g is injective, hence it is a bijection from ran f to some subset D of α. By Lemma 2.1, D is finite; therefore so is ran f. ■

Next, we lay down a few definitions.

2.3. *Definition*

Let $<$ be a partial order on a class A. A member a of A is said to be *maximal in A with respect to* $<$ if there is no $x \in A$ such that $a < x$.

2.4. *Remarks*

(i) When there is no risk of confusion, we shall omit the phrase 'in A with respect to $<$'.

(ii) In general, A may not have a maximal member; or it may have more than one.

(iii) Do not confuse *maximal* with *greatest*. However, if $<$ is a *total* order on A and a is maximal in A then a is also the *greatest* member of A, in the sense that $x < a$ for any other $x < A$ for any other $x \in A$. In this case it is clear that A cannot have more than one maximal member.

2.5. *Definition*

If $\mathcal{A}$ is any class of sets, we put

$$\subset_{\mathcal{A}} =_{\text{df}} \{\langle X, Y\rangle \in \mathcal{A}^2 : X \subset Y\}.$$

$\subset_{\mathcal{A}}$ is called the *restriction of* $\subset$ *to* $\mathcal{A}$.

2.6. *Remarks*

(i) We can also characterize the relation $\subset_{\mathcal{A}}$ by saying that, for any X and Y,

$$X \subset_{\mathcal{A}} Y \Leftrightarrow X \in \mathcal{A} \text{ and } Y \in \mathcal{A} \text{ and } X \subset Y.$$

(ii) As noted in Ex. 2.3.8, if $\mathcal{A}$ is any class of sets, $\subset_{\mathcal{A}}$ is a [sharp] partial order on $\mathcal{A}$.

2.7. Definition

A class $\mathcal{A}$ of sets is *of finite character* if, for any set X,

$$X \in \mathcal{A} \Leftrightarrow Y \in \mathcal{A} \text{ for every } \textit{finite } Y \subseteq X.$$

We shall use the WOT to prove the following useful result.

2.8. Theorem (Tukey–Teichmüller Lemma).

If $\mathcal{A}$ is a set of finite character, then for every $A \in \mathcal{A}$ there exists an $M \in \mathcal{A}$ such that $A \subseteq M$ and M is maximal in $\mathcal{A}$ w.r.t. $\subset_{\mathcal{A}}$.

PROOF

By the WOT, $\mathcal{A}$ is equipollent to some ordinal α. Let G be a bijection from α to $\mathcal{A}$. Thus

$$\mathcal{A} = \{G\xi : \xi < \alpha\}.$$

Take any $A \in \mathcal{A}$; we shall hold A fixed for the rest of the proof. Without loss of generality, we may assume that $A = G\varnothing$ – otherwise, we could compose G with the bijection from $\mathcal{A}$ to itself that interchanges A with $G\varnothing$ and leaves all other members of $\mathcal{A}$ alone.

Using transfinite recursion restricted to α (see Rem. 4.6.8(ii)), we define a map F on α such that, for every $\xi < \alpha$,

$$F\xi = \begin{cases} G\xi & \text{if } \bigcup\{F\eta : \eta < \xi\} \subseteq G\xi, \\ \bigcup\{F\eta : \eta < \xi\} & \text{otherwise.} \end{cases}$$

(Note that $\{F\eta : \eta < \xi\} = \operatorname{ran}(F \restriction \xi)$, so that here $F\xi$ is indeed being determined in terms of $F \restriction \xi$, as required in transfinite recursion.)

It is clear that F is *monotone* in the sense that whenever $\eta \leqslant \xi < \alpha$ then $F\eta \subseteq F\xi$.

We claim that $F\xi \in \mathcal{A}$ for every $\xi < \alpha$. We shall prove this claim by strong transfinite induction restricted to α. Let $\xi < \alpha$; our induction hypothesis is that $F\eta \in \mathcal{A}$ for every $\eta < \xi$.

Now, $F\xi$ is $G\xi$ or $\bigcup\{F\eta : \eta < \xi\}$. Since certainly $G\xi \in \mathcal{A}$, we need only prove that the union $\bigcup\{F\eta : \eta < \xi\}$ belongs to $\mathcal{A}$. But $\mathcal{A}$ is a set of finite character. So it is enough to show that every finite subset of $\bigcup\{F\eta : \eta < \xi\}$ belongs to $\mathcal{A}$. We need only deal with non-empty subsets, since $\varnothing$ is a finite subset of A, and as such must in any case belong to $\mathcal{A}$.

Let B be a non-empty finite subset of $\bigcup\{F\eta : \eta < \xi\}$. Then for each $b \in B$ there exists some $\eta < \xi$ such that $b \in F\eta$. Define a map f from

B to ξ by putting, for each $b \in B$,

$$fb =_{\text{df}} \text{the least } \eta < \xi \text{ such that } b \in F\eta.$$

By Lemma 2.2, ran f is a *finite* non-empty set of ordinals $< \xi$. Hence by Prob. 4.3.17(ii) ran f has a greatest member, say η^*. This means that for every $b \in B$ we have $fb \leqslant \eta^*$; and, since F is monotone, it follows that $F(fb) \subseteq F(\eta^*)$. But by the definition of f we have $b \in F(fb)$; hence

$$b \in F(fb) \subseteq F(\eta^*) \text{ for every } b \in B.$$

Thus $B \subseteq F(\eta^*)$. But $\eta^* < \xi$, so by our induction hypothesis $F(\eta^*)$ belongs to $\mathcal{A}$; and since $\mathcal{A}$ is of finite character B, as a finite subset of $F(\eta^*)$, must also belong to $\mathcal{A}$. This completes the proof that $F\xi \in \mathcal{A}$ for every $\xi \leqslant \alpha$.

We now put $M = \bigcup\{F\eta : \eta < \alpha\}$. We shall show that M has the properties claimed by our theorem. The fact that $M \in \mathcal{A}$ is proved by showing, exactly as before, that every finite subset of M belongs to $\mathcal{A}$. Also, it is easy to see that $F\varnothing = G\varnothing = A$, hence $A \subseteq M$.

It remains to show that M is maximal w.r.t. $\subset_{\mathcal{A}}$. Suppose this were not so. Then there would be some $X \in \mathcal{A}$ such that $M \subset X$. Now, X must be $G\xi$ for some $\xi < \alpha$, so the assumption $M \subset X$ means that $\bigcup\{F\eta : \eta < \alpha\} \subset G\xi$. Hence, *a fortiori*,

$$\bigcup\{F\eta : \eta < \xi\} \subset G\xi.$$

But in this case the definition of F says that $F\xi = G\xi$. It would then follow that $\bigcup\{F\eta : \eta < \alpha\} \subset F\xi$ – which is impossible. ■

2.9. *Definition*

Let $\langle A, < \rangle$ be a poset. A *chain in* $\langle A, < \rangle$ is any subset C of A such that, for all x and y in C, $x < y$ or $x = y$ or $y < x$.

2.10. *Remark*

In other words, a chain in $\langle A, < \rangle$ is a subset of A that is totally ordered by the restriction of $<$ to it.

We shall use the Tukey–Teichmüller (TT) Lemma to prove:

2.11. Theorem (Hausdorff Maximality Principle)

Let $\langle A, <\rangle$ be a poset and let $\mathcal{C}$ be the set of all chains in $\langle A, <\rangle$. Then every member of $\mathcal{C}$ is included in some member of $\mathcal{C}$ that is maximal w.r.t. $\subset_{\mathcal{C}}$.

PROOF

The condition for C being a chain in $\langle A, <\rangle$ (see Def. 2.9) involves only two members of C at a time. Hence it is easy to see that the set $\mathcal{C}$ of all chains is of finite character. Therefore the TT Lemma applies to $\mathcal{C}$. ■

The most famous and frequently used of all the maximality principles that are equivalent to AC is generally known as 'Zorn's Lemma' although it is arguably due to Kuratowski, who published a version of it in 1922, thirteen years before Zorn. We shall now deduce it from the Hausdorff Maximality Principle (HMP). (For the meaning of *upper bound*, see Def. 4.2.23.)

2.12. Theorem (Zorn's Lemma)

Let $\langle A, <\rangle$ be a poset such that every chain in it has an upper bound in A. Then for each $a \in A$ there is some $u \in A$ such that u is maximal in A w.r.t. $<$ and such that $a \leqslant u$.

PROOF

As before, let $\mathcal{C}$ be the set of all chains in $\langle A, <\rangle$, and consider the poset consisting of $\mathcal{C}$ with the partial order $\subset_{\mathcal{C}}$ on it.

The singleton $\{a\}$ is, trivially, a chain in $\langle A, <\rangle$. Hence by the HMP $\{a\}$ is included in a chain C that is maximal in $\mathcal{C}$ w.r.t. $\subset_{\mathcal{C}}$. By hypothesis, C has an upper bound u in A. Since $a \in C$, it follows that $a \leqslant u$.

It remains to show that u is maximal in A. Suppose it were *not* maximal. Then there would exist some v such that $u < v$. Since u is an upper bound for C, it would follow that $x < v$ for all $x \in C$. But then $C \cup \{v\}$ would be a chain that properly includes C – contradicting the maximality of C in $\mathcal{C}$. ■

We have shown that

$$\text{AC} \Rightarrow \text{WOT} \Rightarrow \text{TT Lemma} \Rightarrow \text{HMP} \Rightarrow \text{Zorn's Lemma}.$$

Now we shall complete the cycle:

2.13. Theorem

AC follows from Zorn's Lemma.

PROOF

Let $\mathcal{S}$ be a set of non-empty sets. We must show that there exists a choice function on $\mathcal{S}$.

If $\mathcal{S}$ is empty then $\varnothing$ is the required choice function. So from now on we may assume that $\mathcal{S}$ is non-empty.

Let us say that f is a *partial choice function (pcf)*, if f is a choice function on a subset of $\mathcal{S}$. Such creatures do exist: for example, if A is any member of $\mathcal{S}$ and a is any member of A then $\{\langle A, a\rangle\}$ is a choice function on $\{A\}$ and hence a pcf. Let $\mathcal{F}$ be the set of all pcfs. (It is easy to verify that $\mathcal{F}$ is indeed a set; DIY.) As we have just seen, $\mathcal{F}$ is non-empty.

We now consider the poset $\langle \mathcal{F}, \subset_{\mathcal{F}}\rangle$. Note that if f and g are pcfs, then $f \subset g$ means that $\operatorname{dom} f \subset \operatorname{dom} g$ and $fX = gX$ for each $X \in \operatorname{dom} f$.

We shall show that $\langle \mathcal{F}, \subset_{\mathcal{F}}\rangle$ satisfies the condition of Zorn's Lemma. To this end, let us consider any chain $\mathcal{C}$ in this poset. We claim that its union, $\bigcup \mathcal{C}$, is an upper bound for $\mathcal{C}$ in $\mathcal{F}$.

For any $f \in \mathcal{C}$ we obviously have $f \subseteq \bigcup \mathcal{C}$. So it only remains to show that $\bigcup \mathcal{C}$ belongs to $\mathcal{F}$; in other words, that $\bigcup \mathcal{C}$ is a pcf.

Since every member of $\mathcal{C}$, being a pcf, is a set of ordered pairs $\langle X, x\rangle$ such that $x \in X \in \mathcal{S}$, it is clear that $\bigcup \mathcal{C}$ likewise is a set of ordered pairs of this kind. It only remains to show that $\bigcup \mathcal{C}$ is a function.

Now, if both f and g are members of $\mathcal{C}$ then, since $\mathcal{C}$ is a chain, we must have $f \subseteq g$ or $g \subseteq f$. Therefore $X \in \operatorname{dom} f \cap \operatorname{dom} g$ then $fX = gX$. Thus the coherence condition is fulfilled, showing that $\bigcup \mathcal{C}$ is indeed a function (see Prob. 2.4.8).

We can now apply Zorn's Lemma to the poset $\langle \mathcal{F}, \subset_{\mathcal{F}}\rangle$. Since $\mathcal{F}$ is non-empty, it follows from the Lemma that there exists some $g \in \mathcal{F}$ that is maximal w.r.t. $\subset_{\mathcal{F}}$. Such g is a pcf – a choice function on a subset of $\mathcal{S}$. However, if $\operatorname{dom} g$ were not the whole of $\mathcal{S}$, we could take any $A \in \mathcal{S} - \operatorname{dom} g$ and any $a \in A$, and put

$$f = g \cup \{\langle A, a\rangle\}.$$

Then f would be a pcf such that $g \subset f$, contradicting the maximality of g. Therefore g must be a choice function on the *whole* of $\mathcal{S}$. ■

2.14. Remarks

(i) We have now established

$$\text{AC} \Rightarrow \text{WOT} \Rightarrow \text{TT Lemma} \Rightarrow \text{HMP} \Rightarrow \text{Zorn's Lemma} \Rightarrow \text{AC},$$

hence these five principles are mutually equivalent.

(ii) These principles can be deduced *directly* from each other, without going round the cycle. Some of these deductions are quite easy. For example, to deduce AC directly from the WOT, let $\mathcal{S}$ be any set of non-empty sets. Note that if $X \in \mathcal{S}$ then $X \subseteq \bigcup \mathcal{S}$. By the WOT, there exists a well-ordering $<$ on $\bigcup \mathcal{S}$. Then a choice function g on $\mathcal{S}$ is obtained by putting, for all $X \in \mathcal{S}$,

$$gX = \text{the least member of } X \text{ w.r.t. } <.$$

It is also not difficult to deduce the TT Lemma directly from Zorn's Lemma (DIY!). However, the only direct routes I know from AC to the three maximality principles (TT Lemma, HMP and Zorn's Lemma) are quite rocky.

6
Finite cardinals and alephs

§ 1. Finite cardinals

1.1. Preview

In this chapter we will complete the definition of *cardinal* and *cardinality*, which has so far been left open (see Rem. 3.1.4), and derive some important results about cardinals. In the present section we confine ourselves to finite sets and cardinals; here we shall not invoke AC.

Recall that by Def. 4.3.15 a set is finite iff it is equipollent to a finite ordinal (that is, an ordinal $< \omega$); moreover, by Thm. 4.3.14 this ordinal is unique. Hence the following definition is legitimate.

1.2. Definition

For any finite set A, the *cardinality* $|A|$ *of* A is the (necessarily unique and finite) ordinal α such that $A \approx \alpha$. A *finite cardinal* is an ordinal α such that $|A| = \alpha$ for some finite set A.

1.3. Remarks

(i) Clearly, if A and B are finite sets then $|A| = |B|$ iff $A \approx B$, as required by the incomplete Def. 3.1.3.

(ii) By Def. 1.2, a finite cardinal is a finite ordinal. Conversely, if α is a finite ordinal, then obviously $|\alpha| = \alpha$. Thus the finite cardinals are just the finite ordinals by another name.

(iii) Let n be any natural number. By Def. 3.3.1 and Prob. 4.3.18, the corresponding cardinal, $\boldsymbol{n}$, is finite. This result also follows from the next theorem, in which we calculate these cardinals.

1.4. Theorem

(i) $\mathbf{0} = \varnothing$; $\mathbf{1} = \{\varnothing\}$.
(ii) If α is a finite cardinal, $\alpha + \mathbf{1} = \alpha'$; hence $\alpha + \mathbf{1}$ is a finite cardinal.
(iii) If m is a natural number and $n = m + 1$ then $\boldsymbol{n} = \{\mathbf{0}, \mathbf{1}, \ldots, \boldsymbol{m}\}$.

PROOF

(i) By Def. 3.3.1, $\mathbf{0} = |\varnothing|$ and $\mathbf{1} = |\{\varnothing\}|$. But by Thm. 4.3.3 $\varnothing$ as well as $\varnothing'$ – which by Def. 4.2.26 is $\{\varnothing\}$ – are finite ordinals; hence $|\varnothing| = \varnothing$ and $|\{\varnothing\}| = \{\varnothing\}$.
(ii) Here + is the operation of cardinal addition; so by Def. 3.4.4,

$$\alpha + \mathbf{1} = |A \cup B|,$$

where A and B are any disjoint set such that $|A| = \alpha$ and $|B| = \mathbf{1}$.

As A we take α itself. As B we may then take any set equipollent to $\mathbf{1}$ – that is, any singleton – provided it is disjoint from α. We put $B = \{\alpha\}$, which is disjoint from α because an ordinal cannot belong to itself (see Rem. 4.2.19(ii)). Hence

$$\alpha + \mathbf{1} = |\alpha \cup \{\alpha\}|.$$

But by Def. 4.2.26 this is $|\alpha'|$. Moreover, by Thm. 4.3.3(ii), since α is a finite ordinal so is α'. Hence $\alpha + \mathbf{1} = \alpha'$, which (as we have just noted) is a finite ordinal.
(iii) We proceed by weak mathematical induction on m. For $m = 0$ n is 1 and the required result, $\mathbf{1} = \{\mathbf{0}\}$, follows at once from (i).

Now assume, as induction hypothesis, that m is a number for which (iii) holds. Let $p = (m + 1) + 1 = n + 1$. Then

$$\begin{aligned} \boldsymbol{p} &= \boldsymbol{n} + \mathbf{1} && \text{by Thm. 3.4.6,} \\ &= \boldsymbol{n} \cup \{\boldsymbol{n}\} && \text{by (ii),} \\ &= \{\mathbf{0}, \mathbf{1}, \ldots, \boldsymbol{m}\} \cup \{\boldsymbol{n}\} && \text{by ind. hyp.,} \\ &= \{\mathbf{0}, \mathbf{1}, \ldots, \boldsymbol{n}\}. \end{aligned}$$ ■

1.5. Theorem

For any finite cardinals α and β, $\alpha + \beta$ is a finite cardinal. Moreover, $\alpha + \mathbf{0} = \alpha$ and $\alpha + \beta' = (\alpha + \beta)'$.

PROOF

By Prob. 3.4.7(iii), the equality $\alpha + \mathbf{0} = \alpha$ holds for *all* cardinals α, not just for finite ones.

To prove that $\alpha + \beta$ is a finite cardinal, we apply to β induction on finite ordinals.

For $\beta = \varnothing$, the sum $\alpha + \beta$ is $\alpha + \mathbf{0}$ by Thm. 1.4(i), and we have just seen that this is the finite cardinal α.

Now assume, as induction hypothesis, that β is a finite cardinal such that $\alpha + \beta$ is also a finite cardinal. Then

$$\begin{aligned} \alpha + \beta' &= \alpha + (\beta + \mathbf{1}) && \text{by Thm. 1.4(ii),} \\ &= (\alpha + \beta) + \mathbf{1} && \text{by Prob. 3.4.7(i).} \end{aligned}$$

By our induction hypothesis, $\alpha + \beta$ is a finite cardinal; hence by Thm. 1.4(ii) so is $(\alpha + \beta) + \mathbf{1}$. This shows that $\alpha + \beta'$ is a finite cardinal, and completes the induction on β.

Finally, we have just shown, for any finite cardinals α and β, that $\alpha + \beta' = (\alpha + \beta) + \mathbf{1}$. By Thm. 1.4(ii) this equals $(\alpha + \beta)'$. ■

1.6. Theorem

For any finite cardinals α and β, $\alpha \cdot \beta$ is a finite cardinal. Moreover, $\alpha \cdot \mathbf{0} = \mathbf{0}$ and $\alpha \cdot \beta' = \alpha \cdot \beta + \alpha$.

PROOF

DIY: proceed as in the proof of Thm. 1.5, using Prob. 3.5.5. ■

1.7. Problem

Prove that if $<$ is a [sharp] total order on a finite set A, then $<$ is a well-ordering on A. (Apply induction on finite ordinals to $|A|$. For any non-empty subset B of A you must show that B has a least member. If $B \subset A$, use Lemma 5.2.1. If B is A itself, let a be any member of A and apply the induction hypothesis to $A - \{a\}$.)

1.8. Remark

In 1889, Peano proposed an axiomatization of the theory of natural numbers.[1] In addition to some purely logical axioms (which must be

[1] A translation of his paper, 'The principles of arithmetic, presented by a new method', is in van Heijenoort, *From Frege to Gödel*.

satisfied by any system whatever) he proposed five postulates which we now state, with some inessential modifications.

(1) 0 is a natural number.
(2) Every natural number m has a unique *successor*, $s(m)$.
(3) If m and n are distinct natural numbers, then $s(m) \neq s(n)$.
(4) For every natural number m, $s(m) \neq 0$.
(5) (Principle of Mathematical Induction.) Let K be any set such that K contains 0 and such that if it contains any natural number m then it also contains $s(m)$; then K contains every natural number.

The operations of addition and multiplication of natural numbers can then be introduced by means of four further postulates that assert, for any natural numbers n and m:

(6) $m + 0 = m$.
(7) $m + s(n) = s(m + n)$.
(8) $m \cdot 0 = 0$.
(9) $m \cdot s(n) = m \cdot n + m$.

Intuitively speaking, it is clear that these nine postulates express truths about the system of natural numbers. And in fact they are adequate for an informal axiomatic development of the arithmetic of natural numbers.

Now, speaking more formally, in ZF we have proved for the finite cardinals (a.k.a. finite ordinals) theorems that are exact counterparts of Peano's postulates. To be precise: if in the statement of these postulates we replace the words 'natural number' by 'finite cardinal', and the symbols '0' and 's' respectively by '$\varnothing$' and '$'$' (writing the latter to the right of its argument instead of to its left) and if we understand the symbols for addition and multiplication as denoting respectively addition and multiplication of cardinals, then all nine postulates become theorems of ZF. In this sense, the system consisting of the set ω of finite cardinals together with the operations of succession, addition and multiplication on these cardinals, provides in ZF a *model* for Peano's postulates.

Moreover, this model is *structurally unique* in the following sense. In ZF it is not difficult to prove that *any* system of objects and operations satisfying the appropriate re-interpretation of Peano's postulates must be structurally identical, an exact structural replica of (technically speaking: *isomorphic to*) the system of the finite cardinals.

In this sense, the finite cardinals play within ZF the role of natural numbers. And mathematicians developing (or simulating) various branches of mathematics within set theory are justified in identifying the finite cardinals with the natural numbers, for the purpose of this activity (cf. Rem. 4.3.8).

1.9. Warning

All this does not quite answer the question whether the ZF system of finite cardinals is a faithful and correct representation of the (informal) system of natural numbers, which mathematicians had studied long before the invention of set theory.

Note that for any natural number n, we can prove that the corresponding cardinal $\boldsymbol{n}$ is a finite cardinal (Thm. 1.4, or Def. 3.3.1 and Prob. 4.3.18). But we have *not* proved that

(*) *Every finite cardinal has the form* $\boldsymbol{n}$ *for some natural number* n.

At first sight it seems easy to prove (*) by applying induction on finite ordinals (Cor. 4.3.7) to the 'set'

$$\{\alpha \in \omega : \alpha = \boldsymbol{n} \text{ for some natural number } n\}.$$

But in order to be able to do so, we must first prove that such a set exists as an object of set theory. This, in turn, requires the property *being a natural number*, in terms of which this would-be set is defined, to be a set-theoretic concept (see discussion at the end of § 2 and beginning of § 3 of Ch. 1). But we have taken the notion of *natural number* as given in advance, prior to the development of set theory (cf. Rem. 3.3.6); and without begging the question we cannot presuppose that it is also a set-theoretic notion.

We have no assurance that the ZF system of finite cardinals is a faithful and correct representation of the pre-ZF informal system of natural numbers, so long as the status of (*) is in question. We shall see in the Appendix that this question has a rather surprising answer.

§ 2. Cardinals in general

To extend the definition of cardinality to infinite sets, we invoke AC, via the WOT (Thm. 5.1.5). According to this theorem, every set A is equipollent to some ordinal, and hence by the Least Ordinal Principle (Thm. 4.4.1) there is a unique *least* ordinal to which A is equipollent.

2.1. Definition

For any set A, the *cardinality* $|A|$ *of* A is the least ordinal α such that $A \approx \alpha$. A *cardinal* is an ordinal α such that $|A| = \alpha$ for some set A.

2.2. Remarks

(i) This definition obviously agrees with Def. 1.2 when A is a finite set.

(ii) Def. 2.1 clearly satisfies the condition imposed in Def. 3.1.3: for any sets A and B, $|A| = |B|$ iff $A \approx B$.

(iii) From Def. 2.1 it follows at once that a cardinal is an ordinal that is not equipollent to any smaller ordinal. Conversely, if an ordinal α is not equipollent to any smaller ordinal, then clearly $|\alpha| = \alpha$, so that α is a cardinal.

(iv) If λ and μ are cardinals, then the statement '$\lambda \leqslant \mu$' is apparently ambiguous, because we can interpret '$\leqslant$' according to Def. 4.2.21 (that is, as denoting the order on the class of ordinals) or according to Def. 3.2.1. In the next lemma we shall prove that these two interpretations are in fact equivalent. In the formulation and proof of this lemma we shall use the symbol '$\leqslant$' in the sense of Def. 3.2.1 only, so as not to prejudge the issue. Thereafter, we shall revert to using '$\leqslant$' in either sense, as it will make no difference.

2.3. Lemma

For any cardinals λ and μ, $\lambda \leqslant \mu$ (in the sense of Def. 3.2.1) iff $\lambda \in \mu$ or $\lambda = \mu$.

PROOF

Suppose $\lambda \in \mu$ or $\lambda = \mu$. Since ordinals are transitive sets, it follows that $\lambda \subseteq \mu$. Hence by Thm. 3.2.3 $|\lambda| \leqslant |\mu|$. But λ and μ are cardinals, so $|\lambda| = \lambda$ and $|\mu| = \mu$. Thus $\lambda \leqslant \mu$.

Conversely, suppose that $\lambda \notin \mu$ and $\lambda \neq \mu$. Then, since the class of ordinals is $\in$-well-ordered, we must have $\mu \in \lambda$. In the same way as before, it now follows that $\mu \leqslant \lambda$. Hence we cannot have $\lambda \leqslant \mu$, as by the Schröder–Bernstein Theorem 3.2.7 it would then follow that $\lambda = \mu$, contrary to hypothesis. ■

2.4. Problem

Prove that if α is an infinite ordinal then $|\alpha| = |\alpha'|$; hence α' cannot be a cardinal. (Let f be the map such that $\operatorname{dom} f = \alpha'$, $f\xi = \xi'$ for all finite ξ, $f\xi = \xi$ for all infinite $\xi < \alpha$, and $f\alpha = \varnothing$. Show that f is a bijection from α' to α.)

2.5. Theorem

ω is the least infinite cardinal. ■

2.6. Theorem

If A is a set of cardinals, then $\bigcup A$ is the lub of A in the class of all cardinals, that is, the least cardinal λ such that $\xi \leqslant \lambda$ for all $\xi \in A$.

PROOF

For each $\xi \in A$ we have $\xi \subseteq \bigcup A$ by Def. 1.3.11, hence $|\xi| \leqslant |\bigcup A|$ by Thm. 3.2.3. But ξ is a cardinal, hence $|\xi| = \xi$. Thus $\xi \leqslant |\bigcup A|$ for all $\xi \in A$. This shows that the cardinal $|\bigcup A|$ is an upper bound for A.

Note that $|\bigcup A|$, being a cardinal, is *a fortiori* an ordinal. But by Thm. 4.2.25 $\bigcup A$ itself is the *least* upper bound of A in the class of all ordinals, hence $\bigcup A \leqslant |\bigcup A|$.

On the other hand, from Def. 2.1 it is clear that $|\alpha| \leqslant \alpha$ for every ordinal α. Since by Thm. 4.2.25 $\bigcup A$ is an ordinal, it follows that $|\bigcup A| \leqslant \bigcup A$. Hence $|\bigcup A|$ is $\bigcup A$ itself, and is the lub of A in the class of cardinals. ■

2.7. Theorem

For any set A of cardinals there exists a cardinal (and, in particular, an infinite cardinal) greater than all the members of A.

PROOF

Let λ be the lub of A obtained in Thm. 2.6. By Cantor's Theorem 3.6.8, there exists a cardinal μ such that $\lambda < \mu$, and hence also $\xi < \mu$ for all $\xi \in A$. If μ is infinite, there is nothing further to prove. If μ is finite, then ω is an infinite cardinal such that $\mu < \omega$ and hence also $\xi < \omega$ for all $\xi \in A$. ■

2.8. Corollary

The class of all cardinals is a proper class. ■

2.9. Lemma

We can define a (necessarily unique) function F such that dom $F = W$ *and for every ordinal* α,

$$F\alpha = \textit{the least infinite cardinal not belonging to } \operatorname{ran}(F \restriction \alpha).$$

PROOF

This follows from Thm. 4.6.7 (definition by transfinite recursion). We only need to take as the C of that theorem a function such that whenever x is a set that is also a function, Cx is the least cardinal not belonging to ran x. (Note that ran x is a set by AR, hence by Thm. 2.7 there exists an infinite cardinal not belonging to it; so by the Least Ordinal Principle 4.4.1 there is a *least* such cardinal.) ■

2.10. Definition

For any ordinal α,

$$\aleph_\alpha =_{\text{df}} F\alpha,$$

where F is the function of Lemma 2.9.

2.11. Remarks

(i) '$\aleph$' is *aleph*, the first letter in the Hebrew alphabet. It is also the first letter of the Hebrew word **'אינסוף'** (*einsoph*, meaning infinity), which is a cabbalistic appellation of the deity. The notation is due to Cantor, who was deeply interested in mysticism.

(ii) Combining Def. 2.10 with the characterization of F in Lemma 2.9, we obtain:

$$\aleph_\alpha = \text{the least infinite cardinal not belonging to the set } \{\aleph_\xi : \xi < \alpha\}.$$

2.12. Theorem

(i) *For any* α, $\aleph_\alpha$ *is an infinite cardinal.*
(ii) *For any ordinals* α *and* β, $\alpha < \beta \Rightarrow \aleph_\alpha < \aleph_\beta$.
(iii) $\aleph_0 = \omega$.

PROOF

All three statements follow easily from Rem. 2.11(ii). ■

2.13. Theorem

Every infinite cardinal is $\aleph_\alpha$ *for some ordinal* α.

PROOF

From Thm. 2.12(ii) it follows that $\alpha \neq \beta \Rightarrow \aleph_\alpha \neq \aleph_\beta$. This means that the function F of Lemma 2.9 is a bijection from the class W of all ordinals to the class $\{\aleph_\alpha : \alpha \in W\}$ of all alephs. Since W is a proper class (Cor. 4.2.18), it follows from Prob. 2.4.5 that the class of all alephs must likewise be a proper class.

Now let λ be any infinite cardinal. Then λ, being an ordinal, is a set. Hence there must be some α such that $\aleph_\alpha \notin \lambda$ – otherwise the set λ would include the class of all alephs, and by AS the latter would be a set, contrary to what we have just shown.

Since both λ and $\aleph_\alpha$ are ordinals, the fact that $\aleph_\alpha \notin \lambda$ implies that $\lambda \leqslant \aleph_\alpha$. If $\lambda = \aleph_\alpha$, then there is nothing further to prove. On the other hand, if $\lambda < \aleph_\alpha$ then by Rem. 2.11(ii) it follows that λ belongs to the set $\{\aleph_\xi : \xi < \alpha\}$. Hence $\lambda = \aleph_\xi$ for some $\xi < \alpha$. ■

2.14. Remarks

(i) By Thms. 2.12 and 2.13, the alephs are just the infinite cardinals by another name. Moreover, each infinite cardinal is an $\aleph_\alpha$ for some *unique* ordinal α.
(ii) The theory of real numbers, as other branches of mathematics, can be developed within set theory. In doing so, one identifies the finite cardinals with the natural numbers (see Rem. 1.8). It is then not difficult to show that $\mathsf{P}\aleph_0$ ($= \mathsf{P}\omega$ by Thm. 2.12(iii)) is equipollent to the *continuum* – the set of all real numbers. (It is

also equipollent to the set of all real numbers lying in any given interval, for example, between 0 and 1.) The cardinal $|P\aleph_0|$ is therefore *the cardinality of the continuum*.

Cantor conjectured (but was unable to prove) that $|P\aleph_0| = \aleph_1$. This conjecture is known as the *Continuum Hypothesis (CH)*.

More generally, the *Generalized Continuum Hypothesis (GCH)* is the conjecture that $|P\aleph_\alpha| = \aleph_{\alpha'}$ for every α.

(iii) In 1938 Gödel proved that GCH is consistent relative to the commonly accepted postulates of set theory, in the sense that if they are consistent, then the addition of GCH does not result in inconsistency. In 1963 P. J. Cohen proved that the same holds also for the negation of CH (and hence GCH).

§3. Arithmetic of the alephs

3.1. Preview

In this section we shall present some important results in the arithmetic of the alephs. Some of the proofs are given in a slightly abbreviated form, omitting a few details. We present separately an outline of the proof of Thm. 3.2, although it is a special case of Thm. 3.3. This is done as a dry run, in order to display more clearly, in a simpler context, the idea of the proof.

3.2. Theorem

$\aleph_0 \cdot \aleph_0 = \aleph_0$.

PROOF (OUTLINE)

According to Def. 3.5.1, $\aleph_0 \cdot \aleph_0$ is the cardinality of the set $A \times B$, where A and B are any sets whose cardinality is $\aleph_0$. We shall take both A and B to be $\aleph_0$ itself.

Recall that by Thm. 2.12(iii) $\aleph_0 = \omega$, which is the set of finite ordinals (as well as the set of finite cardinals). Thus we must show that the set $\omega \times \omega$ of all ordered pairs of finite ordinals is equipollent to ω itself.

For any ordinals ξ and η, we let $\max(\xi, \eta)$ be the greater of ξ and η. (If $\xi = \eta$ then $\max(\xi, \eta)$ is equal to both of them.)

We define an order $<$ on the set $\omega \times \omega$ as follows. For any finite ordinals ξ, η, φ and ψ we stipulate that $\langle \xi, \eta \rangle < \langle \varphi, \psi \rangle$ iff one of the following three conditions holds:

(1) $\max(\xi, \eta) < \max(\varphi, \psi)$,
(2) $\max(\xi, \eta) = \max(\varphi, \psi)$ and $\xi < \varphi$,
(3) $\max(\xi, \eta) = \max(\varphi, \psi)$ and $\xi = \varphi$ and $\eta < \psi$.

To make this clearer, here are the first few members of $\omega \times \omega$, listed according to the order $<$:

$$\langle \mathbf{0}, \mathbf{0} \rangle,$$
$$\langle \mathbf{0}, \mathbf{1} \rangle, \langle \mathbf{1}, \mathbf{0} \rangle, \langle \mathbf{1}, \mathbf{1} \rangle,$$
$$\langle \mathbf{0}, \mathbf{2} \rangle, \langle \mathbf{1}, \mathbf{2} \rangle, \langle \mathbf{2}, \mathbf{0} \rangle, \langle \mathbf{2}, \mathbf{1} \rangle, \langle \mathbf{2}, \mathbf{2} \rangle,$$
$$\langle \mathbf{0}, \mathbf{3} \rangle, \langle \mathbf{1}, \mathbf{3} \rangle, \langle \mathbf{2}, \mathbf{3} \rangle, \langle \mathbf{3}, \mathbf{0} \rangle, \langle \mathbf{3}, \mathbf{1} \rangle, \langle \mathbf{3}, \mathbf{2} \rangle, \langle \mathbf{3}, \mathbf{3} \rangle, \ldots.$$

It is not difficult to see that $\omega \times \omega$ with this order on it is similar to ω itself with its $\in$-well-ordering. In particular, $\omega \times \omega$ is equipollent to ω. ■

3.3. *Theorem*

$\aleph_\alpha \cdot \aleph_\alpha = \aleph_\alpha$ *for any ordinal* α.

PROOF

We proceed by transfinite induction. As induction hypothesis we assume that $\aleph_\beta \cdot \aleph_\beta = \aleph_\beta$ for all $\beta < \alpha$.

As in the proof of Thm. 3.2, we define an order $<$ on $\aleph_\alpha \times \aleph_\alpha$ by stipulating, for any ordinals ξ, η, φ and ψ smaller than $\aleph_\alpha$, that $\langle \xi, \eta \rangle < \langle \varphi, \psi \rangle$ iff one of the conditions (1), (2) and (3) listed there holds.

It is easy to verify that $<$ is a well-ordering on $\aleph_\alpha \times \aleph_\alpha$. Hence, by the Representation Theorem 4.5.12, there exists a similarity map f from $\aleph_\alpha \times \aleph_\alpha$ to an ordinal δ. Since F is a bijection from $\aleph_\alpha \times \aleph_\alpha$ to δ, it follows that $\aleph_\alpha \cdot \aleph_\alpha = |\delta|$. We shall show that this δ is in fact $\aleph_\alpha$ itself.

First, note that $\aleph_\alpha = \mathbf{1} \cdot \aleph_\alpha \leqslant \aleph_\alpha \cdot \aleph_\alpha = |\delta| \leqslant \delta$. Now suppose that $\aleph_\alpha < \delta$. This means that $\aleph_\alpha \in \delta = \operatorname{ran} f$; so for some ξ and η, both smaller than $\aleph_\alpha$, we have $f\langle \xi, \eta \rangle = \aleph_\alpha$.

Since ξ and η are smaller than $\aleph_\alpha$, their cardinalities are certainly smaller than $\aleph_\alpha$. Let $\zeta = \max(\xi, \eta)$. Then $|\zeta|$ is either a finite cardinal or some $\aleph_\beta$ such that $\beta < \alpha$.

Let us put $A = \{\langle \varphi, \psi \rangle : \langle \varphi, \psi \rangle < \langle \xi, \eta \rangle\}$. Then, by the definition of $<$, for each $\langle \varphi, \psi \rangle \in A$ we must have $\varphi \leqslant \zeta$ and $\psi \leqslant \zeta$. Therefore A is a subset of $\zeta' \times \zeta'$, hence $|A| \leqslant |\zeta'| \cdot |\zeta'|$.

If ζ is finite then ζ' is finite as well and hence, by Thm. 1.6, so is $|A|$.

If $|\zeta| = \aleph_\beta$ for some $\beta < \alpha$, then by Prob. 2.4 $|\zeta'| = \aleph_\beta$ as well, so by the induction hypothesis $|A| \leqslant \aleph_\beta$. Thus in any case $|A|$ is smaller than $\aleph_\alpha$.

However, since $f\langle \xi, \eta \rangle = \aleph_\alpha$, it follows that $f \restriction A$ is a bijection from A to $\aleph_\alpha$ and hence $|A| = \aleph_\alpha$ – contrary to what we have just shown. This contradiction shows that δ must be equal to $\aleph_\alpha$. ■

3.4. Remark

In view of Thm. 2.13, Thm. 3.3 means simply that $\lambda\lambda = \lambda$ for any infinite cardinal λ.

3.5. Theorem

If μ is an infinite cardinal and λ is any cardinal such that $\mathbf{1} \leqslant \lambda \leqslant \mu$, then $\lambda\mu = \mu$.

PROOF

Using Prob. 3.5.5 and Thm. 3.3, we have:

$$\mu = \mathbf{1}\mu \leqslant \lambda\mu \leqslant \mu\mu = \mu.$$

Thus both $\mu \leqslant \lambda\mu$ and $\lambda\mu \leqslant \mu$. ■

3.6. Theorem

If μ is an infinite cardinal and λ is any cardinal such that $\lambda \leqslant \mu$, then $\lambda + \mu = \mu$.

PROOF

Using Probs. 3.4.7 and 3.5.5 and Thm. 3.3, we have:

$$\mu = \mathbf{0} + \mu \leqslant \lambda + \mu \leqslant \mu + \mu = \mathbf{1}\mu + \mathbf{1}\mu = (\mathbf{1} + \mathbf{1})\mu = \mathbf{2}\mu \leqslant \mu\mu = \mu.$$

Thus both $\mu \leqslant \lambda + \mu$ and $\lambda + \mu \leqslant \mu$. ■

3.7. Theorem

If λ is an infinite cardinal and α is any finite cardinal other than $\mathbf{0}$, then $\lambda^\alpha = \lambda$.

PROOF

DIY, using induction on the finite ordinal α. ■

3.8. Definition

Let A be a class. A map from an ordinal α to A is called an *A-string of length* α. A map from a finite ordinal to A is called a *finite A-string*.

3.9. Theorem

Let A be an infinite set and let S be the class of all finite A-strings. Then S is a set and $|S| = |A|$.

PROOF

If α is a finite ordinal then $\alpha \subseteq \omega$. Hence every finite A-string is a subset of $\omega \times A$. It follows that $S \subseteq \mathsf{P}(\omega \times A)$; so S is a set by AP and AS.

For each finite ordinal α, consider the set $S_\alpha = \mathrm{map}(\alpha, A)$ of all A-strings of length α (see Def. 3.6.1 and Rem. 3.6.2). Clearly, the S_α are pairwise disjoint and

$$S = \bigcup\{S_\alpha : \alpha < \omega\}.$$

Hence it is easy to see that

$$(*) \qquad |S| = \sum\{|S_\alpha| \mid \alpha < \omega\}.$$

Let $|A| = \lambda$. Since $S_\alpha = \mathrm{map}(\alpha, A)$, it follows from Def. 3.6.3 that

$$|S_\alpha| = \lambda^\alpha,$$

which by Thm. 3.7 is equal to λ itself, except when $\alpha = \mathbf{0}$, in which case $\lambda^\alpha = \mathbf{1}$. Therefore by $(*)$ $|S|$ is the sum of $\mathbf{1}$ and $\aleph_0$ times λ:

$$|S| = \mathbf{1} + \aleph_0 \cdot \lambda.$$

Since $\aleph_0$ is the *least* infinite cardinal (Thms. 2.5 and 2.12(iii)), it follows that $\aleph_0 \leqslant \lambda$. Hence by Thms. 3.5 and 3.6 $|S| = \mathbf{1} + \lambda = \lambda$. ■

7

Propositional logic

§ 1. Basic syntax

We shall describe a formal language $\mathcal{L}$. This will be our *object language*, an object of our discussion. It must be distinguished from our *metalanguage*, the language *in which* the discussion is conducted: ordinary English augmented by a special technical vocabulary.

1.1. Specification

The *primitive symbols* of $\mathcal{L}$ fall into two mutually exclusive categories:

(i) an infinite set of *propositional symbols*;
(ii) two distinct *connectives*, $\neg$ and $\rightarrow$, called *negation symbol* and *implication symbol* respectively.

1.2. Warning

The statement just made does *not* mean that, for example, the implication symbol of $\mathcal{L}$ is a boldface arrow-shaped figure. (In fact, for all we care $\mathcal{L}$ may not have a written form at all!) Rather, the boldface arrow is a *syntactic constant*, a symbol in our *metalanguage*, used as a name for the implication symbol of $\mathcal{L}$.

1.3. Definition

If l is a natural number and $\mathbf{s}_1, \mathbf{s}_2, \ldots, \mathbf{s}_l$ are primitive symbols of $\mathcal{L}$, not necessarily distinct, then the concatenation $\mathbf{s}_1\mathbf{s}_2 \ldots \mathbf{s}_l$ is called an *$\mathcal{L}$-string* and the number l is called its *length*. (More formally, an $\mathcal{L}$-string of length l can be defined as map from the set $\{1, 2, \ldots, l\}$ to the set of primitive symbols of $\mathscr{L}$.) In particular, the empty $\mathscr{L}$-string has length 0.

We shall usually omit the prefix '$\mathscr{L}$-', and say simply 'string' rather than '$\mathscr{L}$-string'. Similar ellipses will be used, when there is no risk of confusion, in connection with other bits of terminology introduced later on.

1.4. Definition

$\mathscr{L}$-formulas are strings constructed according to the following three rules.

(1) A string consisting of a single occurrence of a propositional symbol is an *$\mathscr{L}$-formula*.
(2) If $\boldsymbol{\beta}$ is an *$\mathscr{L}$-formula* then $\neg\boldsymbol{\beta}$ (the string obtained by concatenating a single occurrence of $\neg$ and the string $\boldsymbol{\beta}$, in this order) is an *$\mathscr{L}$-formula*.
(3) If $\boldsymbol{\beta}$ and $\boldsymbol{\gamma}$ are *$\mathscr{L}$-formulas* then $\rightarrow\boldsymbol{\beta}\boldsymbol{\gamma}$ (the string obtained by concatenating a single occurrence of $\rightarrow$, the string $\boldsymbol{\beta}$ and the string $\boldsymbol{\gamma}$, in this order) is an *$\mathscr{L}$-formula*.

A formula constructed according to (1) – a single occurrence of a propositional symbol – is called a *prime* formula.

A formula constructed according to (2) is called a *negation* formula; here $\neg\boldsymbol{\beta}$ is the negation *of* $\boldsymbol{\beta}$.

A formula constructed according to (3) is called an *implication* formula; here $\boldsymbol{\beta}$ is the *antecedent* and $\boldsymbol{\gamma}$ the *consequent of* $\rightarrow\boldsymbol{\beta}\boldsymbol{\gamma}$.

1.5. Warnings

(i) In some books, particularly older ones, what we call 'strings' are referred to as 'formulas', whereas what we call 'formulas' are referred to as 'well-formed formulas' ('wffs').
(ii) Def. 1.4 does *not* mean that boldface lower-case Greek letters are $\mathscr{L}$-formulas. Rather, they are *syntactic variables*, symbols in our *metalanguage* used to range over $\mathscr{L}$-formulas.

1.6. Definition

A propositional symbol occurring in a formula $\boldsymbol{\alpha}$ is called a *prime component* of $\boldsymbol{\alpha}$.

1.7. Definition

The *degree of complexity* of a formula $\boldsymbol{\alpha}$ – briefly, $\deg \boldsymbol{\alpha}$ – is the total number of occurrences of connectives ($\neg$ and $\rightarrow$) in $\boldsymbol{\alpha}$.

1.8. Remark

We shall often wish to prove that all formulas $\boldsymbol{\alpha}$ have some property P – briefly, $\forall \boldsymbol{\alpha} P \boldsymbol{\alpha}$. This may be done by *[strong] induction on* $\deg \boldsymbol{\alpha}$, as follows. Define a property Q of natural numbers by stipulating that Q holds for a given number n iff P holds for all formulas $\boldsymbol{\alpha}$ such that $\deg \boldsymbol{\alpha} = n$. Then clearly $\forall \boldsymbol{\alpha} P \boldsymbol{\alpha}$ is equivalent to $\forall n Q n$. As we know (see § 3 of Ch. 0), to prove $\forall n Q n$ by strong induction we deduce Qn (for arbitrary n) from the induction hypothesis $\forall m < n Q m$.

Stated in terms of P rather than Q, this is tantamount to saying: if we deduce $P\boldsymbol{\alpha}$ (for arbitrary $\boldsymbol{\alpha}$) from the induction hypothesis that $P\boldsymbol{\beta}$ holds for all formulas $\boldsymbol{\beta}$ such that $\deg \boldsymbol{\beta} < \deg \boldsymbol{\alpha}$, then it follows that $\forall \boldsymbol{\alpha} P \boldsymbol{\alpha}$.

1.9. Problem

Assign to each primitive symbol $\mathbf{s}$ of $\mathcal{L}$ a *weight* $\mathrm{w}(\mathbf{s})$ by stipulating: if $\mathbf{s}$ is a propositional symbol then $\mathrm{w}(\mathbf{s}) = -1$, while $\mathrm{w}(\neg) = 0$ and $\mathrm{w}(\rightarrow) = 1$. If $\mathbf{s}_1, \mathbf{s}_2, \ldots, \mathbf{s}_l$ are primitive symbols, we assign to the string $\mathbf{s}_1\mathbf{s}_2 \ldots \mathbf{s}_l$ *weight*

$$\mathrm{w}(\mathbf{s}_1\mathbf{s}_2 \ldots \mathbf{s}_l) = \mathrm{w}(\mathbf{s}_1) + \mathrm{w}(\mathbf{s}_2) + \cdots + \mathrm{w}(\mathbf{s}_l).$$

Thus, the weight of a string is the sum obtained by adding -1 for each occurrence of a propositional symbol and $+1$ for each occurrence of $\rightarrow$ in the string (occurrences of $\neg$ make no contribution to the weight). Since a formula is also a string, every formula $\boldsymbol{\alpha}$ has now been assigned a weight $\mathrm{w}(\boldsymbol{\alpha})$. Show that, for any formula $\boldsymbol{\alpha}$,

(i) $\mathrm{w}(\boldsymbol{\alpha}) = -1$;
(ii) if $\boldsymbol{\alpha}$ is the string $\mathbf{s}_1\mathbf{s}_2 \ldots \mathbf{s}_l$ and $k < l$, then $\mathrm{w}(\mathbf{s}_1\mathbf{s}_2 \ldots \mathbf{s}_k) \geqslant 0$.

In other words, (ii) states that any string which is a *proper initial segment* of $\boldsymbol{\alpha}$ (an initial part of $\boldsymbol{\alpha}$ short of the whole of $\boldsymbol{\alpha}$) has non-negative weight. (Prove (i) and (ii) by strong induction on $\deg \boldsymbol{\alpha}$.)

(iii) Show that if $\boldsymbol{\alpha}$ is an implication formula, $\boldsymbol{\alpha} = \rightarrow\boldsymbol{\beta}\boldsymbol{\gamma}$, then $\rightarrow\boldsymbol{\beta}$ is the shortest non-empty initial segment of $\boldsymbol{\alpha}$ whose weight is 0.

§ 2. Notational conventions

In Def. 1.4 we stipulated that in forming an implication formula an implication symbol is placed *before* the antecedent. The advantage of this so-called *Polish notation* (invented by the Polish logician Jan Łukasiewicz) is that $\mathcal{L}$ has no need for brackets or other punctuation marks for indicating grouping of symbols. Thus, in an implication formula (a formula whose initial symbol is $\rightarrow$) the antecedent and the consequent are uniquely determined (see Prob. 1.9). This economy is both elegant and technically useful.

So far we have mimicked this Polish system also in our metalanguage: thus in '$\boldsymbol{\rightarrow\beta\gamma}$' the boldface arrow is placed to the left. However, in practice this metalinguistic notation is difficult to read, partly because it does not conform to common usage. The Polish notation in $\mathcal{L}$ itself causes us no inconvenience, because we do not actually *use* that language, only talk *about* it. But in our metalanguage, which we do use continually, we shall trade off elegance for legibility and conformity to common usage.

2.1. Definition

$(\boldsymbol{\alpha\rightarrow\beta}) =_{\text{df}} \boldsymbol{\rightarrow\alpha\beta}$.

This definition changes nothing in $\mathcal{L}$; as far as $\mathcal{L}$ is concerned Def. 1.4 remains in force. The change is purely in the metalanguage: our metalinguistic notation will no longer mimic the structure of $\mathcal{L}$-formulas, because we shall write '$(\boldsymbol{\alpha\rightarrow\beta})$' instead of '$\boldsymbol{\rightarrow\alpha\beta}$'. For the sake of easier legibility, we use parentheses and brackets of various styles and sizes. In this context, we refer to all of them simply as *brackets*. The brackets are now needed to prevent ambiguity. For example,

$$[(\boldsymbol{\alpha\rightarrow\beta})\boldsymbol{\rightarrow\gamma}] = \boldsymbol{\rightarrow\rightarrow\alpha\beta\gamma}, \text{ but } [\boldsymbol{\alpha\rightarrow}(\boldsymbol{\beta\rightarrow\gamma})] = \boldsymbol{\rightarrow\alpha\rightarrow\beta\gamma}.$$

Here the new notation (introduced in Def. 2.1) is used on the left-hand side, while the old notation for the same formulas is used on the right-hand side.

We now hit a new snag: in long metalinguistic expressions of this kind, written in the new style, the proliferation of brackets can hinder legibility. We therefore abbreviate such expressions by omitting as many pairs of brackets as convenient. Of course, in order to prevent ambiguity such omissions must be governed by certain rules, so that the brackets can be restored to yield a unique unabbreviated expression. We shall need three such rules. The first rule is very simple:

2.2. Rule (Omission of outermost brackets)

A pair of brackets such that no part of the expression lies outside it may be omitted.

For example, $(\alpha\to\beta)\to\gamma = [(\alpha\to\beta)\to\gamma]$ and $\alpha\to(\beta\to\gamma) = [\alpha\to(\beta\to\gamma)]$.

The second rule is easier to formulate as a rule about how to restore omitted brackets. (So a pair of brackets may be omitted if it could then be restored according to this rule.)

2.3. Rule (Association to the right)

If there are two or more occurrences of '→' all enclosed in exactly the same pairs of brackets (or all not enclosed in any brackets) then you may add a new pair of brackets that enclose only the rightmost of these occurrences.

For example,

$$\alpha\to\beta\to\gamma\to\delta = \alpha\to\beta\to(\gamma\to\delta) = \alpha\to[\beta\to(\gamma\to\delta)] = \{\alpha\to[\beta\to(\gamma\to\delta)]\},$$

$$(\alpha\to\beta\to\gamma)\to\delta = [\alpha\to(\beta\to\gamma)]\to\delta = \{[\alpha\to(\beta\to\gamma)]\to\delta\},$$

$$\alpha\to(\beta\to\gamma)\to\delta = \alpha\to[(\beta\to\gamma)\to\delta] = \{\alpha\to[(\beta\to\gamma)\to\delta]\},$$

$$(\alpha\to\beta)\to\gamma\to\delta = (\alpha\to\beta)\to(\gamma\to\delta) = [(\alpha\to\beta)\to(\gamma\to\delta)],$$

$$[(\alpha\to\beta)\to\gamma]\to\delta = \{[(\alpha\to\beta)\to\gamma]\to\delta\}.$$

The third rule is

2.4. Rule (Adhesion of '¬')

Do not omit a pair of brackets whose left member is immediately preceded by an occurrence of '¬'. Equivalently: *In restoring brackets, do not add a new pair of brackets whose left member immediately follows an occurrence of '¬'.*

For example, $\alpha\to\neg\beta\to\gamma = [\alpha\to(\neg\beta\to\gamma)]$ but $\alpha\to\neg(\beta\to\gamma) = [\alpha\to\neg(\beta\to\gamma)]$.

For reasons of economy, we allowed $\mathcal{L}$ to have only two connectives, ¬ and →. Other connectives can however be introduced *metalinguistically*, by definition.

2.5. Definition

(i) $(\alpha \wedge \beta) =_{df} \neg(\alpha \rightarrow \neg\beta)$,
(ii) $(\alpha \vee \beta) =_{df} \neg\alpha \rightarrow \beta$,
(iii) $(\alpha \leftrightarrow \beta) =_{df} (\alpha \rightarrow \beta) \wedge (\beta \rightarrow \alpha)$.

$(\alpha \wedge \beta)$ is called a *conjunction formula* and α and β its *first conjunct* and *second conjunct* respectively; $(\alpha \vee \beta)$ is called a *disjunction formula* and α and β its *first disjunct* and *second disjunct* respectively; $(\alpha \leftrightarrow \beta)$ is called a *bi-implication formula* and α and β its *left-hand side* and *right-hand side* respectively.

2.6. Warning

The metalinguistic symbol '$\wedge$' does not denote anything; strictly speaking it has no meaning on its own – only the *package* '$(\alpha \wedge \beta)$' *as a whole* has been defined as an abbreviation for '$\neg(\alpha \rightarrow \neg\beta)$'. This is an example of a *contextual* definition. Similar remarks apply to the other two clauses of Def. 2.5.

In view of Def. 2.5 we need to modify our procedure for omitting and restoring brackets in metalinguistic expressions. We leave Rules 2.2 and 2.4 as they are, but we replace Rule 2.3 by the following more comprehensive rule for restoring brackets, which takes into account not only '$\rightarrow$' but also the newly introduced metalinguistic symbols '$\wedge$', '$\vee$' and '$\leftrightarrow$'.

2.7. Rule (Ranks and association to the right)

If there are occurrences of '$\leftrightarrow$', '$\rightarrow$', '$\vee$' and '$\wedge$' – at least two occurrences in total – all enclosed in exactly the same pairs of brackets (or all not enclosed in any pair of brackets), order all these occurrences by rank *as follows. Occurrences of '$\leftrightarrow$' have higher* ranks *than those of '$\rightarrow$'; the latter have higher* ranks *than those of '$\vee$'; and occurrences of '$\wedge$' have lowest* ranks. *Moreover, of two occurrences of the same symbol, the one further to the left has the higher* rank. *Then you may add a new pair of brackets that encloses only the symbol-occurrence with the lowest rank.*

For example,

$$\alpha \rightarrow \beta \wedge \gamma \rightarrow \beta \rightarrow \gamma = \alpha \rightarrow (\beta \wedge \gamma) \rightarrow \beta \rightarrow \gamma = \alpha \rightarrow (\beta \wedge \gamma) \rightarrow (\beta \rightarrow \gamma)$$
$$= \alpha \rightarrow [(\beta \wedge \gamma) \rightarrow (\beta \rightarrow \gamma)] = \{\alpha \rightarrow [(\beta \wedge \gamma) \rightarrow (\beta \rightarrow \gamma)]\};$$

$$\begin{aligned}\boldsymbol{\alpha}\wedge\boldsymbol{\beta}\rightarrow\boldsymbol{\gamma}\leftrightarrow\neg\boldsymbol{\alpha}\rightarrow\boldsymbol{\beta}\vee\boldsymbol{\gamma} &= (\boldsymbol{\alpha}\wedge\boldsymbol{\beta})\rightarrow\boldsymbol{\gamma}\leftrightarrow\neg\boldsymbol{\alpha}\rightarrow\boldsymbol{\beta}\vee\boldsymbol{\gamma}\\ &= (\boldsymbol{\alpha}\wedge\boldsymbol{\beta})\rightarrow\boldsymbol{\gamma}\leftrightarrow\neg\boldsymbol{\alpha}\rightarrow(\boldsymbol{\beta}\vee\boldsymbol{\gamma})\\ &= (\boldsymbol{\alpha}\wedge\boldsymbol{\beta})\rightarrow\boldsymbol{\gamma}\leftrightarrow[\neg\boldsymbol{\alpha}\rightarrow(\boldsymbol{\beta}\vee\boldsymbol{\gamma})] = [(\boldsymbol{\alpha}\wedge\boldsymbol{\beta})\rightarrow\boldsymbol{\gamma}]\leftrightarrow[\neg\boldsymbol{\alpha}\rightarrow(\boldsymbol{\beta}\vee\boldsymbol{\gamma})]\\ &= \{[(\boldsymbol{\alpha}\wedge\boldsymbol{\beta})\rightarrow\boldsymbol{\gamma}]\leftrightarrow[\neg\boldsymbol{\alpha}\rightarrow(\boldsymbol{\beta}\vee\boldsymbol{\gamma})]\}.\end{aligned}$$

The idea behind Rule 2.7 is that – in the absence of brackets that indicate otherwise – a symbol-occurrence of higher rank *separates more strongly* than one of lower rank, in much the same way as in English punctuation a full stop separates more strongly than a semicolon, and the latter separates more strongly than a comma.

It must be stressed that the definitions and conventions introduced in this section are metalinguistic devices used in discussing $\mathscr{L}$ and do not change $\mathscr{L}$ itself in any way.

§3. Propositional combinations

A formula $\boldsymbol{\alpha}$ is said to be a *propositional combination* of k formulas $\boldsymbol{\beta}_1$, $\boldsymbol{\beta}_2, \ldots, \boldsymbol{\beta}_k$, if $\boldsymbol{\alpha}$ can be constructed from the $\boldsymbol{\beta}_i$ using $\neg$ and $\rightarrow$. The following definition puts this more precisely.

3.1. Definition

Let $\boldsymbol{\beta}_1, \boldsymbol{\beta}_2, \ldots, \boldsymbol{\beta}_k$ be any formulas. A *propositional combination of* $\boldsymbol{\beta}_1, \boldsymbol{\beta}_2, \ldots, \boldsymbol{\beta}_k$ is any formula constructed according to the following three rules.

(1) Each $\boldsymbol{\beta}_i$ (where $1 \leqslant i \leqslant k$) is a *propositional combination* of $\boldsymbol{\beta}_1, \boldsymbol{\beta}_2, \ldots, \boldsymbol{\beta}_k$.
(2) If $\boldsymbol{\gamma}$ is a *propositional combination* of $\boldsymbol{\beta}_1, \boldsymbol{\beta}_2, \ldots, \boldsymbol{\beta}_k$, then $\neg\boldsymbol{\gamma}$ is a *propositional combination* of $\boldsymbol{\beta}_1, \boldsymbol{\beta}_2, \ldots, \boldsymbol{\beta}_k$.
(3) If $\boldsymbol{\gamma}$ and $\boldsymbol{\delta}$ are *propositional combinations* of $\boldsymbol{\beta}_1, \boldsymbol{\beta}_2, \ldots, \boldsymbol{\beta}_k$, then $\boldsymbol{\gamma}\rightarrow\boldsymbol{\delta}$ is a *propositional combination* of $\boldsymbol{\beta}_1, \boldsymbol{\beta}_2, \ldots, \boldsymbol{\beta}_k$.

For brevity, we shall usually say 'combination', omitting the adjective 'propositional'.

3.2. Warnings

(i) In forming a combination of $\boldsymbol{\beta}_1, \boldsymbol{\beta}_2, \ldots, \boldsymbol{\beta}_k$, not all the $\boldsymbol{\beta}_i$ need actually be used. For example, according to Def. 3.1, both $\boldsymbol{\beta}_2$ and $\boldsymbol{\beta}_1\rightarrow\boldsymbol{\beta}_2$ are combinations of $\boldsymbol{\beta}_1, \boldsymbol{\beta}_2, \boldsymbol{\beta}_3$.

(ii) The $\boldsymbol{\beta}_i$ need not be mutually independent: for example, one of them may be a combination of the others. (Indeed, the $\boldsymbol{\beta}_i$ need not be distinct: some of them may coincide with each other.) For this reason one and the same formula may be obtainable from the $\boldsymbol{\beta}_i$ in more than one way. For example, if $\boldsymbol{\beta}_3 = \neg\boldsymbol{\beta}_1$ then $\neg\boldsymbol{\beta}_1 \rightarrow \boldsymbol{\beta}_2$, obtained from $\boldsymbol{\beta}_1$, $\boldsymbol{\beta}_2$, $\boldsymbol{\beta}_3$ by using clause (1) of Def. 3.1 twice, then clause (2) and clause (3), is the same formula as $\boldsymbol{\beta}_3 \rightarrow \boldsymbol{\beta}_2$, which can be obtained from $\boldsymbol{\beta}_1$, $\boldsymbol{\beta}_2$, $\boldsymbol{\beta}_3$ without using clause (2) of Def. 3.1.

It is clear that every formula is a combination of its prime components (see Def. 1.6). The following problem goes a bit further.

3.3. *Problem*

Let $\boldsymbol{\beta}_1, \boldsymbol{\beta}_2, \ldots, \boldsymbol{\beta}_k$ be distinct prime formulas, among which are all the prime components of a formula $\boldsymbol{\alpha}$. Prove that $\boldsymbol{\alpha}$ can be obtained as a combination of $\boldsymbol{\beta}_1, \boldsymbol{\beta}_2, \ldots, \boldsymbol{\beta}_k$ in exactly one way. (Use induction on $\deg \boldsymbol{\alpha}$, distinguishing three cases corresponding to the three clauses of Def. 1.4.)

§4. Basic semantics

In classical two-valued logic – which is what we are studying here – we admit two distinct *truth values*, namely *truth* and *untruth* (a.k.a. *falsehood*). For brevity, we shall denote them by '$\top$' and '$\bot$' respectively.

4.1. *Remark*

From a purely technical point of view, it does not matter what the truth values $\top$ and $\bot$ are, so long as they are two distinct objects. But intuitively it is best to think of them as abstract entities standing outside the language $\mathcal{L}$.

4.2. *Definition*

(i) A *truth valuation on* $\mathcal{L}$ is a mapping σ from the set of all prime $\mathcal{L}$-formulas to the set $\{\top, \bot\}$ of truth values. For any truth valuation σ and any prime formula $\boldsymbol{\alpha}$ we denote by '$\boldsymbol{\alpha}^\sigma$' the truth value assigned by σ to $\boldsymbol{\alpha}$.

(ii) Given a truth valuation σ, we now extend the definition of $\boldsymbol{\alpha}^\sigma$, the truth value assigned by σ to $\boldsymbol{\alpha}$, to cover *every* $\mathcal{L}$-formula $\boldsymbol{\alpha}$. We proceed by induction on $\deg \boldsymbol{\alpha}$, defining $\boldsymbol{\alpha}^\sigma$ in terms of the truth values assigned by σ to formulas whose degrees are smaller than that of $\boldsymbol{\alpha}$. We distinguish three cases, corresponding to the three clauses of Def. 1.4.

(1) If $\boldsymbol{\alpha}$ is a prime formula, then $\boldsymbol{\alpha}^\sigma$ is already defined.

(2) $(\neg\boldsymbol{\beta})^\sigma = \begin{cases} \bot & \text{if } \boldsymbol{\beta}^\sigma = \top, \\ \top & \text{if } \boldsymbol{\beta}^\sigma = \bot. \end{cases}$

(3) $(\boldsymbol{\beta}\rightarrow\boldsymbol{\gamma})^\sigma = \begin{cases} \bot & \text{if } \boldsymbol{\beta}^\sigma = \top \text{ and } \boldsymbol{\gamma}^\sigma = \bot, \\ \top & \text{otherwise.} \end{cases}$

(iii) Let $\boldsymbol{\alpha}$ be a formula and σ a truth valuation. If $\boldsymbol{\alpha}^\sigma = \top$ we say that $\boldsymbol{\alpha}$ is *true under* σ, whereas if $\boldsymbol{\alpha}^\sigma = \bot$ we say that $\boldsymbol{\alpha}$ is *untrue* (or *false*) *under* σ.

4.3. Remarks

(i) Strictly speaking, in Def. 4.2(ii) we defined a *new* mapping, which *extends* σ: whereas $\operatorname{dom}\sigma$ is the set of *prime* formulas, the domain of the new mapping is the set of *all* formulas, but it agrees with σ on prime formulas. Sacrificing absolute rigour to convenience, we denote by 'σ' this extension as well as the original mapping itself.

(ii) Note that $\boldsymbol{\alpha}^\sigma$ is a truth value rather than an expression in $\mathcal{L}$. (Of course, both '$\boldsymbol{\alpha}$' and '$\boldsymbol{\alpha}^\sigma$' are expressions in our metalanguage.)

4.4. Definition

(i) If $\boldsymbol{\varphi}$ is a formula and σ is a truth valuation such that $\boldsymbol{\varphi}^\sigma = \top$, we say that σ *satisfies* $\boldsymbol{\varphi}$ and write '$\sigma \models \boldsymbol{\varphi}$'.

(ii) If σ is a truth valuation that satisfies every member of a set $\boldsymbol{\Phi}$ of formulas, we say that σ *satisfies* $\boldsymbol{\Phi}$ and write '$\sigma \models \boldsymbol{\Phi}$'.

(iii) If a formula $\boldsymbol{\alpha}$ is satisfied by every truth valuation, we say that $\boldsymbol{\alpha}$ is a *tautology* and write '$\models_0 \boldsymbol{\alpha}$'.

(iv) If $\boldsymbol{\Phi}$ is a set of formulas and $\boldsymbol{\alpha}$ is a formula such that every truth valuation satisfying $\boldsymbol{\Phi}$ also satisfies $\boldsymbol{\alpha}$, we say that $\boldsymbol{\alpha}$ is a *tautological consequence of* $\boldsymbol{\Phi}$ and write '$\boldsymbol{\Phi} \models_0 \boldsymbol{\alpha}$'.

(v) If a set $\boldsymbol{\Phi}$ of formulas is not satisfied by any truth valuation, we say that $\boldsymbol{\Phi}$ is *[propositionally] unsatisfiable* and write '$\boldsymbol{\Phi} \models_0$'.

4.5. Remarks

(i) According to Def. 4.4(ii), a truth valuation σ *fails* to satisfy a set $\mathbf{\Phi}$ of formulas, iff $\mathbf{\Phi}$ has a member that fails to be satisfied by σ. Therefore if σ is any truth valuation, then $\sigma \models \varnothing$. Indeed, $\varnothing$ does not have a member that fails to be satisfied by σ, because it has no members at all.

(ii) By Def. 4.4(iv), $\varnothing \models_0 \boldsymbol{\alpha}$ means that every truth valuation satisfies $\boldsymbol{\alpha}$ (because, as we have just seen, every truth valuation satisfies the empty set $\varnothing$); by Def. 4.4(iii) this means that $\boldsymbol{\alpha}$ is a tautology. Thus, a formula is a tautology iff it is a tautological consequence of the empty set.

(iii) In connection with '$\models_0$' we employ certain notational simplifications that ought to be self-explanatory. Thus, for example, we write '$\mathbf{\Phi}, \boldsymbol{\alpha} \models_0 \boldsymbol{\beta}$' instead of '$\mathbf{\Phi} \cup \{\boldsymbol{\alpha}\} \models_0 \boldsymbol{\beta}$'.

4.6. Problem

(i) For any set $\mathbf{\Phi}$ of formulas and any two formulas $\boldsymbol{\alpha}$ and $\boldsymbol{\beta}$, prove that $\mathbf{\Phi}, \boldsymbol{\alpha} \models_0 \boldsymbol{\beta}$ iff $\mathbf{\Phi} \models_0 \boldsymbol{\alpha} \to \boldsymbol{\beta}$.

(ii) Prove that $\{\boldsymbol{\alpha}_1, \boldsymbol{\alpha}_2, \ldots, \boldsymbol{\alpha}_k\} \models_0 \boldsymbol{\beta}$ iff $\models_0 \boldsymbol{\alpha}_1 \to \boldsymbol{\alpha}_2 \to \cdots \to \boldsymbol{\alpha}_k \to \boldsymbol{\beta}$.

4.7. Warning

Never, never get $\to$ and $\models_0$ confused with each other. (I was *not* referring just now to the *symbols* '$\to$' and '$\models_0$'. You are not likely to get *them* confused, because you can *see* they are different: the former is a boldface arrow-shaped figure, while the latter is shaped like a double-barred turnstile with a little ring on its lower right-hand side. Rather, I was referring to what these symbols *denote*.) Much can be written about this, but the following should help you to avoid the most common errors.

Suppose $\boldsymbol{\alpha}$ and $\boldsymbol{\beta}$ are $\mathcal{L}$-formulas. Then $\boldsymbol{\alpha} \to \boldsymbol{\beta}$ is another such formula. '$\boldsymbol{\alpha} \to \boldsymbol{\beta}$' is a *nominal phrase*: if you write it on its own, you would not be making any statement, but only referring to that formula – just as when I say 'my income-tax statement' and no more I am not making a statement but merely referring to my income-tax statement.[1]

[1] We must exclude here cases of ellipsis, such as when, in reply to the question 'What were you doing last night?', I say 'My income-tax statement.' as an ellipsis for the sentence 'I was doing my income-tax statement.'

On the other hand, if you write '$\boldsymbol{\alpha} \vDash_0 \boldsymbol{\beta}$' on its own, you would be stating that $\boldsymbol{\beta}$ is a tautological consequence of $\boldsymbol{\alpha}$ (or, more precisely, of the singleton $\{\boldsymbol{\alpha}\}$); and if you write '$\vDash_0 \boldsymbol{\alpha} \rightarrow \boldsymbol{\beta}$' on its own, you would be stating that the implication formula $\boldsymbol{\alpha} \rightarrow \boldsymbol{\beta}$ is a tautology. By Prob. 4.6, these two statements are equivalent.

§5. Truth tables

Conditions (2) and (3) of Def. 4.2(ii) may be summarized in *truth tables*:

$\boldsymbol{\beta}$	$\neg\boldsymbol{\beta}$
$\top$	$\bot$
$\bot$	$\top$

$\boldsymbol{\beta}$	$\boldsymbol{\gamma}$	$\boldsymbol{\beta} \rightarrow \boldsymbol{\gamma}$
$\top$	$\top$	$\top$
$\top$	$\bot$	$\bot$
$\bot$	$\top$	$\top$
$\bot$	$\bot$	$\top$

The idea here is that any truth valuation that assigns to $\boldsymbol{\beta}$ (or to $\boldsymbol{\beta}$ and $\boldsymbol{\gamma}$) the truth value(s) shown in the first column (or the first two columns) at a given row must assign to $\neg\boldsymbol{\beta}$ (or to $\boldsymbol{\beta} \rightarrow \boldsymbol{\gamma}$) the truth value shown in the last column at the same row.

This idea can be applied more generally. In the following definition the formula $\boldsymbol{\alpha}$ is any combination of formulas $\boldsymbol{\beta}_1, \boldsymbol{\beta}_2, \ldots, \boldsymbol{\beta}_k$. The definition prescribes how to construct a *truth table for* $\boldsymbol{\alpha}$ *in terms of* $\boldsymbol{\beta}_1, \boldsymbol{\beta}_2, \ldots, \boldsymbol{\beta}_k$. It proceeds by induction on $\deg \boldsymbol{\alpha}$: the induction hypothesis is that if $\boldsymbol{\gamma}$ is any combination of $\boldsymbol{\beta}_1, \boldsymbol{\beta}_2, \ldots, \boldsymbol{\beta}_k$ and $\deg \boldsymbol{\gamma} < \deg \boldsymbol{\alpha}$ then we can construct a truth table for $\boldsymbol{\gamma}$ in terms of $\boldsymbol{\beta}_1, \boldsymbol{\beta}_2, \ldots, \boldsymbol{\beta}_k$; and using this hypothesis the definition tells us how to construct a truth table for $\boldsymbol{\alpha}$ in terms of $\boldsymbol{\beta}_1, \boldsymbol{\beta}_2, \ldots, \boldsymbol{\beta}_k$.

5.1. Definition

Let the formula $\boldsymbol{\alpha}$ be a combination of formulas $\boldsymbol{\beta}_1, \boldsymbol{\beta}_2, \ldots, \boldsymbol{\beta}_k$. A *truth value for* $\boldsymbol{\alpha}$ *in terms of* $\boldsymbol{\beta}_1, \boldsymbol{\beta}_2, \ldots, \boldsymbol{\beta}_k$ is constructed as follows. First, set up a rectangular table with k columns – headed '$\boldsymbol{\beta}_1$', '$\boldsymbol{\beta}_2$', ..., '$\boldsymbol{\beta}_k$' respectively – and 2^k rows. In each of the $k \cdot 2^k$ spaces enter '$\top$' or '$\bot$', so that no two rows are filled out in the same way. Thus each of the 2^k different strings of length k made up of '$\top$'s and '$\bot$'s should appear in exactly one row. (For the sake of definiteness, regard these strings as 'words' in an alphabet consisting of the two letters '$\top$'

and '$\perp$' in this order, and enter the 2^k different strings in lexicographic order.)

Next, add a new last column, headed '$\boldsymbol{\alpha}$', and – proceeding by induction on $\deg \boldsymbol{\alpha}$ – fill it out with '$\top$'s and '$\perp$'s according to the following three rules corresponding to the three clauses of Def. 3.1.

(1) If $\boldsymbol{\alpha} = \boldsymbol{\beta}_i$ (where $1 \leqslant i \leqslant k$), copy the entries of the i-th column (the one headed '$\boldsymbol{\beta}_i$') into the last column, headed '$\boldsymbol{\alpha}$'.

(2) If $\boldsymbol{\alpha} = \neg\boldsymbol{\gamma}$, where $\boldsymbol{\gamma}$ is a combination of $\boldsymbol{\beta}_1, \boldsymbol{\beta}_2, \ldots, \boldsymbol{\beta}_k$, then by the induction hypothesis we already know how to construct a '$\boldsymbol{\gamma}$' column. Now, in the '$\boldsymbol{\alpha}$' column put '$\top$' in each row where the '$\boldsymbol{\gamma}$' column has '$\perp$', and '$\perp$' in each row where the '$\boldsymbol{\gamma}$' column has '$\top$'.

(3) If $\boldsymbol{\alpha} = \boldsymbol{\gamma} \to \boldsymbol{\delta}$, where $\boldsymbol{\gamma}$ and $\boldsymbol{\delta}$ are combinations of $\boldsymbol{\beta}_1, \boldsymbol{\beta}_2, \ldots, \boldsymbol{\beta}_k$, then by the induction hypothesis we already know how to construct '$\boldsymbol{\gamma}$' and '$\boldsymbol{\delta}$' columns. Now, in the '$\boldsymbol{\alpha}$' column put '$\perp$' in each row where the '$\boldsymbol{\gamma}$' column has '$\top$' whereas the '$\boldsymbol{\delta}$' column has '$\perp$'; and '$\top$' elsewhere, that is, in each row where the '$\boldsymbol{\gamma}$' column has '$\perp$' as well as in each row where the '$\boldsymbol{\delta}$' column has '$\top$'.

5.2. *Warning*

Since in general the same $\boldsymbol{\alpha}$ may be obtained as a combination of formulas $\boldsymbol{\beta}_1, \boldsymbol{\beta}_2, \ldots, \boldsymbol{\beta}_k$ in more than one way – see Warning 3.2(ii) – Def. 5.1 may not yield a unique result: $\boldsymbol{\alpha}$ may have more than one truth table in terms of $\boldsymbol{\beta}_1, \boldsymbol{\beta}_2, \ldots, \boldsymbol{\beta}_k$.

5.3. *Problem*

Construct truth tables in terms of $\boldsymbol{\alpha}$, $\boldsymbol{\beta}$ for:

(i) $\boldsymbol{\alpha} \wedge \boldsymbol{\beta}$,
(ii) $\boldsymbol{\alpha} \vee \boldsymbol{\beta}$,
(iii) $\boldsymbol{\alpha} \leftrightarrow \boldsymbol{\beta}$.

(See Def. 2.5.)

5.4. *Problem*

In a truth table in terms of two formulas $\boldsymbol{\alpha}$, $\boldsymbol{\beta}$ there are four ($= 2^2$) rows; thus the last column can be filled out with '$\top$'s and '$\perp$'s in 16 ($= 2^4$) different ways. Find 16 combinations of $\boldsymbol{\alpha}$, $\boldsymbol{\beta}$ whose truth tables in terms of $\boldsymbol{\alpha}$, $\boldsymbol{\beta}$ yield all these 16 different last columns.

5.5. Lemma

Let $\boldsymbol{\alpha}$ *be a combination of* $\boldsymbol{\beta}_1, \boldsymbol{\beta}_2, \ldots, \boldsymbol{\beta}_k$. *Consider a given row in a truth table for* $\boldsymbol{\alpha}$ *in terms of* $\boldsymbol{\beta}_1, \boldsymbol{\beta}_2, \ldots, \boldsymbol{\beta}_k$. *Let* σ *be any truth valuation such that for every i (where* $i = 1, 2, \ldots, k$*)* $\boldsymbol{\beta}_i{}^\sigma$ *is the truth value indicated in the given row at the i-th column (the one headed '*$\boldsymbol{\beta}_i$*'). Then* $\boldsymbol{\alpha}^\sigma$ *is the truth value indicated in the given row at the last column (headed '*$\boldsymbol{\alpha}$*').*

PROOF

Immediate from Def. 5.1 and Def. 4.2(ii), by induction on deg $\boldsymbol{\alpha}$. ■

5.6. Theorem (Semantic soundness of truth tables)

Let $\boldsymbol{\alpha}$ *be a propositional combination of* $\boldsymbol{\beta}_1, \boldsymbol{\beta}_2, \ldots, \boldsymbol{\beta}_k$. *If in a truth table for* $\boldsymbol{\alpha}$ *in terms of* $\boldsymbol{\beta}_1, \boldsymbol{\beta}_2, \ldots, \boldsymbol{\beta}_k$ *all the entries in the last ('*$\boldsymbol{\alpha}$*') column are '*$\top$*', then* $\boldsymbol{\alpha}$ *is a tautology.*

PROOF

Let σ be any truth valuation. Clearly, the truth values $\boldsymbol{\beta}_1{}^\sigma, \boldsymbol{\beta}_2{}^\sigma, \ldots, \boldsymbol{\beta}_k{}^\sigma$ must be respectively the same as those indicated in one particular row of the given truth table. Hence by Lemma 5.5 $\boldsymbol{\alpha}^\sigma$ is the truth value indicated in the same row in the last column. But by assumption this truth value is $\top$. Thus $\boldsymbol{\alpha}^\sigma = \top$ for all σ. ■

5.7. Problem

Verify that for any $\boldsymbol{\alpha}$, $\boldsymbol{\beta}$ and $\boldsymbol{\gamma}$:

(i) $\models_0 \boldsymbol{\alpha} \to \boldsymbol{\beta} \to \boldsymbol{\alpha}$ (Law of Affirmation of the Consequent),

(ii) $\models_0 (\boldsymbol{\alpha} \to \boldsymbol{\beta} \to \boldsymbol{\gamma}) \to (\boldsymbol{\alpha} \to \boldsymbol{\beta}) \to \boldsymbol{\alpha} \to \boldsymbol{\gamma}$ (Self-distributive Law of Implication),

(iii) $\models_0 [(\boldsymbol{\alpha} \to \boldsymbol{\beta}) \to \boldsymbol{\alpha}] \to \boldsymbol{\alpha}$ (Peirce's Law),

(iv) $\models_0 \neg \boldsymbol{\alpha} \to \boldsymbol{\alpha} \to \boldsymbol{\beta}$ (Law of Denial of the Antecedent),

(v) $\models_0 (\boldsymbol{\alpha} \to \neg \boldsymbol{\alpha}) \to \neg \boldsymbol{\alpha}$ (Clavius' Law).

5.8. Warning

The converse of Thm. 5.6 is not generally true. To see this, let $\boldsymbol{\alpha} = \boldsymbol{\beta} \to \boldsymbol{\gamma}$; then a truth table for $\boldsymbol{\alpha}$ in terms of $\boldsymbol{\beta}$, $\boldsymbol{\gamma}$ is shown above (p. 111) and has an '$\bot$' in its last column. Does it follow that $\boldsymbol{\alpha}$ cannot be a tautology? No; this truth table only shows that $\boldsymbol{\alpha}^\sigma = \bot$ *provided* σ is a truth valuation for which $\boldsymbol{\beta}^\sigma = \top$ and $\boldsymbol{\gamma}^\sigma = \bot$. But such a truth valuation may not exist; for example, if $\boldsymbol{\gamma} = \boldsymbol{\beta}$ then of course we cannot have both $\boldsymbol{\beta}^\sigma = \top$ and $\boldsymbol{\gamma}^\sigma = \bot$. Or if $\boldsymbol{\gamma} = \boldsymbol{\delta} \rightarrow \boldsymbol{\delta}$, then $\boldsymbol{\gamma}$ is a tautology, and we can never have $\boldsymbol{\gamma}^\sigma = \bot$, irrespective of what $\boldsymbol{\beta}^\sigma$ happens to be. However, the converse of Thm. 5.6 does hold, provided the $\boldsymbol{\beta}_i$ are subjected to special conditions.

5.9. Theorem (Semantic completeness of truth tables)

Let $\boldsymbol{\alpha}$ be a combination of k distinct prime formulas $\boldsymbol{\beta}_1, \boldsymbol{\beta}_2, \ldots, \boldsymbol{\beta}_k$. If $\boldsymbol{\alpha}$ is a tautology, then in the truth table for $\boldsymbol{\alpha}$ in terms of $\boldsymbol{\beta}_1, \boldsymbol{\beta}_2, \ldots, \boldsymbol{\beta}_k$ all the entries in the last ('$\boldsymbol{\alpha}$') column are '$\top$'.

PROOF

Consider an arbitrary row in this truth table. Since $\boldsymbol{\beta}_1, \boldsymbol{\beta}_2, \ldots, \boldsymbol{\beta}_k$ are prime and distinct, there exists a truth valuation σ such that the truth values $\boldsymbol{\beta}_1{}^\sigma, \boldsymbol{\beta}_2{}^\sigma, \ldots, \boldsymbol{\beta}_k{}^\sigma$ are respectively the same as those indicated in this particular row of the truth table. By Lemma 5.5, $\boldsymbol{\alpha}^\sigma$ is the truth value indicated in the same row at the last column. But $\boldsymbol{\alpha}^\sigma = \top$ since $\boldsymbol{\alpha}$ is a tautology. Thus the entry at the last column in this row is '$\top$'. ■

5.10. Remark

Thms. 5.6 and 5.9 together provide us with an algorithm (a mechanically performable procedure) whereby we can test any formula $\boldsymbol{\alpha}$ and decide whether or not it is a tautology: construct the truth table for $\boldsymbol{\alpha}$ in terms of its prime components (or in terms of any distinct prime formulas among which are all the prime components of $\boldsymbol{\alpha}$; see Prob. 3.3).

Using Prob. 4.6, this algorithm also enables us to decide, for any finite set $\boldsymbol{\Phi}$ of formulas and any formula $\boldsymbol{\alpha}$, whether or not $\boldsymbol{\Phi} \models_0 \boldsymbol{\alpha}$.

5.11. Definition

If $\boldsymbol{\alpha}$ and $\boldsymbol{\beta}$ are formulas satisfied by exactly the same truth valuations (that is, both $\boldsymbol{\alpha} \models_0 \boldsymbol{\beta}$ and $\boldsymbol{\beta} \models_0 \boldsymbol{\alpha}$) we say that $\boldsymbol{\alpha}$ and $\boldsymbol{\beta}$ are *tautologically equivalent* and write '$\boldsymbol{\alpha} \equiv_0 \boldsymbol{\beta}$'.

5.12. Remarks

(i) From Prob. 5.3(iii) it is easy to see that $\boldsymbol{\alpha} \equiv_0 \boldsymbol{\beta}$ iff $\models_0 \boldsymbol{\alpha} \leftrightarrow \boldsymbol{\beta}$.
(ii) An argument similar to the one used in the proof of Thm. 5.6

shows that if $\boldsymbol{\alpha}$ and $\boldsymbol{\beta}$ are combinations of $\boldsymbol{\beta}_1, \boldsymbol{\beta}_2, \ldots, \boldsymbol{\beta}_k$ and if the '$\boldsymbol{\alpha}$' and '$\boldsymbol{\beta}$' columns respectively in truth tables for $\boldsymbol{\alpha}$ and $\boldsymbol{\beta}$ in terms of $\boldsymbol{\beta}_1, \boldsymbol{\beta}_2, \ldots, \boldsymbol{\beta}_k$ have '$\top$'s and '$\bot$'s in the same places, then $\boldsymbol{\alpha} \equiv_0 \boldsymbol{\beta}$.

(iii) An argument similar to that used in the proof of Thm. 5.9 shows that the converse of (ii) holds, provided $\boldsymbol{\beta}_1, \boldsymbol{\beta}_2, \ldots, \boldsymbol{\beta}_k$ are distinct prime formulas.

5.13. Problem

Verify that for any $\boldsymbol{\alpha}, \boldsymbol{\beta}, \boldsymbol{\gamma}, \boldsymbol{\varphi}_1, \boldsymbol{\varphi}_2, \ldots, \boldsymbol{\varphi}_k$:

(i) $\boldsymbol{\alpha} \vee \boldsymbol{\beta} \equiv_0 (\boldsymbol{\alpha} \to \boldsymbol{\beta}) \to \boldsymbol{\beta}$,

(ii) $\boldsymbol{\alpha} \to \boldsymbol{\beta} \equiv_0 \neg\boldsymbol{\beta} \to \neg\boldsymbol{\alpha}$ (Law of Contraposition),

(iii) $\neg(\boldsymbol{\varphi}_1 \wedge \boldsymbol{\varphi}_2 \wedge \ldots \wedge \boldsymbol{\varphi}_k) \equiv_0 \neg\boldsymbol{\varphi}_1 \vee \neg\boldsymbol{\varphi}_2 \vee \ldots \vee \neg\boldsymbol{\varphi}_k$

(iv) $\neg(\boldsymbol{\varphi}_1 \vee \boldsymbol{\varphi}_2 \vee \ldots \vee \boldsymbol{\varphi}_k) \equiv_0 \neg\boldsymbol{\varphi}_1 \wedge \neg\boldsymbol{\varphi}_2 \wedge \ldots \wedge \neg\boldsymbol{\varphi}_k$

(De Morgan's Laws),

(v) $\boldsymbol{\alpha} \wedge \boldsymbol{\beta} \wedge \boldsymbol{\gamma} \equiv_0 (\boldsymbol{\alpha} \wedge \boldsymbol{\beta}) \wedge \boldsymbol{\gamma}$ (Associative Law of Conjunction),

(vi) $\boldsymbol{\alpha} \vee \boldsymbol{\beta} \vee \boldsymbol{\gamma} \equiv_0 (\boldsymbol{\alpha} \vee \boldsymbol{\beta}) \vee \boldsymbol{\gamma}$ (Associative Law of Disjunction),

(vii) $\boldsymbol{\varphi}_1 \wedge \boldsymbol{\varphi}_2 \wedge \ldots \wedge \boldsymbol{\varphi}_k \to \boldsymbol{\alpha} \equiv_0 \boldsymbol{\varphi}_1 \to \boldsymbol{\varphi}_2 \to \cdots \to \boldsymbol{\varphi}_k \to \boldsymbol{\alpha}$.

5.14. Problem

Let $\boldsymbol{\alpha}$ and $\boldsymbol{\beta}$ be any formulas. Let $\boldsymbol{\Phi}$ be the set of all formulas obtainable from $\boldsymbol{\alpha}$ and $\boldsymbol{\beta}$ using negation and conjunction. More precisely,

(1) $\boldsymbol{\alpha}$ and $\boldsymbol{\beta}$ are in $\boldsymbol{\Phi}$;

(2) if $\boldsymbol{\gamma}$ is in $\boldsymbol{\Phi}$ then so is $\neg\boldsymbol{\gamma}$;

(3) if $\boldsymbol{\gamma}$ and $\boldsymbol{\delta}$ are in $\boldsymbol{\Phi}$ then so is $\boldsymbol{\gamma} \wedge \boldsymbol{\delta}$.

Find a formula in $\boldsymbol{\Phi}$ that is tautologically equivalent to $\boldsymbol{\alpha} \to \boldsymbol{\beta}$.

5.15. Problem

The same as Prob. 5.14, but with 'conjunction' and '$\wedge$' replaced by 'disjunction' and '$\vee$' respectively.

5.16. Problem

For any formulas $\boldsymbol{\alpha}$ and $\boldsymbol{\beta}$, put $\boldsymbol{\alpha}|\boldsymbol{\beta} =_{\text{df}} \neg(\boldsymbol{\alpha} \wedge \boldsymbol{\beta})$. The '$|$' here is known as *Sheffer's stroke*. The formula $\boldsymbol{\alpha}|\boldsymbol{\beta}$ is called the *non-conjunction of* $\boldsymbol{\alpha}$

and $\boldsymbol{\beta}$. Let $\boldsymbol{\Phi}$ be the set of all formulas obtainable from $\boldsymbol{\alpha}$ and $\boldsymbol{\beta}$ using non-conjunction. Thus,

(1) $\boldsymbol{\alpha}$ and $\boldsymbol{\beta}$ are in $\boldsymbol{\Phi}$;
(2) if $\boldsymbol{\gamma}$ and $\boldsymbol{\delta}$ are in $\boldsymbol{\Phi}$ then so is $\boldsymbol{\gamma}|\boldsymbol{\delta}$.

Find formulas in $\boldsymbol{\Phi}$ that are tautologically equivalent to $\neg\boldsymbol{\alpha}$ and $\boldsymbol{\alpha}\to\boldsymbol{\beta}$ respectively.

5.17. Problem

Let $\boldsymbol{\alpha}$ and $\boldsymbol{\beta}$ be distinct prime formulas. Let $\boldsymbol{\Phi}$ be defined as in Prob. 5.16, but with 'non-conjunction' and '|' replaced by 'implication' and '→' respectively. Prove that no formula in $\boldsymbol{\Phi}$ is tautologically equivalent to $\boldsymbol{\alpha}\wedge\boldsymbol{\beta}$.

5.18. Problem

Let $\boldsymbol{\alpha}$ and $\boldsymbol{\beta}$ be distinct prime formulas. Let $\boldsymbol{\Phi}$ be defined as in Prob. 5.14, but with 'conjunction' and '∧' replaced by 'bi-implication' and '↔' respectively.

(i) Find eight formulas in $\boldsymbol{\Phi}$ such that every formula in $\boldsymbol{\Phi}$ is tautologically equivalent to exactly one of the eight.
(ii) Prove that no formula in $\boldsymbol{\Phi}$ is tautologically equivalent to $\boldsymbol{\alpha}\to\boldsymbol{\beta}$.

5.19. Remark

Prob. 5.4 means that all binary truth functions are reducible to negation and implication. Prob. 5.14 (Prob. 5.15) means that implication – and hence all binary truth functions – can be reduced to negation and conjunction (negation and disjunction). Prob. 5.16 means that negation and implication – and hence all binary truth functions – can be reduced to non-conjunction. Prob. 5.17 means that conjunction cannot be reduced to implication (although by Prob. 5.13(i) disjunction can be so reduced). Prob. 5.18(ii) means that implication cannot be reduced to negation and bi-implication.

§6. The propositional calculus

The *propositional calculus* (briefly, *Propcal*) presented in this section is a formal mechanism for generating the tautological consequences of any set $\boldsymbol{\Phi}$ of formulas. A central role will be played by *modus ponens*.

6.1. Definition

Modus ponens is the [formal] operation that may be applied to any two formulas of the form $\boldsymbol{\alpha}$ and $\boldsymbol{\alpha}\to\boldsymbol{\beta}$, to yield the formula $\boldsymbol{\beta}$; schematically,

$$\frac{\boldsymbol{\alpha},\ \boldsymbol{\alpha}\to\boldsymbol{\beta}}{\boldsymbol{\beta}}$$

In this connection, $\boldsymbol{\alpha}$ and $\boldsymbol{\alpha}\to\boldsymbol{\beta}$ are called the *minor premiss* and *major premiss* respectively, and $\boldsymbol{\beta}$ is called the *conclusion*.

6.2. Remark

From Def. 4.2 it follows at once that if $\boldsymbol{\alpha}^\sigma = (\boldsymbol{\alpha}\to\boldsymbol{\beta})^\sigma = \top$ then also $\boldsymbol{\beta}^\sigma = \top$. (By Def. 4.4(iv) this amounts to the same thing as $\{\boldsymbol{\alpha}, \boldsymbol{\alpha}\to\boldsymbol{\beta}\} \models_0 \boldsymbol{\beta}$.) We express this by saying that *modus ponens preserves truth* and is therefore *semantically sound* as a *rule of inference*.

We designate as *propositional axioms* all formulas of the following five forms:

6.3. Axiom scheme i. $\boldsymbol{\alpha}\to\boldsymbol{\beta}\to\boldsymbol{\alpha}$,

6.4. Axiom scheme ii. $(\boldsymbol{\alpha}\to\boldsymbol{\beta}\to\boldsymbol{\gamma})\to(\boldsymbol{\alpha}\to\boldsymbol{\beta})\to\boldsymbol{\alpha}\to\boldsymbol{\gamma}$,

6.5. Axiom scheme iii. $[(\boldsymbol{\alpha}\to\boldsymbol{\beta})\to\boldsymbol{\alpha}]\to\boldsymbol{\alpha}$,

6.6. Axiom scheme iv. $\neg\boldsymbol{\alpha}\to\boldsymbol{\alpha}\to\boldsymbol{\beta}$,

6.7. Axiom scheme v. $(\boldsymbol{\alpha}\to\neg\boldsymbol{\alpha})\to\neg\boldsymbol{\alpha}$.

Note that these are not five *single* axioms but *axiom schemes*, each representing infinitely many axioms obtained by all possible choices of formulas $\boldsymbol{\alpha}$, $\boldsymbol{\beta}$, and $\boldsymbol{\gamma}$. We shall refer to them briefly as ‘Ax. i’, ‘Ax. ii’, etc.

6.8. Definition

(i) A *propositional deduction from* a set $\boldsymbol{\Phi}$ of formulas is a non-empty finite sequence of formulas $\boldsymbol{\varphi}_1, \boldsymbol{\varphi}_2, \ldots, \boldsymbol{\varphi}_n$ such that for each k ($k = 1, 2, \ldots, n$) at least one of the following conditions holds:

(1) $\boldsymbol{\varphi}_k$ is a propositional axiom,

(2) $\boldsymbol{\varphi}_k \in \boldsymbol{\Phi}$,

(3) $\boldsymbol{\varphi}_k$ is obtained by *modus ponens* from two earlier formulas in the sequence; that is, there are i and j, both smaller than k, such that $\boldsymbol{\varphi}_j = \boldsymbol{\varphi}_i \rightarrow \boldsymbol{\varphi}_k$.

In this connection $\boldsymbol{\Phi}$ is called a set of *hypotheses*.

(ii) A *propositional proof* is a propositional deduction from the empty set of hypotheses.

Where there is no risk of ambiguity, we shall usually omit the qualification 'propositional' and say simply 'deduction' and 'proof'. Similar ellipses will be used in connection with other bits of terminology introduced below.

6.9. Definition

(i) A deduction (or proof) whose last formula is $\boldsymbol{\alpha}$ is said to be a deduction (or proof, respectively) *of* $\boldsymbol{\alpha}$.

(ii) If there exists a propositional deduction of a formula $\boldsymbol{\alpha}$ from a set $\boldsymbol{\Phi}$ of formulas, we say that $\boldsymbol{\alpha}$ is *[propositionally] deducible* from $\boldsymbol{\Phi}$ and write, briefly, '$\boldsymbol{\Phi} \vdash_0 \boldsymbol{\alpha}$'.

(iii) If there exists a propositional proof of a formula $\boldsymbol{\alpha}$ – that is, a deduction of $\boldsymbol{\alpha}$ from the empty set – we say that $\boldsymbol{\alpha}$ is *[propositionally] provable* and write, briefly, '$\vdash_0 \boldsymbol{\alpha}$'. In this case $\boldsymbol{\alpha}$ is also called a *[propositional] theorem*.

In connection with '$\vdash_0$' we employ notational simplifications like those used in connection with '$\models_0$'. Thus, for example, we write '$\boldsymbol{\Phi}, \boldsymbol{\alpha} \vdash_0 \boldsymbol{\beta}$' instead of '$\boldsymbol{\Phi} \cup \{\boldsymbol{\alpha}\} \vdash_0 \boldsymbol{\beta}$'.

6.10. Remarks

(i) The calculus we have specified here is a *linear* calculus, as distinct from calculi whose deductions have a more complex tree-like branching form rather than being ordinary (linear) sequences as in Def. 6.8. A linear calculus is characterized uniquely by specifying its axioms (by means of axiom-schemes or in some other way) and rules of inference. In the present case the axioms are all instances of Ax. i–Ax. v, and the sole rule of inference is *modus ponens*.

(ii) Many calculi described in the literature, based on other axioms or rules of inference, are equivalent to the one presented here in the sense, roughly speaking,[1] that their relation of deducibility is co-extensive with our $\vdash_0$. (For example, the calculus presented in B&M, Ch. 1, § 10.) All these calculi, including of course the present one, are often referred to collectively as *the [classical] propositional calculus*. Although, strictly speaking, they are distinct calculi, their mutual equivalence makes it possible to regard them as being merely different versions of the same calculus.

(iii) The qualification 'classical' is often omitted; it is however needed sometimes in order to prevent confusion with *non-classical* (a.k.a. *non-standard* or *deviant*) propositional calculi that are broadly similar but not equivalent to the present one; for example, the *intuitionistic* propositional calculus (a version of which is presented in B&M, Ch. 9, § 8).

(iv) We use the term 'theorem' with two quite different meanings, which must be strictly distinguished from each other. A [propositional] theorem in the sense of Def. 6.9(iii) is a *formal* expression, a formula *in* the language $\mathscr{L}$. In this book we never assert such a theorem, since we do not use the language $\mathscr{L}$, only talk *about* it. On the other hand, a theorem such as Thm. 5.6 (which we have asserted above) is a proposition stated in our metalanguage. In order not to get these two kinds of theorem confused with each other, those of the former kind are sometimes called *formal theorems* or *$\mathscr{L}$-theorems* and those of the latter kind *metatheorems*. However, this will rarely be necessary here, as it will usually be clear from the context which meaning of 'theorem' is intended. A similar distinction must be drawn between the two meanings of terms such as 'deduction', 'hypothesis' and 'proof'.

(v) The reason for using the same terms with two alternative meanings is that there is an intended connection between the two sets of meanings. Thus formal deductions are supposed to be stylized and formalized versions or counterparts (or at least analogues) of 'ordinary' deductions in informal or semi-formal axiomatic theories expounded within mathematics and related hypothetico-deductive disciplines. Hypotheses in the sense of Def. 6.8 are supposed to be formal counterparts of the hypotheses or assumptions adopted as a starting point for real (informal or semi-formal) mathematical deductions. (When such hypotheses or

[1] That is, ignoring irrelevant differences between the formal languages in which these various calculi are formulated.

assumptions are adopted as a point of departure for a whole axiomatic theory, rather than for temporary or *ad hoc* ends, they are usually called *postulates* or *[extralogical] axioms*.)

(vi) Formal deductions of the kind studied in Symbolic Logic differ from 'ordinary' mathematical deductions not only in being completely formalized but also in spelling out the logical machinery used. In informal or semi-formal mathematical deductions you are allowed to assert any statement that follows logically from previous ones, but the nature of this relation – being a logical consequence – is not spelt out fully, if at all. In logical calculi, such as Propcal, the purely logical steps in formal deductions are made explicit and formally detailed by specifying logical axioms (such as Ax. i–Ax. v) and rules of inference (such as *modus ponens*).

(vii) In an ordinary mathematical deduction you are allowed to introduce any statement deduced earlier (by a preceding deduction) from the same hypotheses. However, this licence is merely a matter of practical convenience: in principle such a previously deduced statement could be introduced together with its whole deduction, so that every deduction would start from first principles. This latter procedure is mimicked in Def. 6.8.

(viii) Propcal is pitifully inadequate for formalizing any but the most trivial mathematical deductions. Its is however of interest as a sort of pilot project for more powerful and useful systems.

6.11. Example

We show that $\vdash_0 \boldsymbol{\alpha}\to\boldsymbol{\alpha}$ for every $\boldsymbol{\alpha}$. (In other words, we are going to prove a [meta]theorem *about* Propcal, which asserts that, for every formula $\boldsymbol{\alpha}$, $\boldsymbol{\alpha}\to\boldsymbol{\alpha}$ is a propositional theorem, a theorem *of* Propcal.) The following sequence of five formulas is a [propositional] proof of $\boldsymbol{\alpha}\to\boldsymbol{\alpha}$:

$$[\boldsymbol{\alpha}\to(\boldsymbol{\alpha}\to\boldsymbol{\alpha})\to\boldsymbol{\alpha}]\to(\boldsymbol{\alpha}\to\boldsymbol{\alpha}\to\boldsymbol{\alpha})\to\boldsymbol{\alpha}\to\boldsymbol{\alpha}, \qquad \text{(Ax. ii)}$$

$$\boldsymbol{\alpha}\to(\boldsymbol{\alpha}\to\boldsymbol{\alpha})\to\boldsymbol{\alpha}, \qquad \text{(Ax. i)}$$

$$(\boldsymbol{\alpha}\to\boldsymbol{\alpha}\to\boldsymbol{\alpha})\to\boldsymbol{\alpha}\to\boldsymbol{\alpha}, \qquad \text{(m.p.)}$$

$$\boldsymbol{\alpha}\to\boldsymbol{\alpha}\to\boldsymbol{\alpha}, \qquad \text{(Ax. i)}$$

$$\boldsymbol{\alpha}\to\boldsymbol{\alpha}. \qquad \text{(m.p.)}$$

The marginal comments on the right have been added for convenience. Thus the first formula is an instance of Ax. ii, obtained from (6.4) by taking $\boldsymbol{\beta} = \boldsymbol{\alpha} \rightarrow \boldsymbol{\alpha}$ and $\boldsymbol{\gamma} = \boldsymbol{\alpha}$; the second formula is an instance of Ax. i, with $\boldsymbol{\beta} = \boldsymbol{\alpha} \rightarrow \boldsymbol{\alpha}$; the third formula is obtained by *modus ponens* from the preceding two; the fourth formula is an instance of Ax. i, with $\boldsymbol{\beta} = \boldsymbol{\alpha}$; and the fifth formula is again obtained by *modus ponens* from the preceding two. In principle these explanations are redundant, because you can always check whether or not a given formula is an instance of an axiom scheme, or obtainable by *modus ponens* from two earlier formulas.

6.12. Theorem (Semantic soundness of Propcal)

If $\boldsymbol{\Phi} \vdash_0 \boldsymbol{\alpha}$ *then also* $\boldsymbol{\Phi} \models_0 \boldsymbol{\alpha}$. *In particular, if* $\vdash_0 \boldsymbol{\alpha}$ *then also* $\models_0 \boldsymbol{\alpha}$.

PROOF

Let $\boldsymbol{\varphi}_1, \boldsymbol{\varphi}_2, \ldots, \boldsymbol{\varphi}_n$ be a deduction of $\boldsymbol{\alpha}$ from $\boldsymbol{\Phi}$; thus $\boldsymbol{\varphi}_n = \boldsymbol{\alpha}$. We shall prove by [strong] induction on k that $\boldsymbol{\Phi} \models_0 \boldsymbol{\varphi}_k$ for $k = 1, 2, \ldots, n$. Thus, in particular, for $k = n$ it will follow that $\boldsymbol{\Phi} \models_0 \boldsymbol{\alpha}$, as claimed. We distinguish three cases concerning $\boldsymbol{\varphi}_k$, corresponding to the three conditions in Def. 6.8(i).

Case 1: $\boldsymbol{\varphi}_k$ is a propositional axiom. In this case it is easy to verify that $\models_0 \boldsymbol{\varphi}_k$ (see Prob. 5.7); in other words, $\boldsymbol{\varphi}_k$ is satisfied by every truth valuation. Hence *a fortiori* $\boldsymbol{\Phi} \models_0 \boldsymbol{\varphi}_k$.

Case 2: $\boldsymbol{\varphi}_k \in \boldsymbol{\Phi}$. Then obviously $\boldsymbol{\Phi} \models_0 \boldsymbol{\varphi}_k$.

Case 3: $\boldsymbol{\varphi}_k$ is obtained by *modus ponens* from two earlier formulas in the deduction; that is, there are $i, j < k$ such that $\boldsymbol{\varphi}_j = \boldsymbol{\varphi}_i \rightarrow \boldsymbol{\varphi}_k$. In this case, by Rem. 6.2, $\{\boldsymbol{\varphi}_i, \boldsymbol{\varphi}_j\} \models_0 \boldsymbol{\varphi}_k$. But by the induction hypothesis both $\boldsymbol{\Phi} \models_0 \boldsymbol{\varphi}_i$ and $\boldsymbol{\Phi} \models_0 \boldsymbol{\varphi}_j$. Hence clearly $\boldsymbol{\Phi} \models_0 \boldsymbol{\varphi}_k$.

The second claim of our theorem follows from the first by taking $\boldsymbol{\Phi} = \varnothing$. ■

6.13. Theorem (Cut Rule)

If $\boldsymbol{\Phi} \vdash_0 \boldsymbol{\delta}_i$ *for each* $i = 1, 2, \ldots, k$ *and* $\boldsymbol{\Psi} \cup \{\boldsymbol{\delta}_1, \boldsymbol{\delta}_2, \ldots, \boldsymbol{\delta}_k\} \vdash_0 \boldsymbol{\alpha}$ *then* $\boldsymbol{\Phi} \cup \boldsymbol{\Psi} \vdash_0 \boldsymbol{\alpha}$.

PROOF

Take a deduction of $\boldsymbol{\alpha}$ from $\boldsymbol{\Psi} \cup \{\boldsymbol{\delta}_1, \boldsymbol{\delta}_2, \ldots, \boldsymbol{\delta}_k\}$ and whenever $\boldsymbol{\delta}_i$ is used there as an hypothesis replace it by a deduction of $\boldsymbol{\delta}_i$ from $\boldsymbol{\Phi}$. The result is clearly a deduction of $\boldsymbol{\alpha}$ from $\boldsymbol{\Phi} \cup \boldsymbol{\Psi}$. ■

6.14. Remark

The Cut Rule clearly holds for *any* linear calculus, irrespective of its axioms and rules of inference. The strange name of this rule is due to the fact that it allows us to 'cut out the middlemen' $\boldsymbol{\delta}_i$.

We shall often refer to this rule briefly as 'Cut'.

§7. The Deduction Theorem

7.1. Remark

Suppose $\boldsymbol{\Phi} \vdash_0 \boldsymbol{\alpha} \rightarrow \boldsymbol{\beta}$. Since by *modus ponens* we have $\{\boldsymbol{\alpha}, \boldsymbol{\alpha} \rightarrow \boldsymbol{\beta}\} \vdash_0 \boldsymbol{\beta}$, we can apply Cut to the 'middleman' $\boldsymbol{\alpha} \rightarrow \boldsymbol{\beta}$ – see Thm. 6.13 – and get $\boldsymbol{\Phi}, \boldsymbol{\alpha} \vdash_0 \boldsymbol{\beta}$. Thus we have

$$\boldsymbol{\Phi} \vdash_0 \boldsymbol{\alpha} \rightarrow \boldsymbol{\beta} \Rightarrow \boldsymbol{\Phi}, \boldsymbol{\alpha} \vdash_0 \boldsymbol{\beta}.$$

The converse of this result, which we prove next, is of central importance.

7.2. Theorem (Deduction Theorem)

If $\boldsymbol{\Phi}, \boldsymbol{\alpha} \vdash_0 \boldsymbol{\beta}$ *then* $\boldsymbol{\Phi} \vdash_0 \boldsymbol{\alpha} \rightarrow \boldsymbol{\beta}$.

PROOF

Let $\boldsymbol{\varphi}_1, \boldsymbol{\varphi}_2, \ldots, \boldsymbol{\varphi}_n$ be a given deduction of $\boldsymbol{\beta}$ from $\boldsymbol{\Phi} \cup \{\boldsymbol{\alpha}\}$; thus $\boldsymbol{\varphi}_n = \boldsymbol{\beta}$.

We shall prove, by [strong] induction on k, that $\boldsymbol{\Phi} \vdash_0 \boldsymbol{\alpha} \rightarrow \boldsymbol{\varphi}_k$ for $k = 1, 2, \ldots, n$. In particular, for $k = n$ it will follow that $\boldsymbol{\Phi} \vdash_0 \boldsymbol{\alpha} \rightarrow \boldsymbol{\beta}$, as claimed. We distinguish three cases concerning $\boldsymbol{\varphi}_k$, corresponding to the three conditions in Def. 6.8(i).

Case 1: $\boldsymbol{\varphi}_k$ is a propositional axiom. In this case the following sequence of three formulas is a proof of $\boldsymbol{\alpha} \rightarrow \boldsymbol{\varphi}_k$ and hence *a fortiori* a deduction of it from $\boldsymbol{\Phi}$:

$\boldsymbol{\varphi}_k$, (ax.)

$\varphi_k \to \alpha \to \varphi_k$, (Ax. i)

$\alpha \to \varphi_k$. (m.p.)

Case 2: $\varphi_k \in \Phi \cup \{\alpha\}$. Thus $\varphi_k \in \Phi$ or $\varphi_k = \alpha$. Because α plays here a special role, we must split our argument into two subcases.

Subcase 2a: $\varphi_k \in \Phi$. Then the same sequence of three formulas as in Case 1 is a deduction of $\alpha \to \varphi_k$ from Φ, except that now the justification for the presence of φ_k is that it is one of the hypotheses Φ rather than that it is an axiom.

Subcase 2b: $\varphi_k = \alpha$. Then $\alpha \to \varphi_k = \alpha \to \alpha$, so by Ex. 6.11 $\vdash_0 \alpha \to \varphi_k$ and *a fortiori* $\Phi \vdash_0 \alpha \to \varphi_k$.

Case 3: φ_k is obtained by *modus ponens* from two earlier formulas in the given deduction. This means that there are $i, j < k$ such that $\varphi_j = \varphi_i \to \varphi_k$ (so φ_i and φ_j serve as minor and major premiss, respectively, to yield φ_k). By the induction hypothesis, both $\Phi \vdash_0 \alpha \to \varphi_i$ and $\Phi \vdash_0 \alpha \to \varphi_j$ – that is, $\Phi \vdash_0 \alpha \to \varphi_i \to \varphi_k$.

Thanks to Cut, the required result, $\Phi \vdash_0 \alpha \to \varphi_k$, will follow if we show that $\{\alpha \to \varphi_i, \alpha \to \varphi_i \to \varphi_k\} \vdash_0 \alpha \to \varphi_k$. The following sequence of five formulas is a deduction of $\alpha \to \varphi_k$ from $\{\alpha \to \varphi_i, \alpha \to \varphi_i \to \varphi_k\}$:

$\alpha \to \varphi_i$, (hyp.)

$\alpha \to \varphi_i \to \varphi_k$, (hyp.)

$(\alpha \to \varphi_i \to \varphi_k) \to (\alpha \to \varphi_i) \to \alpha \to \varphi_k$, (Ax. ii)

$(\alpha \to \varphi_i) \to \alpha \to \varphi_k$, (m.p.)

$\alpha \to \varphi_k$. (m.p.) ■

7.3. Remarks

(i) We shall refer to the Deduction Theorem briefly as 'DT'.

(ii) In proving DT (and in Ex. 6.11, which is used in the proof) we invoked only Ax. i and Ax. ii. In fact, it is not even necessary for formulas of the forms (6.3) and (6.4) to be axioms: it would have been enough if they were just theorems. More precisely: if $\vdash^*$ is the relation of deducibility in a linear calculus whose sole rule of inference is *modus ponens* and if $\vdash^* \alpha \to \beta \to \alpha$ as well as $\vdash^* (\alpha \to \beta \to \gamma) \to (\alpha \to \beta) \to \alpha \to \gamma$ for all α, β and γ, then DT holds for $\vdash^*$, that is: $\Phi, \alpha \vdash^* \beta \Rightarrow \Phi \vdash^* \alpha \to \beta$.

(iii) Now that we have DT, we shall not need to invoke Ax. i and Ax. ii again. Indeed, the sole purpose of adopting these axiom schemes was to enable us to establish DT.

7.4. Problem

Let $\vdash^*$ be the deducibility relation in a calculus that has *modus ponens* as a – not necessarily sole – rule of inference.

Show that if Cut and DT hold for $\vdash^*$, then $\vdash^* \boldsymbol{\alpha \to \beta \to \alpha}$ and $\vdash^* \boldsymbol{(\alpha \to \beta \to \gamma) \to (\alpha \to \beta) \to \alpha \to \gamma}$ for all $\boldsymbol{\alpha}$, $\boldsymbol{\beta}$ and $\boldsymbol{\gamma}$.

§8. Inconsistency and consistency

8.1. Definition

(i) A set of two formulas $\{\boldsymbol{\alpha}, \neg\boldsymbol{\alpha}\}$, one of which is the negation of the other, is called a *contradictory pair*.

(ii) A set $\boldsymbol{\Phi}$ of formulas is said to be *[propositionally] inconsistent* – in symbols: '$\boldsymbol{\Phi} \vdash_0$' – if both members of some contradictory pair are propositionally deducible from $\boldsymbol{\Phi}$; that is, for some formula $\boldsymbol{\alpha}$ $\boldsymbol{\Phi} \vdash_0 \boldsymbol{\alpha}$ as well as $\boldsymbol{\Phi} \vdash_0 \neg\boldsymbol{\alpha}$. Otherwise, $\boldsymbol{\Phi}$ is said to be *[propositionally] consistent*.

8.2. Warning

Some authors use 'contradictory', 'consistent' and 'inconsistent' as semantic terms; so that, for example, a set $\boldsymbol{\Phi}$ of formulas would be said to be *inconsistent* if $\boldsymbol{\Phi} \vDash_0$, that is, if it is not satisfied by any truth valuation. We shall strictly avoid that semantic usage. Although it will transpire that a set $\boldsymbol{\Phi}$ of formulas is satisfied by some truth valuation iff it is consistent (in the proof-theoretic sense of Def. 8.1), this fact is a far from trivial theorem rather than a mere matter of definition.

8.3. Problem

(i) Prove that if $\boldsymbol{\Psi} \subseteq \boldsymbol{\Phi}$ and $\boldsymbol{\Psi}$ is inconsistent then $\boldsymbol{\Phi}$ is inconsistent.

(ii) Prove that if $\boldsymbol{\Phi}$ is inconsistent then it has an inconsistent *finite* subset.

8.4. Theorem

An inconsistent set of formulas is not satisfied by any truth valuation: if $\boldsymbol{\Phi} \vdash_0$ *then* $\boldsymbol{\Phi} \vDash_0$.

PROOF

Suppose $\mathbf{\Phi} \vdash_0$. Then for some $\boldsymbol{\alpha}$ both $\mathbf{\Phi} \vdash_0 \boldsymbol{\alpha}$ and $\mathbf{\Phi} \vdash_0 \neg\boldsymbol{\alpha}$. By the soundness of Propcal (Thm. 6.12) it follows that both $\mathbf{\Phi} \models_0 \boldsymbol{\alpha}$ and $\mathbf{\Phi} \models_0 \neg\boldsymbol{\alpha}$. Thus any truth valuation satisfying $\mathbf{\Phi}$ would have to satisfy both $\boldsymbol{\alpha}$ and $\neg\boldsymbol{\alpha}$, which is impossible by clause (2) of Def. 4.2(ii). ■

8.5. *Corollary (Consistency of Propcal)*

It is impossible, for any $\boldsymbol{\alpha}$, *that both* $\vdash_0 \boldsymbol{\alpha}$ *and* $\vdash_0 \neg\boldsymbol{\alpha}$.

PROOF

The claim is equivalent to saying that the empty set is consistent; but the empty set is satisfied by every truth valuation (cf. Rem. 4.5(i)). ■

8.6. *Theorem (Inconsistency Effect)*

If $\mathbf{\Phi} \vdash_0$ *then* $\mathbf{\Phi} \vdash_0 \boldsymbol{\beta}$ *for every formula* $\boldsymbol{\beta}$.

PROOF

Assume $\mathbf{\Phi} \vdash_0$. Then for some $\boldsymbol{\alpha}$ both $\mathbf{\Phi} \vdash_0 \boldsymbol{\alpha}$ and $\mathbf{\Phi} \vdash_0 \neg\boldsymbol{\alpha}$. Now, for any $\boldsymbol{\beta}$, the formula $\neg\boldsymbol{\alpha}\to\boldsymbol{\alpha}\to\boldsymbol{\beta}$ is an instance of Ax. iv; hence $\{\boldsymbol{\alpha}, \neg\boldsymbol{\alpha}\} \vdash_0 \boldsymbol{\beta}$. By Cut, $\mathbf{\Phi} \vdash_0 \boldsymbol{\beta}$. ■

8.7. *Remarks*

(i) For brevity, we shall refer to the Inconsistency Effect as 'IE'.

(ii) The converse of Thm. 8.6 is trivial: if all formulas are deducible from $\mathbf{\Phi}$, then in particular both members of *any* contradictory pair are deducible from it.

(iii) Our sole purpose in adopting Ax. iv was to enable us to establish IE. From now on this axiom scheme will not have to be invoked.

8.8. *Problem*

Let $\vdash^*$ be the deducibility relation in a calculus for which both DT and IE hold. Prove that $\vdash^* \neg\boldsymbol{\alpha}\to\boldsymbol{\alpha}\to\boldsymbol{\beta}$ for all $\boldsymbol{\alpha}$ and $\boldsymbol{\beta}$.

8.9. *Theorem (Reductio ad absurdum)*

If $\mathbf{\Phi}, \boldsymbol{\alpha} \vdash_0$ *then* $\mathbf{\Phi} \vdash_0 \neg\boldsymbol{\alpha}$.

PROOF

Assume $\mathbf{\Phi}, \boldsymbol{\alpha} \vdash_0$. Then by IE we have $\mathbf{\Phi}, \boldsymbol{\alpha} \vdash_0 \neg\boldsymbol{\alpha}$ and hence, by DT, $\mathbf{\Phi} \vdash_0 \boldsymbol{\alpha}\rightarrow\neg\boldsymbol{\alpha}$.

Now, $(\boldsymbol{\alpha}\rightarrow\neg\boldsymbol{\alpha})\rightarrow\neg\boldsymbol{\alpha}$ is an instance of Ax. v; hence $\boldsymbol{\alpha}\rightarrow\neg\boldsymbol{\alpha} \vdash_0 \neg\boldsymbol{\alpha}$. Using Cut, we get $\mathbf{\Phi} \vdash_0 \neg\boldsymbol{\alpha}$, as claimed. ■

8.10. Remarks

(i) The converse of *reductio* is immediate: if $\mathbf{\Phi} \vdash_0 \neg\boldsymbol{\alpha}$ then *a fortiori* $\mathbf{\Phi}, \boldsymbol{\alpha} \vdash_0 \neg\boldsymbol{\alpha}$. But clearly also $\mathbf{\Phi}, \boldsymbol{\alpha} \vdash_0 \boldsymbol{\alpha}$; hence $\mathbf{\Phi}, \boldsymbol{\alpha} \vdash_0$.

(ii) The sole purpose of adopting Ax. v was to enable us to prove *reductio*. Henceforth there will be no need to invoke that axiom scheme.

8.11. Problem

Let $\vdash^*$ be the deducibility relation in a calculus that has *modus ponens* as a rule of inference and for which DT and *reductio* hold. Prove that $\vdash^* (\boldsymbol{\alpha}\rightarrow\neg\boldsymbol{\alpha})\rightarrow\neg\boldsymbol{\alpha}$ for all $\boldsymbol{\alpha}$.

8.12. Problem

Prove that $\boldsymbol{\alpha} \vdash_0 \neg\neg\boldsymbol{\alpha}$ for all $\boldsymbol{\alpha}$.

8.13. Remark

All the proof-theoretic results we have obtained so far – Cut, DT, IE and *reductio* – hold also for the intuitionistic propositional calculus (the most important non-classical propositional calculus). But the following result – the inverse of Prob. 8.12 – does not hold for that calculus, so in order to prove it we shall have to invoke Ax. iii, which is not valid in intuitionistic logic.

8.14. Lemma

$\neg\neg\boldsymbol{\alpha} \vdash_0 \boldsymbol{\alpha}$ *for all* $\boldsymbol{\alpha}$.

PROOF

Clearly, $\{\boldsymbol{\alpha}\rightarrow\neg\boldsymbol{\alpha}, \boldsymbol{\alpha}\} \vdash_0 \boldsymbol{\alpha}$; but also $\{\boldsymbol{\alpha}\rightarrow\neg\boldsymbol{\alpha}, \boldsymbol{\alpha}\} \vdash_0 \neg\boldsymbol{\alpha}$, by *modus ponens*. Therefore $\{\boldsymbol{\alpha}\to\neg\boldsymbol{\alpha}, \boldsymbol{\alpha}\} \vdash_0$ and by *reductio* we get[1]

(1) $$\boldsymbol{\alpha}\to\neg\boldsymbol{\alpha} \vdash_0 \neg\boldsymbol{\alpha}.$$

Now, $\{\neg\boldsymbol{\alpha}, \neg\neg\boldsymbol{\alpha}\}$ is a contradictory pair, so it follows from (1) that $\{\neg\neg\boldsymbol{\alpha}, \boldsymbol{\alpha}\to\neg\boldsymbol{\alpha}\} \vdash_0$. Hence by IE we have $\{\neg\neg\boldsymbol{\alpha}, \boldsymbol{\alpha}\to\neg\boldsymbol{\alpha}\} \vdash_0 \boldsymbol{\alpha}$, and by DT

(2) $$\neg\neg\boldsymbol{\alpha} \vdash_0 (\boldsymbol{\alpha}\to\neg\boldsymbol{\alpha})\to\boldsymbol{\alpha}.$$

Next, $[(\boldsymbol{\alpha}\to\neg\boldsymbol{\alpha})\to\boldsymbol{\alpha}]\to\boldsymbol{\alpha}$ is an instance of Ax. iii, therefore $(\boldsymbol{\alpha}\to\neg\boldsymbol{\alpha})\to\boldsymbol{\alpha} \vdash_0 \boldsymbol{\alpha}$. From this and (2) we get by Cut $\neg\neg\boldsymbol{\alpha} \vdash_0 \boldsymbol{\alpha}$, as claimed. ■

8.15. Theorem (Principle of Indirect Proof)

If $\boldsymbol{\Phi}, \neg\boldsymbol{\alpha} \vdash_0$ *then* $\boldsymbol{\Phi} \vdash_0 \boldsymbol{\alpha}$.

PROOF

Assume $\boldsymbol{\Phi}, \neg\boldsymbol{\alpha} \vdash_0$. By *reductio*, $\boldsymbol{\Phi} \vdash_0 \neg\neg\boldsymbol{\alpha}$; hence, using Lemma 8.14 and Cut, $\boldsymbol{\Phi} \vdash_0 \boldsymbol{\alpha}$, as claimed. ■

8.16. Remarks

(i) For brevity, we shall refer to the Principle of Indirect Proof as 'PIP'.
(ii) Lemma 8.14 is a special case of PIP, for clearly $\{\neg\neg\boldsymbol{\alpha}, \neg\boldsymbol{\alpha}\} \vdash_0$.
(iii) The converse of PIP is immediate.
(iv) The sole purpose of adopting Ax. iii was to enable us to prove PIP. Henceforth it will no longer be necessary to invoke this axiom scheme.
(v) Indeed, from now on we shall not invoke *any* propositional axiom, because the four proof-theoretic principles – DT, IE, *reductio* and PIP – jointly contain all the information that the choice of axioms was designed to provide (cf. Probs. 7.4, 8.8, 8.11 and 8.18). We use these four principles even where, as in the proof of Lemma 8.14, it would have been quicker to invoke an axiom. The reason for this apparent perversity is that the axioms are forgettable, mere scaffolding, whereas the four principles (together with *modus ponens* and Cut) encapsulate the most

[1] We could have got (1) more directly, as in the proof of Thm. 8.9; but see Rem. 8.16(v).

important inherent structural facts about the propositional calculus.

8.17. Warning

Do not commit the solecism of confusing PIP with *reductio*. The two principles, though formally similar to each other, are quite distinct. (Intuitionistic logic rejects the former and upholds the latter.)

8.18. Problem

Let $\vdash^*$ be the deducibility relation in a calculus that has *modus ponens* as a rule of inference and for which Cut, DT, IE and PIP hold. Prove that $\vdash^* [(\boldsymbol{\alpha}\to\boldsymbol{\beta})\to\boldsymbol{\alpha}]\to\boldsymbol{\alpha}$ for all $\boldsymbol{\alpha}$ and $\boldsymbol{\beta}$.

8.19. Problem

Prove:

(i) $\neg\boldsymbol{\alpha} \vdash_0 \boldsymbol{\alpha}\to\boldsymbol{\beta}$,
(ii) $\boldsymbol{\beta} \vdash_0 \boldsymbol{\alpha}\to\boldsymbol{\beta}$,
(iii) $\{\boldsymbol{\alpha}, \neg\boldsymbol{\beta}\} \vdash_0 \neg(\boldsymbol{\alpha}\to\boldsymbol{\beta})$,
(iv) $\neg(\boldsymbol{\alpha}\to\boldsymbol{\beta}) \vdash_0 \boldsymbol{\alpha}$,
(v) $\neg(\boldsymbol{\alpha}\to\boldsymbol{\beta}) \vdash_0 \neg\boldsymbol{\beta}$.

8.20. Problem

Using Def. 2.5, prove:

(i) $\boldsymbol{\alpha}\wedge\neg\boldsymbol{\alpha} \vdash_0$,
(ii) $\vdash_0 \boldsymbol{\alpha}\vee\neg\boldsymbol{\alpha}$,
(iii) $\boldsymbol{\alpha}\wedge\boldsymbol{\beta} \vdash_0 \boldsymbol{\beta}\wedge\boldsymbol{\alpha}$,
(iv) $\boldsymbol{\alpha}\vee\boldsymbol{\beta} \vdash_0 \boldsymbol{\beta}\vee\boldsymbol{\alpha}$.

8.21. Remark

In Prob. 8.20, (ii) does not depend on the intuitionistically invalid PIP (or Ax. iii), whereas (iv) does. On the other hand, it is well known that in intuitionistic logic the law of excluded middle is invalid, whereas disjunction has a symmetric meaning. This apparent incongruity is due to the fact that in intuitionistic logic Def. 2.5(ii) itself is not acceptable, because disjunction (and, for that matter, conjunction) cannot be reduced to negation and implication.

8.22. Problem

Prove that $\vdash_0 (\neg\alpha\rightarrow\beta)\rightarrow(\neg\alpha\rightarrow\neg\beta)\rightarrow\alpha$.

8.23. Remark

The version of the propositional calculus introduced in B&M Ch. 1, § 10 differs from the present one solely in having the axiom scheme $(\neg\alpha\rightarrow\beta)\rightarrow(\neg\alpha\rightarrow\neg\beta)\rightarrow\alpha$ instead of our last three axiom schemes. Hence by Prob. 8.22 all the axioms of that version are theorems in the present version. On the other hand, since (as shown in B&M, Ch. 1, § 10) IE, *reductio* and PIP hold for the B&M version, it follows from Probs. 8.8, 8.11 and 8.18 that the converse also holds: all axioms of the present version are theorems in the B&M version. The two versions are therefore equivalent.

§ 9. Weak completeness

9.1. Observation

We reproduce below the truth tables for $\neg\beta$ in terms of β and for $\beta\rightarrow\gamma$ in terms of β, γ (cf. p. 111). Alongside these tables we quote some proven results concerning deducibility.

β	$\neg\beta$		
$\top$	$\bot$	$\beta \vdash_0 \neg\neg\beta$	(Prob. 8.12),
$\bot$	$\top$	$\neg\beta \vdash_0 \neg\beta$	(obvious).

β	γ	$\beta\rightarrow\gamma$		
$\top$	$\top$	$\top$	$\{\beta, \gamma\} \vdash_0 \beta\rightarrow\gamma$	(Prob. 8.19(ii)),
$\top$	$\bot$	$\bot$	$\{\beta, \neg\gamma\} \vdash_0 \neg(\beta\rightarrow\gamma)$	(Prob. 8.19(iii)),
$\bot$	$\top$	$\top$	$\{\neg\beta, \gamma\} \vdash_0 \beta\rightarrow\gamma$	(Prob. 8.19(i) or 8.19(ii)),
$\bot$	$\bot$	$\top$	$\{\neg\beta, \neg\gamma\} \vdash_0 \beta\rightarrow\gamma$	(Prob. 8.19(i)).

Observe that there is a systematic relationship between each row in the truth tables and the deducibility statement to its right. The formulas involved in each statement are related to the headings of the columns in the truth tables. Where the entry in the truth table is '$\top$', the corresponding formula on the right is exactly the one indicated at the head of the relevant column; but where the entry in the truth table is '$\bot$', the corresponding formula on the right is the negation of that indicated at the head of the column.

We shall now generalize this observation to *all* truth tables.

9.2. Lemma

Let $\boldsymbol{\alpha}$ *be a combination of formulas* $\boldsymbol{\beta}_1, \boldsymbol{\beta}_2, \ldots, \boldsymbol{\beta}_k$. *Select a given row in a truth table for* $\boldsymbol{\alpha}$ *in terms of* $\boldsymbol{\beta}_1, \boldsymbol{\beta}_2, \ldots, \boldsymbol{\beta}_k$. *For each* $i = 1, 2, \ldots, k$ *let* $\boldsymbol{\beta}_i'$ *be* $\boldsymbol{\beta}_i$ *or* $\neg\boldsymbol{\beta}_i$, *according as the entry in the given row at the i-th column is '*$\top$*' or '*$\bot$*'. Similarly, let* $\boldsymbol{\alpha}'$ *be* $\boldsymbol{\alpha}$ *or* $\neg\boldsymbol{\alpha}$, *according as the entry in the given row at the last column is '*$\top$*' or '*$\bot$*'.*

Then $\{\boldsymbol{\beta}_1', \boldsymbol{\beta}_2', \ldots, \boldsymbol{\beta}_k'\} \vdash_0 \boldsymbol{\alpha}'$.

PROOF

For brevity, we put $\boldsymbol{\Phi} = \{\boldsymbol{\beta}_1', \boldsymbol{\beta}_2', \ldots, \boldsymbol{\beta}_k'\}$, so we must prove $\boldsymbol{\Phi} \vdash_0 \boldsymbol{\alpha}'$. We proceed by induction on $\deg \boldsymbol{\alpha}$ and distinguish three cases, according to which of the three rules in Def. 5.1 was used to construct the last column in the truth table in question.

Case 1: $\boldsymbol{\alpha} = \boldsymbol{\beta}_i$ (where $1 \leqslant i \leqslant k$) and Rule (1) of Def. 5.1 was used. In this case the entry in the given row and last column is a copy of the one in the i-th column. Then $\boldsymbol{\alpha}' = \boldsymbol{\beta}_i' \in \boldsymbol{\Phi}$ and obviously $\boldsymbol{\Phi} \vdash_0 \boldsymbol{\alpha}'$.

Case 2: $\boldsymbol{\alpha} = \neg\boldsymbol{\gamma}$, where $\boldsymbol{\gamma}$ is a combination of $\boldsymbol{\beta}_1, \boldsymbol{\beta}_2, \ldots, \boldsymbol{\beta}_k$ and Rule (2) of Def. 5.1 was used. By the induction hypothesis, $\boldsymbol{\Phi} \vdash_0 \boldsymbol{\gamma}'$, so the required result, $\boldsymbol{\Phi} \vdash_0 \boldsymbol{\alpha}'$, will follow (thanks to Cut) if we show that $\boldsymbol{\gamma}' \vdash_0 \boldsymbol{\alpha}'$. We distinguish two subcases.

Subcase 2a: $\boldsymbol{\gamma}' = \boldsymbol{\gamma}$. Then according to Rule (2) we get $\boldsymbol{\alpha}' = \neg\boldsymbol{\alpha} = \neg\neg\boldsymbol{\gamma}$. Thus $\boldsymbol{\gamma}' \vdash_0 \boldsymbol{\alpha}'$ is the same as $\boldsymbol{\gamma} \vdash_0 \neg\neg\boldsymbol{\gamma}$, which holds by Prob. 8.12.

Subcase 2b: $\boldsymbol{\gamma}' = \neg\boldsymbol{\gamma}$. Then according to Rule (2) we get $\boldsymbol{\alpha}' = \boldsymbol{\alpha} = \neg\boldsymbol{\gamma}$; and $\boldsymbol{\gamma}' \vdash_0 \boldsymbol{\alpha}'$ is the same as $\neg\boldsymbol{\gamma} \vdash_0 \neg\boldsymbol{\gamma}$, which is obvious.

Case 3: $\boldsymbol{\alpha} = \boldsymbol{\gamma} \to \boldsymbol{\delta}$, where $\boldsymbol{\gamma}$ and $\boldsymbol{\delta}$ are combinations of $\boldsymbol{\beta}_1, \boldsymbol{\beta}_2, \ldots, \boldsymbol{\beta}_k$ and Rule (3) of Def. 5.1 was used. By the induction hypothesis, $\boldsymbol{\Phi} \vdash_0 \boldsymbol{\gamma}'$ and $\boldsymbol{\Phi} \vdash_0 \boldsymbol{\delta}'$, so the required result, $\boldsymbol{\Phi} \vdash_0 \boldsymbol{\alpha}'$, will follow (thanks to Cut) if we show that $\{\boldsymbol{\gamma}', \boldsymbol{\delta}'\} \vdash_0 \boldsymbol{\alpha}'$. We distinguish three subcases (the first two of which are not mutually exclusive).

Subcase 3a: $\boldsymbol{\gamma}' = \neg\boldsymbol{\gamma}$. Then according to Rule (3) we get $\boldsymbol{\alpha}' = \boldsymbol{\alpha} = \boldsymbol{\gamma} \to \boldsymbol{\delta}$. So $\boldsymbol{\gamma}' \vdash_0 \boldsymbol{\alpha}'$ is the same as $\neg\boldsymbol{\gamma} \vdash_0 \boldsymbol{\gamma} \to \boldsymbol{\delta}$, which holds by Prob. 8.19(i). Therefore *a fortiori* $\{\boldsymbol{\gamma}', \boldsymbol{\delta}'\} \vdash_0 \boldsymbol{\alpha}'$.

Subcase 3b: $\boldsymbol{\delta}' = \boldsymbol{\delta}$. According to Rule (3) we have again $\boldsymbol{\alpha}' = \boldsymbol{\alpha} = \boldsymbol{\gamma} \rightarrow \boldsymbol{\delta}$. So $\boldsymbol{\delta}' \vdash_0 \boldsymbol{\alpha}'$ is the same as $\boldsymbol{\delta} \vdash_0 \boldsymbol{\gamma} \rightarrow \boldsymbol{\delta}$, which holds by Prob. 8.19(ii). Therefore *a fortiori* $\{\boldsymbol{\gamma}', \boldsymbol{\delta}'\} \vdash_0 \boldsymbol{\alpha}'$.

Subcase 3c: neither of the previous two subcases holds; so $\boldsymbol{\gamma}' = \boldsymbol{\gamma}$ and $\boldsymbol{\delta}' = \neg\boldsymbol{\delta}$. Then by Rule (3) $\boldsymbol{\alpha}' = \neg\boldsymbol{\alpha} = \neg(\boldsymbol{\gamma} \rightarrow \boldsymbol{\delta})$. So $\{\boldsymbol{\gamma}', \boldsymbol{\delta}'\} \vdash_0 \boldsymbol{\alpha}'$ is the same as $\{\boldsymbol{\gamma}, \neg\boldsymbol{\delta}\} \vdash_0 \neg(\boldsymbol{\gamma} \rightarrow \boldsymbol{\delta})$, which holds by Prob. 8.19(iii). ■

9.3. Lemma

Let $\boldsymbol{\alpha}$ be a combination of $\boldsymbol{\beta}_1, \boldsymbol{\beta}_2, \ldots, \boldsymbol{\beta}_k$, and suppose that in some truth table for $\boldsymbol{\alpha}$ in terms of $\boldsymbol{\beta}_1, \boldsymbol{\beta}_2, \ldots, \boldsymbol{\beta}_k$ all the entries in the last column are '⊤'. For each $i = 1, 2, \ldots, k$ let $\boldsymbol{\beta}_i'$ be chosen arbitrarily as $\boldsymbol{\beta}_i$ or $\neg\boldsymbol{\beta}_i$ – the choice being made independently for different i. Then $\{\boldsymbol{\beta}_1', \boldsymbol{\beta}_2', \ldots, \boldsymbol{\beta}_{k-p}'\} \vdash_0 \boldsymbol{\alpha}$ for every $p = 0, 1, \ldots, k$. In particular, for $p = k$, $\vdash_0 \boldsymbol{\alpha}$.

PROOF

By induction on p. For $p = 0$ the claim is that $\{\boldsymbol{\beta}_1', \boldsymbol{\beta}_2', \ldots, \boldsymbol{\beta}_k'\} \vdash_0 \boldsymbol{\alpha}$. This holds by Lemma 9.2, because according to our present assumption the formula $\boldsymbol{\alpha}'$ (defined there) is always $\boldsymbol{\alpha}$ itself.

For the induction step, let $p < k$. We must show that $\boldsymbol{\Phi} \vdash_0 \boldsymbol{\alpha}$, where $\boldsymbol{\Phi} = \{\boldsymbol{\beta}_1', \boldsymbol{\beta}_2', \ldots, \boldsymbol{\beta}_{k-(p+1)}'\}$. (If $p = k - 1$ then $\boldsymbol{\Phi} = \emptyset$.)

The induction hypothesis is that $\boldsymbol{\Phi}, \boldsymbol{\beta}_{k-p}' \vdash_0 \boldsymbol{\alpha}$. But we are free to choose $\boldsymbol{\beta}_{k-p}'$ in two ways: as $\boldsymbol{\beta}_{k-p}$ or as $\neg\boldsymbol{\beta}_{k-p}$. So we have

$$\boldsymbol{\Phi}, \boldsymbol{\beta}_{k-p} \vdash_0 \boldsymbol{\alpha} \quad \text{and} \quad \boldsymbol{\Phi}, \neg\boldsymbol{\beta}_{k-p} \vdash_0 \boldsymbol{\alpha}.$$

Hence

$$\boldsymbol{\Phi}, \neg\boldsymbol{\alpha}, \boldsymbol{\beta}_{k-p} \vdash_0 \quad \text{and} \quad \boldsymbol{\Phi}, \neg\boldsymbol{\alpha}, \neg\boldsymbol{\beta}_{k-p} \vdash_0.$$

By *reductio* and PIP respectively, we get

$$\boldsymbol{\Phi}, \neg\boldsymbol{\alpha} \vdash_0 \neg\boldsymbol{\beta}_{k-p} \quad \text{and} \quad \boldsymbol{\Phi}, \neg\boldsymbol{\alpha} \vdash_0 \boldsymbol{\beta}_{k-p}.$$

This shows that $\boldsymbol{\Phi}, \neg\boldsymbol{\alpha} \vdash_0$. So by PIP $\boldsymbol{\Phi} \vdash_0 \boldsymbol{\alpha}$, as required. ■

We are now in a position to prove a partial converse of Thm. 6.12. The converse is only partial because of the restriction to *finite* sets of formulas; hence also the qualification 'weak' in the name of the theorem:

9.4. Theorem (Weak semantic completeness of Propcal)

For any formula $\boldsymbol{\alpha}$, *if* $\models_0 \boldsymbol{\alpha}$ *then* $\vdash_0 \boldsymbol{\alpha}$. *More generally, if* $\boldsymbol{\Phi}$ *is a finite set of formulas and* $\boldsymbol{\Phi} \models_0 \boldsymbol{\alpha}$, *then* $\boldsymbol{\Phi} \vdash_0 \boldsymbol{\alpha}$.

PROOF

Suppose $\models_0 \boldsymbol{\alpha}$. Then by Thm. 5.9 the truth table of $\boldsymbol{\alpha}$ in terms of all its prime components satisfies the assumption of Lemma 9.3; hence by that lemma $\vdash_0 \boldsymbol{\alpha}$.

To prove the second part of the theorem, assume that $\boldsymbol{\Phi} \models_0 \boldsymbol{\alpha}$, where $\boldsymbol{\Phi}$ is a finite set of formulas. Let $\boldsymbol{\varphi}_1, \boldsymbol{\varphi}_2, \ldots, \boldsymbol{\varphi}_k$ be all the members of $\boldsymbol{\Phi}$; then $\boldsymbol{\Phi} = \{\boldsymbol{\varphi}_1, \boldsymbol{\varphi}_2, \ldots, \boldsymbol{\Phi}_k\}$ and we have $\{\boldsymbol{\varphi}_1, \boldsymbol{\varphi}_2, \ldots, \boldsymbol{\varphi}_k\} \models_0 \boldsymbol{\alpha}$.

By Prob. 4.6(ii) we get $\models_0 \boldsymbol{\varphi}_1 \rightarrow \boldsymbol{\varphi}_2 \rightarrow \cdots \rightarrow \boldsymbol{\varphi}_k \rightarrow \boldsymbol{\alpha}$. Therefore, by the first part of the present theorem, $\vdash_0 \boldsymbol{\varphi}_1 \rightarrow \boldsymbol{\varphi}_2 \rightarrow \cdots \rightarrow \boldsymbol{\varphi}_k \rightarrow \boldsymbol{\alpha}$. Hence, by k applications of *modus ponens*, we obtain $\{\boldsymbol{\varphi}_1, \boldsymbol{\varphi}_2, \ldots, \boldsymbol{\varphi}_k\} \vdash_0 \boldsymbol{\alpha}$, that is, $\boldsymbol{\Phi} \vdash_0 \boldsymbol{\alpha}$. ■

A partial converse of Thm. 8.4 can now be proved.

9.5. Theorem

A finite unsatisfiable set of formulas is inconsistent: if $\boldsymbol{\Phi}$ *is finite and* $\boldsymbol{\Phi} \models_0$, *then* $\boldsymbol{\Phi} \vdash_0$.

PROOF

Suppose $\boldsymbol{\Phi} \models_0$. Then trivially $\boldsymbol{\Phi} \models_0 \boldsymbol{\alpha}$ for any formula $\boldsymbol{\alpha}$. If $\boldsymbol{\Phi}$ is finite, then by Thm. 9.4 it follows that $\boldsymbol{\Phi} \vdash_0 \boldsymbol{\alpha}$ for any $\boldsymbol{\alpha}$; hence clearly (cf. Rem. 8.7(ii)) $\boldsymbol{\Phi} \vdash_0$. ■

9.6. Remarks

(i) Thm. 9.5 has been formulated contrapositively. An equivalent positive formulation is: *A finite consistent set of formulas is satisfiable [by some truth valuation].*

(ii) Thms. 9.4 and 9.5 are equivalent. We have just seen that the latter follows from the former, but the converse also holds. Indeed, if $\boldsymbol{\Phi}$ is finite and $\boldsymbol{\Phi} \models_0 \boldsymbol{\alpha}$, then clearly $\boldsymbol{\Phi} \cup \{\neg\boldsymbol{\alpha}\}$ is finite and unsatisfiable; hence by Thm. 9.5 $\boldsymbol{\Phi}, \neg\boldsymbol{\alpha} \vdash_0$, and by PIP $\boldsymbol{\Phi} \vdash_0 \boldsymbol{\alpha}$.

§ 10. Hintikka sets

10.1 Preview

Our final task in propositional logic will be to prove the *full* converse of Thm. 6.12 – the *strong* semantic completeness of the propositional calculus. From Rem. 9.6 it should be clear that this task can be accomplished by proving first the full converse of Thm. 8.4: A *consistent – finite or infinite – set of formulas is satisfiable*. We shall do so in three easy stages.

First, we shall show that certain special sets of formulas, called *Hintikka* sets, are satisfiable. This will be quite easy, because the definition of these sets is rigged for this very purpose.

Second, we shall introduce the even more special *maximal consistent* sets of formulas and show that each such set is a Hintikka set, and hence satisfiable. In fact, it will transpire that there is a one-to-one correspondence between maximal consistent sets and truth valuations.

Finally, using a simple but powerful result from set theory, we shall show that every consistent set of formulas is a subset of a maximal consistent set, and is therefore automatically satisfied by the truth valuation that satisfies the latter.

10.2. Definition

A *[propositional] Hintikka* set *[in $\mathcal{L}$]* is a set $\boldsymbol{\Phi}$ of formulas satisfying the following four conditions for all formulas $\boldsymbol{\alpha}$ and $\boldsymbol{\beta}$:

(1) If $\boldsymbol{\alpha}$ is prime and $\boldsymbol{\alpha} \in \boldsymbol{\Phi}$, then $\neg\boldsymbol{\alpha} \notin \boldsymbol{\Phi}$.
(2) If $\neg\neg\boldsymbol{\alpha} \in \boldsymbol{\Phi}$, then also $\boldsymbol{\alpha} \in \boldsymbol{\Phi}$.
(3) If $\boldsymbol{\alpha}\to\boldsymbol{\beta} \in \boldsymbol{\Phi}$ then $\neg\boldsymbol{\alpha} \in \boldsymbol{\Phi}$ or $\boldsymbol{\beta} \in \boldsymbol{\Phi}$.
(4) If $\neg(\boldsymbol{\alpha}\to\boldsymbol{\beta}) \in \boldsymbol{\Phi}$ then $\boldsymbol{\alpha} \in \boldsymbol{\Phi}$ and $\neg\boldsymbol{\beta} \in \boldsymbol{\Phi}$.

10.3. Theorem

If $\boldsymbol{\Phi}$ is a Hintikka set, it is satisfied by some truth valuation.

PROOF

Let $\boldsymbol{\Phi}$ be a Hintikka set. Define a truth valuation σ by stipulating that $\boldsymbol{\alpha}^\sigma = \top$ for every prime formula $\boldsymbol{\alpha}$ belonging to $\boldsymbol{\Phi}$, and $\boldsymbol{\alpha}^\sigma = \bot$ for any other prime $\boldsymbol{\alpha}$. We claim that, for any formula $\boldsymbol{\varphi}$,

$$\text{(a) } \boldsymbol{\varphi} \in \boldsymbol{\Phi} \Rightarrow \boldsymbol{\varphi}^\sigma = \top, \qquad \text{(b) } \neg\boldsymbol{\varphi} \in \boldsymbol{\Phi} \Rightarrow \boldsymbol{\varphi}^\sigma = \bot.$$

We shall prove this double claim simultaneously[1] by induction on $\deg \varphi$. We distinguish three cases, corresponding to the three clauses of Def. 1.4 and those of Def. 4.2(ii).

Case 1: φ is prime.

(1a) $\varphi \in \Phi \Rightarrow \varphi^\sigma = \top$ by the definition of σ.

(1b) $\neg\varphi \in \Phi \Rightarrow \varphi \notin \Phi$ by clause (1) of Def. 10.2,
$\Rightarrow \varphi^\sigma = \bot$ by the definition of σ.

Case 2: φ is a negation formula; say $\varphi = \neg\alpha$.

(2a) $\varphi \in \Phi \Rightarrow \neg\alpha \in \Phi$
$\Rightarrow \alpha^\sigma = \bot$ by part (b) of ind. hyp.,
$\Rightarrow \varphi^\sigma = \top$ by clause (2) of Def. 4.2(ii).

(2b) $\neg\varphi \in \Phi \Rightarrow \neg\neg\alpha \in \Phi$
$\Rightarrow \alpha \in \Phi$ by clause (2) of Def. 10.2,
$\Rightarrow \alpha^\sigma = \top$ by part (a) of ind. hyp.,
$\Rightarrow \varphi^\sigma = \bot$ by clause (2) of Def. 4.2(ii).

Case 3: φ is an implication formula; say $\varphi = \alpha \rightarrow \beta$.

(3a) $\varphi \in \Phi \Rightarrow \alpha \rightarrow \beta \in \Phi$
$\Rightarrow \neg\alpha \in \Phi$ or $\beta \in \Phi$ by clause (3) of Def. 10.2,
$\Rightarrow \alpha^\sigma \in \bot$ or $\beta^\sigma = \top$ by ind. hyp.,
$\Rightarrow \varphi^\sigma = \top$ by clause (3) of Def. 4.2(ii).

(3b) $\neg\varphi \in \Phi \Rightarrow \alpha \in \Phi \;\&\; \neg\beta \in \Phi$ by clause (4) of Def. 10.2,
$\Rightarrow \alpha^\sigma = \top$ and $\beta^\sigma = \bot$ by ind. hyp.,
$\Rightarrow \varphi^\sigma = \bot$ by clause (3) of Def. 4.2(ii). ■

§ 11. The ambient metatheory

Let us pause to consider the mathematical presuppositions that underlie our study of propositional logic. This study is being conducted in mathematical fashion: we frame precise definitions and prove [meta]-theorems concerning the object language $\mathcal{L}$, its syntax and semantics. The mathematical theory in which this study is conducted is our

[1] Note that (a) by itself is sufficient for proving our theorem; and once (a) is established for all φ then (b) would follow automatically. But if you try to prove (a) on its own, you will find out that the inductive argument runs into snags.

metatheory. (The prefix *meta* is used here to distinguish this theory, which is developed in our metalanguage, from formal *object* theories that may be constructed in the object language and serve as objects of our study.)

As any other mathematical theory, our metatheory must start from a launching pad of presuppositions: some underlying concepts, regarded as known, in terms of which further concepts of the theory are defined; and certain fundamental propositions, on the basis of which the theorems of our theory can be rigorously proved.

Set theory – in the form of ZF or some broadly similar codification – is certainly strong enough to underpin our study of logic. Indeed, the entire technical development in the Logic part of this book can be read as occurring within set theory. Interpreted in this way, not only the term *set* but also other mathematical terms such as *natural number* and *finite*, must be understood in the appropriate technical sense: a *natural number* as an ordinal belonging to ω, and a *finite* set as a set equipollent to a natural number (cf. Rem. 6.1.8).

But all that we have done so far in this chapter does not really require such a strong ambient theory. Set theory is vital only where there is need to regard infinite pluralities as single objects: sets that in turn can themselves be members of classes. So far we have hardly had any need for positing such *completed* (or *actual*) infinities. Though we have used set-theoretic terminology, this was not essential. For example, although in Specification 1.1(i) we refer to the totality of propositional symbols of $\mathcal{L}$ as a *set*, we have never had to regard this totality as a single object that can be a member of a class. We only need the stock of propositional symbols to be *potentially* infinite; so everything we have done works just as well if we replace the word 'set' here by 'collection' or by one of its synonyms such as 'plurality' or 'class'. The same applies to other places where the term 'set' has been used.

There was one context that seems to be an exception to what we have just said and where we did refer to infinite entities as objects. In Def. 4.4(iii) a formula $\boldsymbol{\alpha}$ was defined to be a tautology, $\models_0 \boldsymbol{\alpha}$, just in case $\boldsymbol{\alpha}^\sigma = \top$ for *every* truth valuation σ. This definition refers (at least implicitly) to the class of *all* truth valuations. Now, by Def. 4.2(i), a truth valuation is a map with an infinite domain, and hence is itself infinite.

However, this reliance on infinite objects can be avoided by a simple device. Clearly, the truth value $\boldsymbol{\alpha}^\sigma$ (defined in Def. 4.2(ii)) depends only on the values assigned by σ to the prime components of $\boldsymbol{\alpha}$.

Therefore, instead of truth valuations proper we may consider *partial* truth valuations, whose domain is a *finite* set of propositional symbols, among which are all the prime components of $\boldsymbol{\alpha}$. These partial truth valuations themselves are finite objects; and the notion of *tautology* can be redefined by referring to the class of these objects rather than of truth valuations proper.

A similar device can be used in connection with the definition of the notion of *tautological consequence*, $\boldsymbol{\Phi} \models_0 \boldsymbol{\alpha}$, provided the collection $\boldsymbol{\Phi}$ is finite. (It is enough to consider partial truth valuations whose domains are finite sets of propositional symbols, among which are all the prime components of $\boldsymbol{\alpha}$ and all the members of $\boldsymbol{\Phi}$.)

Thus, provided we restrict parts (iv) and (v) of Def. 4.4 to finite sets $\boldsymbol{\Phi}$, the rest of the development of propositional logic so far does not require the framework of set theory. Looked at in this way, terms such as *natural number* and *finite* are to be understood informally rather than in their technical set-theoretic sense. To be sure, some – relatively modest – mathematical presuppositions are still needed as underpinning. We shall not attempt to specify these presuppositions in detail, but merely point out that among them the Principle of Mathematical Induction takes pride of place.

But such modest mathematical underpinning is no longer adequate for the development in the following sections of this chapter. Here, particularly in the proof of Thm. 13.1, some set-theoretic machinery must be used. So this development must be understood as taking place within a sufficiently strong ambient set theory. (See, however, Rem. 13.3(i).)

§ 12. Maximal consistent sets

12.1. Definition

A *maximal [propositionally] consistent* set is a consistent set of formulas that is not a proper subset of any consistent set of formulas.

12.2. Remarks

(i) In other words, a set $\boldsymbol{\Phi}$ of formulas is maximal consistent iff $\boldsymbol{\Phi}$ is consistent, but by adding to $\boldsymbol{\Phi}$ even a single new formula (that is, one not already belonging to it) we obtain an inconsistent set.

(ii) Maximal consistency is an instance of a general set-theoretic concept. Let $\mathcal{K}$ be the class of all consistent sets of formulas. The relation $\subset_{\mathcal{K}}$ is then a partial order on $\mathcal{K}$ (see Def. 5.2.5 and Rem. 5.2.6). A maximal consistent set is just a maximal member of $\mathcal{K}$ with respect to the partial order $\subset_{\mathcal{K}}$ (see Def. 5.2.3).

The following theorem shows that a maximal consistent set is *saturated with respect to deducibility*.

12.3. Theorem

If $\mathbf{\Phi}$ *is maximal consistent and* $\mathbf{\Phi} \vdash_0 \boldsymbol{\alpha}$, *then* $\boldsymbol{\alpha} \in \mathbf{\Phi}$.

PROOF

Let $\mathbf{\Phi}$ be maximal consistent and $\mathbf{\Phi} \vdash_0 \boldsymbol{\alpha}$. Suppose it were the case that $\boldsymbol{\alpha} \notin \mathbf{\Phi}$. Then, by the maximality of $\mathbf{\Phi}$, we would get $\mathbf{\Phi}, \boldsymbol{\alpha} \vdash_0$ (cf. Rem. 12.2(i)). Hence by *reductio* we would have $\mathbf{\Phi} \vdash_0 \neg\boldsymbol{\alpha}$, showing that $\mathbf{\Phi}$ itself is inconsistent, contrary to hypothesis. ■

The following theorem provides an alternative characterization of maximal consistent sets.

12.4. Theorem

A consistent set $\mathbf{\Phi}$ *is maximal consistent iff for every formula* $\boldsymbol{\alpha}$ *either* $\boldsymbol{\alpha} \in \mathbf{\Phi}$ *or* $\neg\boldsymbol{\alpha} \in \mathbf{\Phi}$.

First, assume $\mathbf{\Phi}$ is maximal consistent. If $\boldsymbol{\alpha} \notin \mathbf{\Phi}$ then by the maximality of $\mathbf{\Phi}$ it follows that $\mathbf{\Phi}, \boldsymbol{\alpha} \vdash_0$. Hence by *reductio* $\mathbf{\Phi} \vdash_0 \neg\boldsymbol{\alpha}$, and by Thm. 12.3 $\neg\boldsymbol{\alpha} \in \mathbf{\Phi}$.

Conversely, assume $\mathbf{\Phi}$ is consistent and satisfies the condition in question. Let $\boldsymbol{\alpha}$ be any formula that does not belong to $\mathbf{\Phi}$; so by the assumed condition $\neg\boldsymbol{\alpha} \in \mathbf{\Phi}$. It follows that $\mathbf{\Phi}, \boldsymbol{\alpha} \vdash_0$. Thus we see that by adding to $\mathbf{\Phi}$ even a single new formula we get an inconsistent set. Thus (cf. Rem. 12.2(i)) $\mathbf{\Phi}$ is maximal consistent. ■

12.5. Theorem

Every maximal consistent set is a Hintikka set.

PROOF

Let $\mathbf{\Phi}$ be maximal consistent. We shall show that $\mathbf{\Phi}$ fulfils the four conditions of Def. 10.2.

Condition (1) of that definition is obviously satisfied, since $\mathbf{\Phi}$ is consistent.

Now suppose $\neg\neg\boldsymbol{\alpha} \in \mathbf{\Phi}$. Then by Lemma 8.14 $\mathbf{\Phi} \vdash_0 \boldsymbol{\alpha}$, hence by Thm. 12.3 $\boldsymbol{\alpha} \in \mathbf{\Phi}$. Thus condition (2) of Def. 10.2 is satisfied.

Next, suppose $\boldsymbol{\alpha} \to \boldsymbol{\beta} \in \mathbf{\Phi}$. If $\neg\boldsymbol{\alpha} \in \mathbf{\Phi}$ then condition (3) of Def. 10.2 is satisfied. On the other hand, if $\neg\boldsymbol{\alpha} \notin \mathbf{\Phi}$ then by Thm. 12.4 $\boldsymbol{\alpha} \in \mathbf{\Phi}$. Since we have assumed that $\boldsymbol{\alpha} \to \boldsymbol{\beta} \in \mathbf{\Phi}$, it now follows that $\mathbf{\Phi} \vdash_0 \boldsymbol{\beta}$ and hence by Thm. 12.3 $\boldsymbol{\beta} \in \mathbf{\Phi}$. Thus condition (3) of Def. 10.2 is satisfied in this case as well.

Finally, suppose $\neg(\boldsymbol{\alpha} \to \boldsymbol{\beta}) \in \mathbf{\Phi}$. By parts (iv) and (v) of Prob. 8.19 we have $\mathbf{\Phi} \vdash_0 \boldsymbol{\alpha}$ and $\mathbf{\Phi} \vdash_0 \neg\boldsymbol{\beta}$. Hence by Thm. 12.3 $\boldsymbol{\alpha} \in \mathbf{\Phi}$ and $\neg\boldsymbol{\beta} \in \mathbf{\Phi}$. Thus condition (4) of Def. 10.2 is satisfied. ■

The following theorem establishes a one-to-one correspondence between truth valuations and maximal consistent sets.

12.6. Theorem

(i) *For any truth valuation* σ, *the set* $\{\boldsymbol{\varphi}: \boldsymbol{\varphi}^\sigma = \top\}$ *is maximal consistent.*

(ii) *Conversely, if* $\mathbf{\Phi}$ *is maximal consistent then* $\mathbf{\Phi} = \{\boldsymbol{\varphi}: \boldsymbol{\varphi}^\sigma = \top\}$ *for some truth valuation* σ. *Moreover, this* σ *is the unique truth valuation satisfying* $\mathbf{\Phi}$.

PROOF

(i) Put $\mathbf{\Psi} = \{\boldsymbol{\varphi}: \boldsymbol{\varphi}^\sigma = \top\}$. $\mathbf{\Psi}$ is evidently satisfiable: it is satisfied by σ. Hence by Thm. 8.4 it must be consistent.

If $\boldsymbol{\alpha}$ is a formula such that $\boldsymbol{\alpha} \notin \mathbf{\Psi}$ then, by the definition of $\mathbf{\Psi}$, it follows that $\boldsymbol{\alpha}^\sigma = \bot$. Hence $(\neg\boldsymbol{\alpha})^\sigma = \top$ and so $\neg\boldsymbol{\alpha} \in \mathbf{\Psi}$. Thus by Thm. 12.4 $\mathbf{\Psi}$ is maximal consistent.

(ii) Conversely, let $\mathbf{\Phi}$ be any maximal consistent set. By Thm. 12.5, $\mathbf{\Phi}$ is a Hintikka set and hence by Thm. 10.3 it is satisfiable. Let σ be a truth valuation satisfying $\mathbf{\Phi}$. Again let us put $\mathbf{\Psi} = \{\boldsymbol{\varphi}: \boldsymbol{\varphi}^\sigma = \top\}$. Now $\mathbf{\Psi}$ is the set of *all* formulas satisfied by σ, so $\mathbf{\Phi} \subseteq \mathbf{\Psi}$. By (i), $\mathbf{\Psi}$ is consistent; but $\mathbf{\Phi}$, being *maximal* consistent, cannot be included in another consistent set. Therefore $\mathbf{\Psi}$ cannot be other than $\mathbf{\Phi}$ itself. Thus $\mathbf{\Phi} = \mathbf{\Psi} = \{\boldsymbol{\varphi}: \boldsymbol{\varphi}^\sigma = \top\}$.

As we have just seen, if σ is *any* truth valuation satisfying $\mathbf{\Phi}$ then $\boldsymbol{\varphi}^\sigma = \top$ holds *just* for formulas $\boldsymbol{\varphi}$ belonging to $\mathbf{\Phi}$ *and for no others*. This means that σ is uniquely determined by $\mathbf{\Phi}$. ■

12.7. Remark

It is now clear that showing a set of formulas to be satisfiable is tantamount to showing that it is included in a maximal consistent set.

12.8. Problem (The [classical] logic of implication)

An *implicational valuation* is a mapping from the set of all prime formulas and all negation formulas to the set $\{\top, \bot\}$ of truth values. An implicational valuation is then extended to implication formulas as well by imposing condition (3) of Def. 4.2(ii). Let $\models^*$ be the resulting consequence relation; thus $\mathbf{\Phi} \models^* \boldsymbol{\alpha}$ iff every implicational valuation satisfying $\mathbf{\Phi}$ also satisfies $\boldsymbol{\alpha}$.

Let $\vdash^*$ be the relation of deducibility in the *[classical] calculus of implication* – the linear calculus based on Ax. i, Ax. ii and Ax. iii, with *modus ponens* as sole rule of inference.

(i) Verify that the calculus of implication is semantically sound: $\mathbf{\Phi} \vdash^* \boldsymbol{\alpha} \Rightarrow \mathbf{\Phi} \models^* \boldsymbol{\alpha}$.
(ii) Show that $\boldsymbol{\beta}\rightarrow\boldsymbol{\alpha}, (\boldsymbol{\beta}\rightarrow\boldsymbol{\gamma})\rightarrow\boldsymbol{\alpha} \vdash^* \boldsymbol{\alpha}$ for all $\boldsymbol{\alpha}$, $\boldsymbol{\beta}$ and $\boldsymbol{\gamma}$.
(iii) Let $\boldsymbol{\alpha}$ be a formula and let $\mathbf{\Phi}$ be a set of formulas such that $\mathbf{\Phi} \nvdash^* \boldsymbol{\alpha}$ and which is maximal with this property (that is, $\mathbf{\Phi}$ is not a proper subset of any $\mathbf{\Psi}$ such that $\mathbf{\Psi} \nvdash^* \boldsymbol{\alpha}$). Show that $\mathbf{\Phi}$ is saturated with respect to $\vdash^*$: if $\mathbf{\Phi} \vdash^* \boldsymbol{\beta}$ then $\boldsymbol{\beta} \in \mathbf{\Phi}$.
(iv) Let $\boldsymbol{\alpha}$ and $\mathbf{\Phi}$ be as in (iii). Show that there is a unique implicational valuation that satisfies $\mathbf{\Phi}$ but does not satisfy $\boldsymbol{\alpha}$.

§13. Strong completeness

The road to the strong completeness theorem is now clear.

13.1. Theorem

Every consistent set of formulas is satisfied by a truth valuation.

PROOF

Let $\mathbf{\Phi}$ be any set of formulas. If $\mathbf{\Phi}$ is consistent then clearly every subset of $\mathbf{\Phi}$, and in particular every finite subset, is consistent (cf. Prob. 8.3(i)). Conversely, if every finite subset of $\mathbf{\Phi}$ is consistent then by Prob. 8.3(ii) $\mathbf{\Phi}$ itself is consistent.

Thus the class $\mathcal{K}$ of all consistent sets of formulas is of finite character (see Def. 5.2.7). It is not difficult to see that $\mathcal{K}$ is in fact a set. (The class **S** of all $\mathcal{L}$-strings is a set by Thm. 6.3.9; and $\mathcal{K}$ is included in **PS**.) So if $\mathbf{\Phi}$ is any consistent set, it follows from the TT Lemma (Thm. 5.2.8) that $\mathbf{\Phi}$ is included in some (not necessarily unique) maximal consistent set $\mathbf{\Psi}$. By Thm. 12.6(ii) $\mathbf{\Psi}$ is satisfiable, and hence so is $\mathbf{\Phi}$. ■

13.2. Theorem (Strong semantic completeness of Propcal)

For any set $\mathbf{\Phi}$ *of formulas and any formula* $\boldsymbol{\alpha}$, if $\mathbf{\Phi} \models_0 \boldsymbol{\alpha}$ *then* $\mathbf{\Phi} \vdash_0 \boldsymbol{\alpha}$.

PROOF

If $\mathbf{\Phi} \models_0 \boldsymbol{\alpha}$ then every truth valuation satisfying $\mathbf{\Phi}$ must satisfy $\boldsymbol{\alpha}$ and hence cannot satisfy $\neg\boldsymbol{\alpha}$. Thus $\mathbf{\Phi}, \neg\boldsymbol{\alpha} \not\models_0$. By Thm. 13.1 $\mathbf{\Phi}, \neg\boldsymbol{\alpha} \not\vdash_0$; hence by PIP $\mathbf{\Phi} \vdash_0 \boldsymbol{\alpha}$. ■

13.3. Remarks

(i) If the primitive symbols of $\mathcal{L}$ are given by an explicit enumeration:

$$\{\mathbf{p}_n : n \in N\},$$

then the proof of Thm. 13.1 can be made more elementary and constructive. First, it is easy to define explicitly an enumeration of all $\mathcal{L}$-formulas:

$$\{\boldsymbol{\varphi}_n : n \in N\}.$$

Next, given a consistent set $\mathbf{\Phi}$, we define, by induction on n, sets $\mathbf{\Phi}_n$ as follows. We put $\mathbf{\Phi}_0 = \mathbf{\Phi}$; and

$$\mathbf{\Phi}_{n+1} = \begin{cases} \mathbf{\Phi}_n \cup \{\boldsymbol{\varphi}_n\} & \text{if this set is consistent,} \\ \mathbf{\Phi}_n & \text{otherwise.} \end{cases}$$

It is then quite easy to show that the union $\mathbf{\Psi} = \bigcup\{\mathbf{\Phi}_n : n \in N\}$ is a maximal consistent set; and $\mathbf{\Psi}$ clearly includes $\mathbf{\Phi}$.

(ii) The soundness and completeness theorems (Thms. 6.12 and 13.2) jointly mean that the relations of deducibility and tautological consequence are co-extensive: $\mathbf{\Phi} \vdash_0 \boldsymbol{\alpha}$ iff $\mathbf{\Phi} \models_0 \boldsymbol{\alpha}$. Similarly, Thms. 8.4 and 13.1 jointly mean that consistency and satisfiability are co-extensive: $\mathbf{\Phi} \not\vdash_0$ iff $\mathbf{\Phi} \not\models_0$. Therefore any fact proved for $\vdash_0$ holds also for $\models_0$ and vice versa. An important example is the following result.

13.4. Theorem (Compactness theorem for propositional logic)

If $\mathbf{\Phi}$ *is a set of formulas such that every finite subset of* $\mathbf{\Phi}$ *is satisfiable, then so is* $\mathbf{\Phi}$ *itself.*

PROOF

Immediate from Prob. 8.3(ii). ■

13.5. Problem (The logic of implication – continued)

Let $\models^*$ and $\vdash^*$ be as in Prob. 12.8. Prove the strong completeness of the calculus of implication: if $\mathbf{\Phi} \models^* \boldsymbol{\alpha}$ then $\mathbf{\Phi} \vdash^* \boldsymbol{\alpha}$. (If $\mathbf{\Phi} \nvdash^* \boldsymbol{\alpha}$, show that $\mathbf{\Phi}$ is included in a set $\mathbf{\Psi}$ such that $\mathbf{\Psi} \nvdash^* \boldsymbol{\alpha}$ and such that $\mathbf{\Psi}$ is maximal with this property; then use Prob. 12.8(iv).)

8
First-order logic

§ 1. Basic syntax

From now on, our formal object language $\mathcal{L}$ will be a fixed but (unless stated otherwise) arbitrary *first-order* language. We begin by specifying the *primitive symbols* of such a language.

1.1. Specification

The *primitive symbols* of a *first-order* language $\mathcal{L}$ fall into five mutually exclusive categories:

(i) An infinite sequence of *[individual] variables*:

$$\mathbf{v}_1, \mathbf{v}_2, \mathbf{v}_3, \ldots, \mathbf{v}_n, \ldots.$$

The order of the variables indicated here will be referred to as their *alphabetic* order.

(ii) For each natural number n, a set of *n-ary function symbols*. These sets must be pairwise disjoint and some or all of them may be empty. The 0-ary function symbols (if any) are called *[individual] constants*.

(iii) For each positive natural number n, a set of *n-ary predicate symbols*. These sets must be pairwise disjoint and at least one of them must be non-empty.

(iv) Two distinct *connectives*, $\neg$ and $\rightarrow$, called *negation symbol* and *implication symbol* respectively.

(v) The *universal quantifier* $\forall$.

A particular binary predicate symbol $=$ may be singled out as the *equality symbol*, in which case $\mathcal{L}$ is referred to as a language *with equality*. We further stipulate that if $\mathcal{L}$ has at least one function symbol that is not an individual constant (that is, at least one n-ary function symbol with positive n), then it *must* be a language with equality.

The variables, the connectives, the universal quantifier and the equality symbol (if present) are the *logical* symbols of $\mathcal{L}$. All other primitive symbols (namely, the function symbols and the predicate symbols other than $\boldsymbol{=}$) are *extralogical*.

1.2. Warnings

(i) Specification 1.1 must not be read as *exhibiting* any symbol of the object language $\mathcal{L}$, which indeed may not have a written form. Thus, for example, it must *not* be supposed that '$\mathbf{v}_1$' is a variable of $\mathcal{L}$. Rather, it is a *syntactic constant*, belonging to our metalanguage and denoting the first variable (in alphabetic order) of $\mathcal{L}$. Also, '$\boldsymbol{=}$' should not be taken to be the equality symbol of $\mathcal{L}$. Rather, it is a syntactic constant used to denote the equality symbol of $\mathcal{L}$, if it has one. (Cf. Warning 7.1.2.)

(ii) Note carefully the distinction between '$\boldsymbol{=}$' and '='. Both are symbols in our metalanguage. The former is a name (in the metalanguage) of the equality symbol of the object language (if it has one); the latter is the equality symbol of the metalanguage, an abbreviation of the phrase 'is the same as'.

The similarity of shape between '$\boldsymbol{=}$' and '=' – which may be confusing at first – is an intended pun and a mnemonic device; see Rem. 4.3(iii) below.

1.3. Remark

The difference in the logical symbols between two different first-order languages is clearly inessential, and there would be no real loss of generality if we were to assume that all first-order languages share the same logical symbols. (In the case of the equality symbol this would mean that all first-order languages *with equality* have the same equality symbol.) Two first-order languages are *essentially* different if only one of them is with equality, of if they have different stocks of extralogical symbols.

1.4. Definition

An *$\mathcal{L}$-string* is defined in the same way as in propositional logic (see Def. 7.1.3), namely as a finite sequence of primitive symbols of $\mathcal{L}$.

In propositional logic we had one significant type of string: the formulas. Here we have two types: *terms* as well as formulas.

1.5. Definition

$\mathscr{L}$-terms are strings constructed according to the following two rules.

(1) A string consisting of a single occurrence of a variable is an *$\mathscr{L}$-term*.
(2) If $\mathbf{f}$ is an n-ary function symbol and $\mathbf{t}_1, \mathbf{t}_2, \ldots, \mathbf{t}_n$ are *$\mathscr{L}$-terms* then the string $\mathbf{f}\mathbf{t}_1\mathbf{t}_2 \ldots \mathbf{t}_n$ (obtained by concatenating a single occurrence of $\mathbf{f}$ and $\mathbf{t}_1, \mathbf{t}_2, \ldots, \mathbf{t}_n$, in this order) is an *$\mathscr{L}$-term*.

In a term $\mathbf{f}\mathbf{t}_1\mathbf{t}_2 \ldots \mathbf{t}_n$ constructed according to clause (2), the terms $\mathbf{t}_1$, $\mathbf{t}_2, \ldots, \mathbf{t}_n$ are the *first argument*, *second argument*, ..., *nth argument*, respectively.

For $n = 0$, (2) says that a single occurrence of a constant is an $\mathscr{L}$-term (see Specification 1.1(ii)).

1.6. Definition

The *degree of complexity* of a term $\mathbf{t}$ – briefly, $\deg \mathbf{t}$ – is the total number of occurrences of function symbols in $\mathbf{t}$.

We shall often use induction on $\deg \mathbf{t}$ in order to prove general statements about all terms $\mathbf{t}$.

1.7. Definition

$\mathscr{L}$-formulas are strings constructed according to the following four rules.

(1) If $\mathbf{P}$ is an n-ary predicate symbol and $\mathbf{t}_1, \mathbf{t}_2, \ldots, \mathbf{t}_n$ are $\mathscr{L}$-terms then the string $\mathbf{P}\mathbf{t}_1\mathbf{t}_2 \ldots \mathbf{t}_n$ (obtained by concatenating a single occurrence of $\mathbf{P}$ and $\mathbf{t}_1, \mathbf{t}_2, \ldots, \mathbf{t}_n$, in this order) is an *$\mathscr{L}$-formula*.
(2) If $\boldsymbol{\beta}$ is an *$\mathscr{L}$-formula* then $\neg\boldsymbol{\beta}$ (the string obtained by concatenating a single occurrence of $\neg$ and the string $\boldsymbol{\beta}$, in this order) is an *$\mathscr{L}$-formula*.
(3) If $\boldsymbol{\beta}$ and $\boldsymbol{\gamma}$ are *$\mathscr{L}$-formulas* then $\rightarrow\boldsymbol{\beta}\boldsymbol{\gamma}$ (the string obtained by concatenating a single occurrence of $\rightarrow$, the string $\boldsymbol{\beta}$ and the string $\boldsymbol{\gamma}$, in this order) is an *$\mathscr{L}$-formula*.
(4) If $\mathbf{x}$ is a variable and $\boldsymbol{\beta}$ is an *$\mathscr{L}$-formula* then $\forall\mathbf{x}\boldsymbol{\beta}$ (the string

obtained by concatenating a single occurrence of $\forall$, a single occurrence of $\mathbf{x}$ and the string $\boldsymbol{\beta}$, in this order) is an $\mathcal{L}$-*formula*.

A formula $\mathbf{P t}_1 \mathbf{t}_2 \ldots \mathbf{t}_n$ constructed according to (1) is called an *atomic* formula; the terms $\mathbf{t}_1, \mathbf{t}_2, \ldots, \mathbf{t}_n$ are its *first argument*, *second argument*, ..., *nth argument*, respectively. In the particular case where $\mathbf{P}$ is the equality symbol $=$ (in which case n must be 2) the atomic formula is also called an *equation* and its first and second arguments are called its *left-hand side* and *right-hand side* respectively.

In connection with formulas constructed according to (2) and (3) we use the same terminology as before (see Def. 7.1.4).

A formula $\forall \mathbf{x} \boldsymbol{\beta}$ constructed according to (4) is called a *universal* formula; here $\mathbf{x}$ is the *variable of quantification* and the string $\mathbf{x}\boldsymbol{\beta}$ is the *scope* of the initial occurrence of the universal quantifier.

1.8. Definition

The *degree of complexity* of a formula $\boldsymbol{\alpha}$ – briefly, $\deg \boldsymbol{\alpha}$ – is the total number of occurrences of connectives ($\neg$ and $\rightarrow$) and the universal quantifier $\forall$ in $\boldsymbol{\alpha}$.

1.9. Definition

An $\mathcal{L}$-*expression* is an $\mathcal{L}$-term or an $\mathcal{L}$-formula.

1.10. Remark

We use '$\mathbf{r}$', '$\mathbf{s}$' and '$\mathbf{t}$' (sometimes with subscripts) as syntactic variables ranging over $\mathcal{L}$-terms. Boldface lower-case Greek letters (sometimes with subscripts) are used as syntactic variables ranging over $\mathcal{L}$-formulas. These and other notational conventions of this kind should be self-evident.

§2. Adaptation of previous material

In this section we adapt the notational conventions, definitions and results of Ch. 7 to the new setting. Some of these will be slightly extended to fit this new setting.

The following problem can be solved similarly to Prob. 7.1.9.

2.1. Problem

Assign to each primitive symbol $\mathbf{p}$ of $\mathcal{L}$ a *weight* $\mathrm{w}(\mathbf{p})$ by stipulating that if $\mathbf{x}$ is a variable then $\mathrm{w}(\mathbf{x}) = -1$; if $\mathbf{f}$ is an n-ary function symbol then $\mathrm{w}(\mathbf{f}) = n - 1$; if $\mathbf{P}$ is an n-ary predicate symbol then $\mathrm{w}(\mathbf{P}) = n - 1$; while $\mathrm{w}(\neg) = 0$, $\mathrm{w}(\rightarrow) = 1$ and $\mathrm{w}(\forall) = 1$. If $\mathbf{p}_1, \mathbf{p}_2, \ldots, \mathbf{p}_l$ are primitive symbols, assign to the string $\mathbf{p}_1\mathbf{p}_2 \ldots \mathbf{p}_l$ *weight*

$$\mathrm{w}(\mathbf{p}_1\mathbf{p}_2 \ldots \mathbf{p}_l) = \mathrm{w}(\mathbf{p}_1) + \mathrm{w}(\mathbf{p}_2) + \cdots + \mathrm{w}(\mathbf{p}_l).$$

Thus, the weight of a string is the sum obtained by adding -1 for each occurrence of a variable, $n - 1$ for each occurrence of an n-ary function symbol or predicate symbol, and $+1$ for each occurrence of $\rightarrow$ or $\forall$ in the string (occurrences of $\neg$ make no contribution to the weight). Show that, for any term $\mathbf{t}$,

(i) $\mathrm{w}(\mathbf{t}) = -1$;
(ii) if $\mathbf{t}$ is the string $\mathbf{p}_1\mathbf{p}_2 \ldots \mathbf{p}_l$ and $k < l$, then $\mathrm{w}(\mathbf{p}_1\mathbf{p}_2 \ldots \mathbf{p}_k) \geqslant 0$.
(iii) Show that if $\mathbf{t}$ is a term $\mathbf{f}\mathbf{t}_1\mathbf{t}_2 \ldots \mathbf{t}_n$ formed according to Def. 1.5(2), then for each $k = 0, 1, \ldots, n$, $\mathbf{f}\mathbf{t}_1\mathbf{t}_2 \ldots \mathbf{t}_k$ is the shortest non-empty initial segment of $\mathbf{t}$ whose weight is $n - k - 1$.
(iv) Show that if $\boldsymbol{\alpha}$ is a formula $\mathbf{P}\mathbf{t}_1\mathbf{t}_2 \ldots \mathbf{t}_n$ formed according to Def. 1.7(1), then for each $k = 0, 1, \ldots, n$, $\mathbf{P}\mathbf{t}_1\mathbf{t}_2 \ldots \mathbf{t}_k$ is the shortest non-empty initial segment of $\boldsymbol{\alpha}$ whose weight is $n - k - 1$.
(v) Also show that the results of Prob. 7.1.9 concerning formulas hold for the present language $\mathcal{L}$. (For (i) and (ii) of Prob. 7.1.9, four cases now need to be considered, corresponding to the four clauses of Def. 1.7. In the case where $\boldsymbol{\alpha}$ is atomic, the previous results of the present problem are invoked.)

Prob. 2.1 shows that the Polish notation decreed for $\mathcal{L}$ makes brackets and other punctuation marks unnecessary in that language.[1] However, for reasons explained in § 2 of Ch. 7, we decree:

2.2. Definition

(i) The same as Def. 7.2.1.
(ii) $(\mathbf{r}{=}\mathbf{s}) =_{\mathrm{df}} {=}\mathbf{r}\mathbf{s}$,

[1] The ambiguities that might otherwise arise are illustrated by a piece that appeared in *The Guardian* on 10 October 1985, reporting 'grisly new details of the murder by Lord Lucan in 1974 of one of his children's two nannies'. Did the writer intend to say '. . . of [one of (his children's two nannies)]' or '. . . of [(one of his children)'s two nannies]'? Did Lord Lucan murder one of the two nannies of his children, or did he commit the double murder of two nannies of one of his children?

(iii) $(\mathbf{r}{\neq}\mathbf{s}) =_{\mathrm{df}} \neg(\mathbf{r}{=}\mathbf{s})$.

Also, we introduce by contextual definition surrogates for three additional connectives and the existential quantifier:

2.3. *Definition*

(i)–(iii) The same as Def. 7.2.5(i)–(iii).
(iv) $\exists\mathbf{x}\boldsymbol{\alpha} =_{\mathrm{df}} \neg\forall\mathbf{x}\neg\boldsymbol{\alpha}$.

With this more conventional metalinguistic notation, brackets are needed, and so are rules for omitting and restoring them. We adopt the same rules as before: ommission of outermost brackets (Rule 7.2.2), adhesion of '$\neg$' (Rule 7.2.4), ranks and association to the right (Rule 7.2.7) and add to them one more rule:

2.4. *Rule (Adhesion of '$\forall\mathbf{x}$' and '$\exists\mathbf{x}$')*

Do not omit a pair of brackets whose left member is immediately preceded by an occurrence of '$\forall\mathbf{x}$' or '$\exists\mathbf{x}$'. Equivalently: *In restoring brackets, do not add a new pair of brackets whose left member immediately follows an occurrence of* '$\forall\mathbf{x}$' or '$\exists\mathbf{x}$'. *Similarly with* '$\mathbf{x}$' *replaced by* '$\mathbf{y}$', or '$\mathbf{z}$', *or by any other syntactic variable ranging over $\mathcal{L}$-variables, or by a syntactic constant denoting an $\mathcal{L}$-variable.*

In order to adapt the rest of the material of Ch. 7 to our present setting, we need to redefine the notions *prime* formula and *prime component* of a formula.

2.5. *Definition*

A *prime* formula is a formula that is atomic or universal.

2.6. *Definition*

The set of *prime components* of a formula $\boldsymbol{\alpha}$ is the smallest set of prime formulas from which $\boldsymbol{\alpha}$ can be obtained as a propositional combination. In detail, by induction on $\deg\boldsymbol{\alpha}$:

(1) If $\boldsymbol{\alpha}$ is a prime formula, then the set of *prime components* of $\boldsymbol{\alpha}$ is $\{\boldsymbol{\alpha}\}$.

(2) If $\boldsymbol{\alpha} = \neg\boldsymbol{\beta}$ then the set of *prime components* of $\boldsymbol{\alpha}$ is the same as that of $\boldsymbol{\beta}$.
(3) If $\boldsymbol{\alpha} = \boldsymbol{\beta}\rightarrow\boldsymbol{\gamma}$ then the set of *prime components* of $\boldsymbol{\alpha}$ is the union of those of $\boldsymbol{\beta}$ and $\boldsymbol{\gamma}$.

With these redefinitions, all the material of §§3–13 of Ch. 7 carries over lock, stock and barrel into the present setting. From now on, whenever we use a piece of notation introduced in Ch. 7, or refer to a definition, result or remark in that chapter, we shall interpret that notation, definition, result or remark as relating to the present setting, in which $\mathcal{L}$ is a first-order language rather than the language of Ch. 7.

§3. Mathematical structures

3.1. Preview

Of course, we have not introduced our first-order language $\mathcal{L}$ merely as a vehicle for propositional logic—this would leave the variables, the function symbols, the predicate symbols and the universal quantifier without gainful employment, while only the connectives would be doing a significant job. The point of having a first-order language is that such a language, *when suitably interpreted*, can be used to 'talk about' this or that mathematical structure. In this section we shall explain what a mathematical structure is.

We shall make use of the material presented in Ch. 2; in particular, the notions of *relation* and *property* (Def. 2.1.14) and that of *map* (a.k.a *mapping* or *function*, Def. 2.2.1). We shall also need the following definition.

3.2. Definition

For $n \geqslant 1$, an *n-ary operation on* a class A is a map from A^n to A.

If f is an n-ary operation on A, and $a_1, a_2, \ldots, a_n \in A$, then the value of f at the n-tuple $\langle a_1, a_2, \ldots, a_n\rangle$ is usually denoted by '$f(a_1, a_2, \ldots, a_n)$' with parentheses instead of corner brackets.

3.3. Remark

From Def. 3.2 and the definitions made in Ch. 2 it is not difficult to see that f is an n-ary operation on A iff f is an $(n+1)$-ary relation on A such that for any $a_1, a_2, \ldots, a_n \in A$ there is a unique $a \in A$ for which $\langle a_1, a_2, \ldots, a_n, a\rangle \in f$.

So far we have defined the notion of *n-ary operation* for positive n only. If we were to extend Def. 3.2 directly to $n = 0$, then a 0-*ary operation on* A would be defined as a set of the form $\{\langle \emptyset, a\rangle\}$, with $a \in A$. On the other hand, were we to extend the condition of Rem. 3.3 to the case $n = 0$, then a 0-*ary operation on* A would have to be defined as a set of the form $\{a\}$, with $a \in A$. In either case, there would be a one-to-one correspondence between 0-ary operations on A and members of A. It fact it turns out to be most convenient to take neither of these courses, but – in the spirit of reductionism – simply to *identify* 0-ary operations on a class with its members:

3.4. Definition

A 0-*ary operation on* a class A is a member of A.

We are now ready to lay down the main definition of this section.

3.5. Definition

A *mathematical structure* is a composite entity $\mathfrak{U}$ consisting of the following ingredients.

(i) A non-empty set U called the *domain* or *universe* of $\mathfrak{U}$. The members of the domain are called the *individuals* of $\mathfrak{U}$.
(ii) A set of operations on U, called the *basic operations* of $\mathfrak{U}$.
(iii) A non-empty set of relations on U, called the *basic relations* of $\mathfrak{U}$.

Note that the set of basic operations may be empty. Among the basic operations there may be some 0-ary ones, which by Def. 3.4 are individuals of the structure. Such an object – that is, an individual of the structure which is also among its basic operations – is called a *designated individual* of the structure.

Perhaps the most fundamental structure of classical mathematics is:

3.6. Example

The *elementary* (or *first-order*) *structure of natural numbers* may be defined as the structure $\mathfrak{N}$ having the following ingredients.

(i) Its domain is the set $N = \{0, 1, 2, \ldots\}$ of all natural numbers.
(ii) Its four basic operations are the designated individual 0; the

unary operation s which assigns to each number n its immediate successor; and two binary operations, $+$ and $\times$, which assign to each pair of numbers their sum and product respectively.

(iii) Its only basic relation is the identity relation on N, namely $\mathrm{id}_N = \{\langle n, n\rangle : n \in N\}$.

3.7. Example

A more general notion of structure than that prescribed by Def. 3.5 is obtained by allowing the domain to be a proper class rather than a set, and also admitting a basic relation which is a proper class. The most important example of this liberalized notion is the *structure of sets* $\mathfrak{M}$, having the following ingredients.

(i) Its domain is the class M of all objects, that is sets and individuals (if any) of set theory, a.k.a. the *universal class*.

(ii) No basic operations.

(iii) Its basic relations are the identity relation on M and the relation $\in$ of membership between objects and sets.

3.8. Remark

A great many mathematical statements are, or can be construed as, statements about mathematical structures. The *structuralist* view of mathematics holds that mathematics is essentially the study of such structures.

§4. Basic semantics

4.1. Preview

By itself, $\mathcal{L}$ is meaningless; its expressions express nothing: they are just strings of meaningless symbols, combined according to apparently arbitrary formal syntactic rules. In this section we introduce the basic semantic apparatus needed to endow $\mathcal{L}$-expressions with meaning.

First, we shall define the notion of *$\mathcal{L}$-interpretation* (a.k.a. *$\mathcal{L}$-structure*). Roughly speaking, an $\mathcal{L}$-interpretation is a mathematical structure (cf. Def. 3.5) together with a sort of 'dictionary' that assigns a reference to each function symbol of $\mathcal{L}$, making it a name of some basic operation of the structure; and to each extralogical predicate symbol of $\mathcal{L}$, making it a name of some basic relation of the structure.

Under a given $\mathcal{L}$-interpretation, each *closed* term (a term not containing variables) receives a reference, becoming a name for some individual (a member of the domain of the structure). A term containing variables does not receive any particular reference, but once the variables are assigned values (belonging to the domain) the term itself receives a value (also belonging to the domain).

Certain formulas, known as *sentences*, also receive meaning under an $\mathcal{L}$-interpretation: each sentence expresses a proposition about the mathematical structure concerned, and thus receives a truth value $\top$ or $\bot$, according as that proposition is true or not. A formula that is not a sentence does not express a proposition and thus cannot be said to be true or false outright. Rather, it expresses a *condition* which may or may not be *satisfied* by a given assignment of values (belonging to the domain) to certain variables, the *free* variables of the formula.

In order to deal with *all* terms (including those that contain variables) and *all* formulas (including those that are not sentences), we shall introduce the notion of *$\mathcal{L}$-valuation*, which is an $\mathcal{L}$-interpretation together with an assignment of an individual (member of the domain) as *value* to each variable of $\mathcal{L}$. Under an $\mathcal{L}$-valuation, each term receives a value (belonging to the domain of the structure) and each formula receives a truth value.

4.2. Definition

An *$\mathcal{L}$-interpretation* (or *$\mathcal{L}$-structure*) is a package – that is, a composite entity (or, to be pedantic, an ordered triple) – $\mathfrak{U}$, consisting of the following three components.

(i) A non-empty set U, called the *domain* or *universe [of discourse]* of $\mathfrak{U}$. The members of U are called *individuals* of $\mathfrak{U}$.

(ii) A mapping that assigns to each function symbol $\mathbf{f}$ of $\mathcal{L}$ an operation $\mathbf{f}^{\mathfrak{U}}$ on U, such that if $\mathbf{f}$ is an n-ary function symbol then $\mathbf{f}^{\mathfrak{U}}$ is an n-ary operation on U. In particular, if $\mathbf{c}$ is a constant of $\mathcal{L}$ then $\mathbf{c}^{\mathfrak{U}}$ is an individual of $\mathfrak{U}$. Operations of the form $\mathbf{f}^{\mathfrak{U}}$ are called *basic* operations of $\mathfrak{U}$; individuals of the form $\mathbf{c}^{\mathfrak{U}}$ are called *designated* individuals of $\mathfrak{U}$.

(iii) A mapping that assigns to each predicate symbol $\mathbf{P}$ of $\mathcal{L}$ a relation $\mathbf{P}^{\mathfrak{U}}$ on U, such that if $\mathbf{P}$ is an n-ary predicate symbol then $\mathbf{P}^{\mathfrak{U}}$ is an n-ary relation on U and such that if $\mathcal{L}$ has the equality symbol $=$ then $=^{\mathfrak{U}}$ is the identity (diagonal) relation on U, namely $\mathrm{id}_U = \{\langle u, u\rangle : u \in U\}$.

4.3. Remarks

(i) The requirement that the domain U be non-empty has some technical advantages and is adopted by most modern authors. However, it is not essential and some authors (for example, Wilfrid Hodges, *Logic*, Penguin 1977) do allow structures with empty domain; the resulting treatment differs in some minor points from the conventional one.

(ii) The mappings mentioned in clauses (ii) and (iii) of Def. 4.2 are not assumed to be one-to-one. For example, it is possible to have $\mathbf{c} \neq \mathbf{d}$ with $\mathbf{c}^{\mathfrak{U}} = \mathbf{d}^{\mathfrak{U}}$; in other words, two distinct constants may have the same interpretation. (This is like an object having more than one name in ordinary language.)

(iii) The special role of the equality symbol of $\mathscr{L}$, and the reason why we have denoted it by '=', are made clear in clause (iii) of Def. 4.2. Many authors confine the mapping in this clause to *extra-logical* predicate symbols; and the additional requirement that the equality symbol = of $\mathscr{L}$ be interpreted as denoting the identity relation on U is then introduced separately as part of the Basic Semantic Definition (see, for example, B&M, pp. 49 and 51). In the end it amounts to the same thing.

(iv) We use upper-case German (Fraktur) letters to denote $\mathscr{L}$-structures. We adopt the convention that where a structure is denoted by a given German letter, its domain will be denoted by the corresponding upper-case *italic*, unless specified otherwise.

(v) Note that the meaning of the term 'individual' here (as well as in Def. 3.5) is different from its special meaning in set theory (see 1.1.3).

4.4. Definition

(i) An *$\mathscr{L}$-valuation* is a package (say an ordered pair) σ whose two components are: an $\mathscr{L}$-interpretation $\mathfrak{U}$; and a mapping that assigns to every variable $\mathbf{x}$ of $\mathscr{L}$ a *value* $\mathbf{x}^\sigma \in U$.

(ii) The $\mathscr{L}$-structure that forms part of an $\mathscr{L}$-valuation σ is called the *underlying* structure of σ. We also say that σ is *based* on this structure.

(iii) If σ is an $\mathscr{L}$-valuation with underlying structure $\mathfrak{U}$, then by the *universe of* σ we mean the domain U of $\mathfrak{U}$; and we put $\mathbf{f}^\sigma =_{\text{df}} \mathbf{f}^{\mathfrak{U}}$ for every function symbol $\mathbf{f}$ of $\mathscr{L}$ and $\mathbf{P}^\sigma =_{\text{df}} \mathbf{P}^{\mathfrak{U}}$ for every predicate symbol $\mathbf{P}$ of $\mathscr{L}$.

4.5. Definition

If σ is a valuation and u is an individual in its universe, then $\sigma(\mathbf{x}/u)$ is the valuation that is based on the same structure as σ and assigns the same values as σ to all variables other than $\mathbf{x}$, while $\mathbf{x}^{\sigma(\mathbf{x}/u)} = u$. We say that $\sigma(\mathbf{x}/u)$ is obtained from σ by *revaluing* $\mathbf{x}$ as u.

Thus $\mathbf{f}^{\sigma(\mathbf{x}/u)} = \mathbf{f}^{\sigma}$ for every function symbol $\mathbf{f}$; and $\mathbf{P}^{\sigma(\mathbf{x}/u)} = \mathbf{P}^{\sigma}$ for every predicate symbol $\mathbf{P}$; and $\mathbf{y}^{\sigma(\mathbf{x}/u)} = \mathbf{y}^{\sigma}$ for every variable $\mathbf{y} \neq \mathbf{x}$; while $\mathbf{x}^{\sigma(\mathbf{x}/u)} = u$.

The following definition is of central importance. It was first stated explicitly by Alfred Tarski in 1933, but had been used tacitly long before that. For any valuation σ, the first section of the definition assigns to each term $\mathbf{t}$ a *value* $\mathbf{t}^{\sigma}$ belonging to the universe of σ. This is done by induction on $\deg \mathbf{t}$, in two clauses corresponding to those of Def. 1.5. The second section of the definition assigns to each formula $\boldsymbol{\alpha}$ a truth value $\boldsymbol{\alpha}^{\sigma}$. This is done by induction on $\deg \boldsymbol{\alpha}$, in four clauses corresponding to those of Def. 1.7.

4.6. Basic Semantic Definition (BSD)

Let σ be a valuation with universe U.

(T1) If $\mathbf{x}$ is a variable, then $\mathbf{x}^{\sigma}$ is already defined (see Def. 4.4).

(T2) If $\mathbf{f}$ is an n-ary function symbol and $\mathbf{t}_1, \mathbf{t}_2, \ldots, \mathbf{t}_n$ are terms, then

$$(\mathbf{f}\mathbf{t}_1\mathbf{t}_2 \ldots \mathbf{t}_n)^{\sigma} = \mathbf{f}^{\sigma}(\mathbf{t}_1{}^{\sigma}, \mathbf{t}_2{}^{\sigma}, \ldots, \mathbf{t}_n{}^{\sigma}).$$

(F1) If $\mathbf{P}$ is an n-ary predicate symbol and $\mathbf{t}_1, \mathbf{t}_2, \ldots, \mathbf{t}_n$ are terms, then

$$(\mathbf{P}\mathbf{t}_1\mathbf{t}_2 \ldots \mathbf{t}_n)^{\sigma} = \begin{cases} \top & \text{if } \langle \mathbf{t}_1{}^{\sigma}, \mathbf{t}_2{}^{\sigma}, \ldots, \mathbf{t}_n{}^{\sigma} \rangle \in \mathbf{P}^{\sigma}, \\ \bot & \text{otherwise.} \end{cases}$$

In particular,

$$(\mathbf{s}{=}\mathbf{t})^{\sigma} = \begin{cases} \top & \text{if } \mathbf{s}^{\sigma} = \mathbf{t}^{\sigma}, \\ \bot & \text{otherwise.} \end{cases}$$

(F2)

$$(\neg\boldsymbol{\beta})^{\sigma} = \begin{cases} \bot & \text{if } \boldsymbol{\beta}^{\sigma} = \top, \\ \top & \text{if } \boldsymbol{\beta}^{\sigma} = \bot. \end{cases}$$

(F3)
$$(\boldsymbol{\beta}\rightarrow\boldsymbol{\gamma})^\sigma = \begin{cases} \bot & \text{if } \boldsymbol{\beta}^\sigma = \top \text{ and } \boldsymbol{\gamma}^\sigma = \bot, \\ \top & \text{otherwise.} \end{cases}$$

(F4)
$$(\forall \mathbf{x}\boldsymbol{\beta})^\sigma = \begin{cases} \top & \text{if } \boldsymbol{\beta}^{\sigma(\mathbf{x}/u)} = \top \text{ for every } u \in U, \\ \bot & \text{otherwise.} \end{cases}$$

4.7. Remarks

(i) Strictly speaking, what the BSD defines is a pair of *new* mappings *induced* by the given valuation σ and extending it to larger domains. This is somewhat obscured by the fact that both of these two new mappings are also denoted by 'σ'. The first of these, defined in (T1) and (T2), is a map from the set of all terms to the universe U of σ. The second map induced by σ, defined in (F1)–(F4), maps the set of all formulas to the set $\{\top, \bot\}$ of truth values.

(ii) Clauses (F2) and (F3) are identical with clauses (2) and (3) of Def. 7.4.2(ii), and so ensure that [the second mapping induced by] a valuation assigns truth values to formulas in just the way a truth valuation is required to do in propositional semantics. Thus, as far as its effect on formulas is concerned, a valuation may be regarded as a special case of a truth valuation. Note however that not every truth valuation can be obtained in this way from a valuation. For example, if $\boldsymbol{\alpha}$ is any formula and $\mathbf{x}$ is any variable, then by Def. 2.5 the formula $\forall\mathbf{x}(\boldsymbol{\alpha}\rightarrow\boldsymbol{\alpha})$ is prime; hence there are truth valuations under which $\forall\mathbf{x}(\boldsymbol{\alpha}\rightarrow\boldsymbol{\alpha})$ has the truth value $\bot$. But it is easy to see that $[\forall\mathbf{x}(\boldsymbol{\alpha}\rightarrow\boldsymbol{\alpha})]^\sigma = \top$ for any valuation σ. (If $\mathscr{L}$ is a language with equality, then a simpler counter-example is provided by the equation $\mathbf{x}=\mathbf{x}$, where $\mathbf{x}$ is any variable: this formula is prime, but its truth value under any valuation is $\top$.)

(iii) Due to (F4), the BSD has a strongly *non-effective* character: it does not, in general, provide us with a method whereby the truth value $\boldsymbol{\alpha}^\sigma$ (for given $\boldsymbol{\alpha}$ and σ) might be found in a finite number of steps. For, if the universe U is infinite and $\boldsymbol{\alpha}$ is a universal formula, $\forall\mathbf{x}\boldsymbol{\beta}$, then by (F4) the truth value $\boldsymbol{\alpha}^\sigma$ depends on the infinitely many truth values $\boldsymbol{\beta}^{\sigma(\mathbf{x}/u)}$, for all the infinitely many u in U. Of course, for some *particular* $\mathbf{x}$, $\boldsymbol{\beta}$ and σ with infinite universe U it may be possible to determine, using some theoretical argument, whether or not $\boldsymbol{\beta}^{\sigma(\mathbf{x}/u)} = \top$ for all the infinitely many u in U. But there is no a priori reason to suppose that there is some universally applicable method for arriving at such an argument. (Indeed, we shall see later that there can be no such method.) Compare this to the situation in propositional logic, where the truth value of any given formula under any given truth valuation can be computed mechanically: for example, using a truth table.

4.8. *Problem*

Using Def. 2.3(iv), show that

$$(\exists \mathbf{x}\boldsymbol{\beta})^\sigma = \top \Leftrightarrow \boldsymbol{\beta}^{\sigma(\mathbf{x}/u)} = \top \text{ for some } u \text{ in the universe } U \text{ of } \sigma.$$

4.9. *Definition*

(i) If $\boldsymbol{\varphi}$ is a formula and σ is a valuation such that $\boldsymbol{\varphi}^\sigma = \top$, we say that σ *satisfies* $\boldsymbol{\varphi}$ and write '$\sigma \vDash \boldsymbol{\varphi}$'.

(ii) If $\boldsymbol{\Phi}$ is a set of formulas and σ is a valuation that satisfies every member of $\boldsymbol{\Phi}$, we say that σ *satisfies* $\boldsymbol{\Phi}$ and write '$\sigma \vDash \boldsymbol{\Phi}$'.

4.10. *Definition*

(i) If the formula $\boldsymbol{\alpha}$ is satisfied by every valuation, we say that $\boldsymbol{\alpha}$ is *logically true* (or *logically valid*) and write '$\vDash \boldsymbol{\alpha}$'.

(ii) If $\boldsymbol{\Phi}$ is a set of formulas and $\boldsymbol{\alpha}$ is a formula such that every valuation that satisfies $\boldsymbol{\Phi}$ also satisfies $\boldsymbol{\alpha}$, we say that $\boldsymbol{\alpha}$ is a *logical consequence* of $\boldsymbol{\Phi}$ and write '$\boldsymbol{\Phi} \vDash \boldsymbol{\alpha}$'. In this connection we employ simplified notation similar to that used in connection with '$\vDash_0$'. For example, we write '$\boldsymbol{\Phi}, \boldsymbol{\varphi} \vDash \boldsymbol{\alpha}$' as short for '$\boldsymbol{\Phi} \cup \{\boldsymbol{\varphi}\} \vDash \boldsymbol{\alpha}$'.

(iii) If $\boldsymbol{\Phi}$ is a set of formulas that is satisfied by some valuation, we say that $\boldsymbol{\Phi}$ is *satisfiable*. If $\boldsymbol{\Phi}$ is not satisfied by any valuation we say that it is *unsatisfiable* and write '$\boldsymbol{\Phi} \vDash$'.

(iv) If $\boldsymbol{\alpha} \vDash \boldsymbol{\beta}$ and also $\boldsymbol{\beta} \vDash \boldsymbol{\alpha}$ (that is, $\boldsymbol{\alpha}^\sigma = \boldsymbol{\beta}^\sigma$ for every valuation σ) then we say that $\boldsymbol{\alpha}$ and $\boldsymbol{\beta}$ are *logically equivalent* and write '$\boldsymbol{\alpha} \equiv \boldsymbol{\beta}$'.

4.11. *Theorem*

If $\boldsymbol{\Phi} \vDash_0 \boldsymbol{\alpha}$ *then also* $\boldsymbol{\Phi} \vDash \boldsymbol{\alpha}$. *In particular, if* $\vDash_0 \boldsymbol{\alpha}$ then also $\vDash \boldsymbol{\alpha}$; and if $\boldsymbol{\alpha} \equiv_0 \boldsymbol{\beta}$ then also $\boldsymbol{\alpha} \equiv \boldsymbol{\beta}$.

PROOF

Immediate from Rem. 4.7(ii). ■

The converse of this theorem is of course false. As pointed out in Rem. 4.7(ii), the logically true formula $\forall \mathbf{x}(\boldsymbol{\alpha}\rightarrow\boldsymbol{\alpha})$ is prime, so cannot be a tautology. And this same formula is logically equivalent, but not tautologically equivalent, to the formula $\forall \mathbf{x}(\boldsymbol{\alpha}\leftrightarrow\boldsymbol{\alpha})$.

4.12. Problem

(i) For any set $\boldsymbol{\Phi}$ of formulas and any two formulas $\boldsymbol{\alpha}$ and $\boldsymbol{\beta}$, prove that $\boldsymbol{\Phi}, \boldsymbol{\alpha} \models \boldsymbol{\beta}$ iff $\boldsymbol{\Phi} \models \boldsymbol{\alpha}\rightarrow\boldsymbol{\beta}$.

(ii) Prove that $\{\boldsymbol{\alpha}_1, \boldsymbol{\alpha}_2, \ldots, \boldsymbol{\alpha}_k\} \models \boldsymbol{\beta}$ iff $\models \boldsymbol{\alpha}_1\rightarrow\boldsymbol{\alpha}_2\rightarrow\cdots\rightarrow\boldsymbol{\alpha}_k\rightarrow\boldsymbol{\beta}$.

(iii) Prove that $\boldsymbol{\alpha} \equiv \boldsymbol{\beta}$ iff $\models \boldsymbol{\alpha}\leftrightarrow\boldsymbol{\beta}$.

4.13. Remark

We say that $\boldsymbol{\beta}$ is a *subformula* of $\boldsymbol{\alpha}$ if the formula $\boldsymbol{\beta}$, regarded as an $\mathscr{L}$-string, occurs as a consecutive part of the formula $\boldsymbol{\alpha}$, where the latter is also regarded as an $\mathscr{L}$-string. (Note that $\boldsymbol{\beta}$ can occur in $\boldsymbol{\alpha}$ more than once; but using Prob. 2.1(v) it is easy to show that two distinct occurrences of $\boldsymbol{\beta}$ in $\boldsymbol{\alpha}$ cannot overlap.)

An obvious feature of the BSD is that if $\boldsymbol{\alpha}$ is a non-atomic formula, then $\boldsymbol{\alpha}^\sigma$ is determined in terms of the truth values of certain subformulas of $\boldsymbol{\alpha}$ under σ itself and (if $\boldsymbol{\alpha}$ is a universal formula) under certain other valuations. Note that it is the *truth values* of these subformulas that matter, not the subformulas themselves.

This has the following consequence. Suppose that $\boldsymbol{\beta}'$ is a formula such that $\boldsymbol{\beta}' \equiv \boldsymbol{\beta}$ and let $\boldsymbol{\alpha}'$ result from $\boldsymbol{\alpha}$ when an occurrence of a subformula $\boldsymbol{\beta}$ in $\boldsymbol{\alpha}$ is replaced by $\boldsymbol{\beta}'$. Then $\boldsymbol{\alpha}' \equiv \boldsymbol{\alpha}$. This rather obvious result can be proved rigorously by a simple but tedious induction on $\deg \boldsymbol{\alpha}$.

4.14. Remark

Let us pause to consider the issue raised in § 11 of Ch. 7: that of the ambient metatheory. While the mathematical presuppositions required for the first three sections of this chapter are rather modest, the Tarskian semantics presented in this section is quite another matter. This is mainly due to Def. 4.10, which refers (albeit implicitly) to the class of all $\mathscr{L}$-valuations, and thereby requires these valuations to be objects.

Now, a valuation is in general an infinite entity, for two reasons. First, because a valuation must assign interpretations to all extralogical symbols of $\mathscr{L}$, of which there *may* be infinitely many; and values to all variables of $\mathscr{L}$, of which there *must* be infinitely many. Second, because one component of a valuation is its universe, which may be an infinite set.

The first reason is not essential, at least if we are prepared to confine our semantic treatment to a single formula (or to finitely many formulas) at a time. We can employ a device similar to that described in §11 of Ch. 7 in connection with truth valuations: instead of using full valuations, we may use *partial* valuations, which assign interpretations and values to finitely many extralogical symbols and variables, including all those that occur in the given formula(s). But the possible infinitude of the universe of a valuation cannot be circumvented in this way, because clause (F4) of the BSD makes the truth value of a universal formula, $(\forall \mathbf{x} \boldsymbol{\beta})^\sigma$, dependent on the *whole* universe U of σ.

For this reason, those parts of our investigation that depend on concepts defined in Def. 4.10 will generally presuppose the existence of infinite sets as objects, and must be viewed as taking place in an ambient theory that incorporates a sufficiently rich set theory.

§5. Free and bound occurrences of variables

5.1. Preview

The value $\mathbf{t}^\sigma$ of a term and the truth value $\boldsymbol{\alpha}^\sigma$ of a formula under a valuation σ clearly ought not to depend on the whole of σ but only on its 'relevant' parts. For example, if $\mathbf{f}$ is a function symbol that does not occur in $\mathbf{t}$ (or in $\boldsymbol{\alpha}$) then surely $\mathbf{t}^\sigma$ (or $\boldsymbol{\alpha}^\sigma$) ought not to depend on $\mathbf{f}^\sigma$. We shall soon state this proposition more precisely, and prove that it is indeed correct. However, when it comes to variables, we must distinguish two ways in which they occur in formulas: an occurrence of a variable in a formula can be either *free* or *bound*. It will transpire that $\boldsymbol{\alpha}^\sigma$ does not depend on $\mathbf{x}^\sigma$ even if the variable $\mathbf{x}$ does occur in $\boldsymbol{\alpha}$, provided that all its occurrences are bound.

5.2. Definition

We say that valuations σ and τ *agree* on a variable $\mathbf{x}$ (or function symbol $\mathbf{f}$, or extralogical predicate symbol $\mathbf{P}$) if σ and τ have the same universe and $\mathbf{x}^\sigma = \mathbf{x}^\tau$ (or $\mathbf{f}^\sigma = \mathbf{f}^\tau$, or $\mathbf{P}^\sigma = \mathbf{P}^\tau$, respectively).

5.3. Remark

We can characterize $\sigma(\mathbf{x}/u)$ as the valuation that agrees with σ on all extralogical symbols and all variables other than $\mathbf{x}$, whereas $\mathbf{x}^{\sigma(\mathbf{x}/u)} = u$.

5.4. Theorem

Let $\mathbf{t}$ *be a term and let* σ *and* τ *be valuations that agree on all function symbols and all variables occurring in* $\mathbf{t}$. *Then* $\mathbf{t}^\sigma = \mathbf{t}^\tau$.

PROOF

Easy, by induction on $\deg \mathbf{t}$. DIY or see B&M, p. 54. ■

5.5. Remark

In particular, if $\mathbf{t}$ contains no variables (and is therefore made up entirely of constants and other function symbols) and the valuations σ and τ are based on the same $\mathscr{L}$-structure then $\mathbf{t}^\sigma = \mathbf{t}^\tau$.

5.6. Definition

A term $\mathbf{t}$ is *closed* if it contains no variables. If $\mathbf{t}$ is such a term and $\mathfrak{U}$ is an $\mathscr{L}$-structure, we put $\mathbf{t}^{\mathfrak{U}} =_{\text{df}} \mathbf{t}^\sigma$, where σ is some valuation based on $\mathfrak{U}$. (By Rem. 5.5 it makes no difference which valuation based on $\mathfrak{U}$ is chosen.)

An occurrence of a variable $\mathbf{x}$ in a formula $\boldsymbol{\alpha}$ is *bound* if it falls inside the scope of a quantifier that has $\mathbf{x}$ as its variable of quantification. Any other occurrence of $\mathbf{x}$ in $\boldsymbol{\alpha}$ is *free*. More precisely, we define these concepts by induction on $\deg \boldsymbol{\alpha}$.

5.7. Definition

The occurrences of a variable $\mathbf{x}$ in a formula $\boldsymbol{\alpha}$ are classified into two mutually exclusive kinds, *free* occurrences of $\mathbf{x}$ in $\boldsymbol{\alpha}$ and *bound* occurrences of $\mathbf{x}$ in $\boldsymbol{\alpha}$, as follows:

(1) If $\boldsymbol{\alpha}$ is atomic, then all occurrences of $\mathbf{x}$ in $\boldsymbol{\alpha}$ are *free* in $\boldsymbol{\alpha}$.
(2) If $\boldsymbol{\alpha} = \neg\boldsymbol{\beta}$ then an occurrence of $\mathbf{x}$ in $\boldsymbol{\alpha}$ is *free* in $\boldsymbol{\alpha}$ iff it is *free* in $\boldsymbol{\beta}$.
(3) If $\boldsymbol{\alpha} = \boldsymbol{\beta} \rightarrow \boldsymbol{\gamma}$ then an occurrence of $\mathbf{x}$ in $\boldsymbol{\alpha}$ is *free* in $\boldsymbol{\alpha}$ iff it is *free* in $\boldsymbol{\beta}$ or in $\boldsymbol{\gamma}$.

(4) If $\boldsymbol{\alpha} = \forall \mathbf{x}\boldsymbol{\beta}$ then all occurrences of $\mathbf{x}$ in $\boldsymbol{\alpha}$ are *bound* in $\boldsymbol{\alpha}$. But if $\boldsymbol{\alpha} = \forall \mathbf{y}\boldsymbol{\beta}$, where $\mathbf{y}$ is a variable other than $\mathbf{x}$, then an occurrence of $\mathbf{x}$ in $\boldsymbol{\alpha}$ is *free* in $\boldsymbol{\alpha}$ iff it is *free* in $\boldsymbol{\beta}$.

A variable $\mathbf{x}$ is *free in* a formula $\boldsymbol{\alpha}$ if $\mathbf{x}$ has a free occurrence in $\boldsymbol{\alpha}$. The *free variables of* a formula are those that are free in it.

5.8. Theorem

Let $\boldsymbol{\alpha}$ be a formula and let σ and τ be valuations that have the same universe and agree on all the extralogical symbols and free variables of $\boldsymbol{\alpha}$. Then $\boldsymbol{\alpha}^\sigma = \boldsymbol{\alpha}^\tau$.

PROOF

By induction on deg $\boldsymbol{\alpha}$. We distinguish four cases.

Case 1: $\boldsymbol{\alpha}$ is an atomic formula, say $\mathbf{P}\mathbf{t}_1\mathbf{t}_2 \ldots \mathbf{t}_n$. Then

$$\begin{aligned} \boldsymbol{\alpha}^\sigma = \top &\Leftrightarrow \langle \mathbf{t}_1{}^\sigma, \mathbf{t}_2{}^\sigma, \ldots, \mathbf{t}_n{}^\sigma \rangle \in \mathbf{P}^\sigma && \text{by BSD F1,} \\ &\Leftrightarrow \langle \mathbf{t}_1{}^\sigma, \mathbf{t}_2{}^\sigma, \ldots, \mathbf{t}_n{}^\sigma \rangle \in \mathbf{P}^\tau && \text{by assumption,} \\ &\Leftrightarrow \langle \mathbf{t}_1{}^\tau, \mathbf{t}_2{}^\tau, \ldots, \mathbf{t}_n{}^\tau \rangle \in \mathbf{P}^\tau && \text{by Thm. 5.4,} \\ &\Leftrightarrow \boldsymbol{\alpha}^\tau = \top && \text{by BSD F1.} \end{aligned}$$

Case 2: $\boldsymbol{\alpha}$ is a negation formula, say $\neg\boldsymbol{\beta}$. Note that σ and τ agree on the extralogical symbols and free variables of $\boldsymbol{\beta}$, since they are the same as those of $\boldsymbol{\alpha}$. Hence

$$\begin{aligned} \boldsymbol{\alpha}^\sigma = \top &\Leftrightarrow \boldsymbol{\beta}^\sigma = \bot && \text{by BSD F2,} \\ &\Leftrightarrow \boldsymbol{\beta}^\tau = \bot && \text{by ind. hyp.,} \\ &\Leftrightarrow \boldsymbol{\alpha}^\tau = \top && \text{by BSD F2.} \end{aligned}$$

Case 3: $\boldsymbol{\alpha}$ is an implication formula. DIY.

Case 4: $\boldsymbol{\alpha}$ is a universal formula, say $\forall \mathbf{x}\boldsymbol{\beta}$. The valuations σ and τ agree on all the extralogical symbols of $\boldsymbol{\beta}$, because they are exactly those of $\boldsymbol{\alpha}$. Every free variable of $\boldsymbol{\alpha}$ is also free in $\boldsymbol{\beta}$; but $\boldsymbol{\beta}$ may have one additional free variable, namely $\mathbf{x}$. Now, σ and τ need not agree on $\mathbf{x}$; but if u is any member of U then $\sigma(\mathbf{x}/u)$ and $\tau(\mathbf{x}/u)$ do agree on $\mathbf{x}$ as well: both assign to it the value u. Hence by the induction hypothesis $\boldsymbol{\beta}^{\sigma(\mathbf{x}/u)} = \boldsymbol{\beta}^{\tau(\mathbf{x}/u)}$. So

$$\begin{aligned} \boldsymbol{\alpha}^{\sigma} = \top &\Leftrightarrow \boldsymbol{\beta}^{\sigma(\mathbf{x}/u)} = \top \text{ for every } u \in U && \text{by BSD F4,} \\ &\Leftrightarrow \boldsymbol{\beta}^{\tau(\mathbf{x}/u)} = \top \text{ for every } u \in U && \text{by ind. hyp.,} \\ &\Leftrightarrow \boldsymbol{\alpha}^{\tau} = \top && \text{by BSD F4.} \end{aligned}$$ ■

5.9. Remark

In particular, if $\boldsymbol{\alpha}$ has no free variables (so that all occurrences of variables in it, if any, are bound) and the valuations σ and τ are based on the same structure then $\boldsymbol{\alpha}^{\sigma} = \boldsymbol{\alpha}^{\tau}$.

5.10. Definition

A *sentence* is a formula without free variables. If $\boldsymbol{\alpha}$ is a sentence and $\mathfrak{U}$ is an $\mathscr{L}$-structure, and $\boldsymbol{\alpha}$ is satisfied by some – and hence (cf. Rem. 5.9) by every – valuation based on $\mathfrak{U}$, then we say that $\boldsymbol{\alpha}$ *holds* (or *is satisfied*) *in* $\mathfrak{U}$, and that $\mathfrak{U}$ is a *model for* $\boldsymbol{\alpha}$, and write '$\mathfrak{U} \vDash \boldsymbol{\alpha}$'.

If $\mathfrak{U} \vDash \boldsymbol{\varphi}$ for every member $\boldsymbol{\varphi}$ of a set $\boldsymbol{\Sigma}$ of sentences, we say that $\mathfrak{U}$ is a *model* for $\boldsymbol{\Sigma}$.

5.11. Problem

Prove: $\vDash \forall \mathbf{x}(\boldsymbol{\alpha} \rightarrow \boldsymbol{\beta}) \rightarrow \forall \mathbf{x}\boldsymbol{\alpha} \rightarrow \forall \mathbf{x}\boldsymbol{\beta}$. (Use Prob. 4.12.)

5.12. Problem

Show that if $\mathbf{x}$ is not free in $\boldsymbol{\alpha}$ then $\forall \mathbf{x}\boldsymbol{\alpha} \equiv \boldsymbol{\alpha} \equiv \exists \mathbf{x}\boldsymbol{\alpha}$.

5.13. Problem

Assuming that $\mathbf{x}$ is not free in $\boldsymbol{\beta}$, show that

(i) $\forall \mathbf{x}(\boldsymbol{\alpha} \wedge \boldsymbol{\beta}) \equiv \forall \mathbf{x}\boldsymbol{\alpha} \wedge \boldsymbol{\beta}$,
(ii) $\exists \mathbf{x}(\boldsymbol{\alpha} \wedge \boldsymbol{\beta}) \equiv \exists \mathbf{x}\boldsymbol{\alpha} \wedge \boldsymbol{\beta}$;
(iii) $\forall \mathbf{x}(\boldsymbol{\alpha} \vee \boldsymbol{\beta}) \equiv \forall \mathbf{x}\boldsymbol{\alpha} \vee \boldsymbol{\beta}$,
(iv) $\exists \mathbf{x}(\boldsymbol{\alpha} \vee \boldsymbol{\beta}) \equiv \exists \mathbf{x}\boldsymbol{\alpha} \vee \boldsymbol{\beta}$;
(v) $\forall \mathbf{x}(\boldsymbol{\alpha} \rightarrow \boldsymbol{\beta}) \equiv \exists \mathbf{x}\boldsymbol{\alpha} \rightarrow \boldsymbol{\beta}$,
(vi) $\exists \mathbf{x}(\boldsymbol{\alpha} \rightarrow \boldsymbol{\beta}) \equiv \forall \mathbf{x}\boldsymbol{\alpha} \rightarrow \boldsymbol{\beta}$;
(vii) $\forall \mathbf{x}(\boldsymbol{\beta} \rightarrow \boldsymbol{\alpha}) \equiv \boldsymbol{\beta} \rightarrow \forall \mathbf{x}\boldsymbol{\alpha}$,
(viii) $\exists \mathbf{x}(\boldsymbol{\beta} \rightarrow \boldsymbol{\alpha}) \equiv \boldsymbol{\beta} \rightarrow \exists \mathbf{x}\boldsymbol{\alpha}$.

5.14. Problem

Construct a sentence $\boldsymbol{\alpha}$ containing only logical symbols (that is, no function symbols and no predicate symbols other than $=$) such that $\boldsymbol{\alpha}$ holds in a structure $\mathfrak{U}$ iff the domain U of $\mathfrak{U}$ has

(i) at least three members,
(ii) at most three members,
(iii) exactly three members.

5.15. Problem

Let $\mathcal{L}$ be a language without $=$ whose only extralogical symbol is a binary predicate symbol $\mathbf{P}$. Construct an $\mathcal{L}$-sentence $\boldsymbol{\alpha}$ such that $\boldsymbol{\alpha}$ has no finite model (that is, $\boldsymbol{\alpha}$ does not hold in any structure whose domain is finite) and such that if U is any infinite set then there is a binary relation P on U such that the $\mathcal{L}$-structure $\mathfrak{U}$ with domain U and with $\mathbf{P}^{\mathfrak{U}} = P$ is a model for $\boldsymbol{\alpha}$. (In writing your solution, do not be tempted to denote the predicate symbol of $\mathcal{L}$ by anything other than '$\mathbf{P}$'. Note that any condition that you wish to impose on the interpretation P of $\mathbf{P}$ must be written into $\boldsymbol{\alpha}$.)

§6. Substitution

Substitution is a purely syntactic operation: occurrences of a variable in a given expression are replaced by [occurrences of] a term. Thus, three $\mathcal{L}$-entities are involved: first, the expression *in* which the substitution is made; second, the variable *for* which a term is substituted; and third, the term which is *substituted* for occurrences of this variable. We start with the straightforward case where the first mentioned entity, the expression in which the substitution is made, is itself a term.

We denote by '$\mathbf{s}(\mathbf{x}/\mathbf{t})$' (read: '$\mathbf{s}$, *with* $\mathbf{x}$ *replaced by* $\mathbf{t}$') the result obtained from the term $\mathbf{s}$ when all occurrences of the variable $\mathbf{x}$ in $\mathbf{s}$ are simultaneously replaced by occurrences of the term $\mathbf{t}$. In detail, $\mathbf{s}(\mathbf{x}/\mathbf{t})$ is defined by induction on $\deg \mathbf{s}$.

6.1. Definition

For any variable $\mathbf{x}$ and any term $\mathbf{t}$,

(1a) $\mathbf{x}(\mathbf{x}/\mathbf{t}) = \mathbf{t}$;
(1b) $\mathbf{y}(\mathbf{x}/\mathbf{t}) = \mathbf{y}$ for any variable $\mathbf{y}$ other than $\mathbf{x}$;

(2) if $\mathbf{s} = \mathbf{f}\mathbf{s}_1\mathbf{s}_2 \ldots \mathbf{s}_n$ where $\mathbf{f}$ is an n-ary function symbol and $\mathbf{s}_1$, $\mathbf{s}_2$, $\ldots$, $\mathbf{s}_n$ are terms, then $\mathbf{s}(\mathbf{x}/\mathbf{t}) = \mathbf{f}\mathbf{s}_1(\mathbf{x}/\mathbf{t})\mathbf{s}_2(\mathbf{x}/\mathbf{t}) \ldots \mathbf{s}_n(\mathbf{x}/\mathbf{t})$.

The most important fact about $\mathbf{s}(\mathbf{x}/\mathbf{t})$ is its semantic behaviour: the way its value under a valuation σ depends on $\mathbf{s}$, $\mathbf{x}$, $\mathbf{t}$ and σ.

We must not expect the value $\mathbf{s}(\mathbf{x}/\mathbf{t})^\sigma$ to be the same as $\mathbf{s}^\sigma$, because $\mathbf{s}(\mathbf{x}/\mathbf{t})$ and $\mathbf{s}$ are, in general, two different terms. However, note that in the former term $\mathbf{t}$ occupies the same positions that $\mathbf{x}$ occupies in the latter. Thus we ought to expect the value $\mathbf{s}(\mathbf{x}/\mathbf{t})^\sigma$ to be the same as the value of $\mathbf{s}$ not under σ itself, but under the valuation obtained from σ by revaluing $\mathbf{x}$ and assigning to it the value that $\mathbf{t}$ has under σ (see Def. 4.5). Thus we ought to have:

$$\mathbf{s}(\mathbf{x}/\mathbf{t})^\sigma = \mathbf{s}^{\sigma(\mathbf{x}/t)}, \text{ where } t = \mathbf{t}^\sigma. \tag{6.2}$$

6.3. Remark

For purely typographical reasons, the printed form of (6.2) is a bit more complicated than it need be. When writing this formula by hand, there is no need to use 't' at all, because the 't' in the main part of the formula can be replaced by '$\mathbf{t}^\sigma$'. The form then taken by (6.2) is shown here:

$$\mathbf{s}(\mathbf{x}/\mathbf{t})^\sigma = \mathbf{s}^{\sigma(\mathbf{x}/\mathbf{t}^\sigma)}. \tag{6.2'}$$

Unfortunately, this requires three levels of print and the third-floor characters have to be smaller than ordinary small print. This is technically difficult to typeset as well hard on the eye. So in print we use the verbose form (6.2); but in hand-written texts it is better to use the more compact (6.2′). A similar remark applies also to (6.6) below.

6.4. Theorem

(6.2) *holds for all* $\mathbf{s}$, $\mathbf{x}$, $\mathbf{t}$ *and* σ.

PROOF

By induction on $\deg \mathbf{s}$. Three cases must be considered, corresponding to the three clauses in Def. 6.1. Throughout, we put $t = \mathbf{t}^\sigma$.

Case 1a: $\mathbf{s}$ is $\mathbf{x}$. Then $\mathbf{s}(\mathbf{x}/\mathbf{t})^\sigma = \mathbf{x}(\mathbf{x}/\mathbf{t})^\sigma = \mathbf{t}^\sigma$, by Def. 6.1. On the other hand, $\mathbf{s}^{\sigma(\mathbf{x}/t)} = \mathbf{x}^{\sigma(\mathbf{x}/t)} = t$, by Def. 4.5. So (6.2) holds in this case.

Case 1b: $\mathbf{s}$ is a variable $\mathbf{y} \neq \mathbf{x}$. Then $\mathbf{s}(\mathbf{x}/\mathbf{t})^\sigma = \mathbf{y}(\mathbf{x}/\mathbf{t})^\sigma = \mathbf{y}^\sigma$, by Def. 6.1. On the other hand, $\mathbf{s}^{\sigma(\mathbf{x}/t)} = \mathbf{y}^{\sigma(\mathbf{x}/t)} = \mathbf{y}^\sigma$, by Def. 4.5. So (6.2) holds also in this case.

Case 2: $\mathbf{s}$ is $\mathbf{f}\mathbf{s}_1\mathbf{s}_2 \ldots \mathbf{s}_n$. Then

$$\begin{aligned}
\mathbf{s}(\mathbf{x}/\mathbf{t})^\sigma &= [\mathbf{f}\mathbf{s}_1\mathbf{s}_2 \ldots \mathbf{s}_n](\mathbf{x}/\mathbf{t})^\sigma \\
&= [\mathbf{f}\mathbf{s}_1(\mathbf{x}/\mathbf{t})\mathbf{s}_2(\mathbf{x}/\mathbf{t}) \ldots \mathbf{s}_n(\mathbf{x}/\mathbf{t})]^\sigma && \text{by Def. 6.1,} \\
&= \mathbf{f}^\sigma(\mathbf{s}_1(\mathbf{x}/\mathbf{t})^\sigma, \mathbf{s}_2(\mathbf{x}/\mathbf{t})^\sigma, \ldots, \mathbf{s}_n(\mathbf{x}/\mathbf{t})^\sigma) && \text{by BSD T2,} \\
&= \mathbf{f}^{\sigma(\mathbf{x}/t)}(\mathbf{s}_1(\mathbf{x}/\mathbf{t})^\sigma, \mathbf{s}_2(\mathbf{x}/\mathbf{t})^\sigma, \ldots, \mathbf{s}_n(\mathbf{x}/\mathbf{t})^\sigma) && \text{by Def. 4.5,} \\
&= \mathbf{f}^{\sigma(\mathbf{x}/t)}(\mathbf{s}_1{}^{\sigma(\mathbf{x}/t)}, \mathbf{s}_2{}^{\sigma(\mathbf{x}/t)}, \ldots, \mathbf{s}_n{}^{\sigma(\mathbf{x}/t)}) && \text{by ind. hyp.,} \\
&= [\mathbf{f}\mathbf{s}_1\mathbf{s}_2 \ldots \mathbf{s}_n]^{\sigma(\mathbf{x}/t)} && \text{by BSD T2,} \\
&= \mathbf{s}^{\sigma(\mathbf{x}/t)}. && \blacksquare
\end{aligned}$$

6.5. Remark

Thm. 6.5 does not tell us anything unexpected about the semantic effect of substitution – on the contrary, the result is what we anticipated. The point of the theorem is that it confirms that Def. 6.1 was correct, in the sense of ensuring the desired effect.

Let us turn to the case where a term $\mathbf{t}$ is to be substituted for a variable $\mathbf{x}$ in a formula $\boldsymbol{\alpha}$. For reasons that should now be clear, we must define the substitution in such a way that

$$\boldsymbol{\alpha}(\mathbf{x}/\mathbf{t})^\sigma = \boldsymbol{\alpha}^{\sigma(\mathbf{x}/t)}, \text{ where } t = \mathbf{t}^\sigma. \tag{6.6}$$

Now however complications arise due to the different roles played by free and bound occurrences of variables. Here we shall only outline the way these complications are resolved. Full technical details can be found in B&M, pp. 57–64.

First, it is clear that when substituting $\mathbf{t}$ for $\mathbf{x}$ in $\boldsymbol{\alpha}$, only *free* occurrences of $\mathbf{x}$ in $\boldsymbol{\alpha}$ should be replaced by $\mathbf{t}$. Intuitively speaking, the reason for this is that the truth value $\boldsymbol{\alpha}^\sigma$ depends on the value $\mathbf{x}^\sigma$ only through the free occurrences of $\mathbf{x}$ in $\boldsymbol{\alpha}$ (see Thm. 5.8). Besides, if we replace *all* occurrences of $\mathbf{x}$ by $\mathbf{t}$, the result may not be a formula at all. Indeed, if $\mathbf{x}$ has bound occurrences in $\boldsymbol{\alpha}$, then at least one of them must immediately follow an occurrence of $\forall$. If such an occurrence of $\mathbf{x}$ is replaced by $\mathbf{t}$, then the result will not be a formula, unless $\mathbf{t}$ itself happens to be a variable, because in a formula each $\forall$ must be followed by a variable.

Can we therefore define $\boldsymbol{\alpha}(\mathbf{x}/\mathbf{t})$ as the result of replacing all *free* occurrences of $\mathbf{x}$ in $\boldsymbol{\alpha}$ by $\mathbf{t}$? Unfortunately, this does not always work. Cases where if *fails* to work are those in which some free occurrence of $\mathbf{x}$ in $\boldsymbol{\alpha}$ occurs within the scope of a $\mathbf{y}$-quantifier, where the variable $\mathbf{y}$ (which must of course be distinct from $\mathbf{x}$) happens to occur in $\mathbf{t}$. If we then simply replace such an occurrence of $\mathbf{x}$ by $\mathbf{t}$, the resulting occurrence of $\mathbf{y}$ in the new formula so obtained will be *captured*: it becomes bound by the $\mathbf{y}$-quantifier. It turns out that when capturing takes place, (6.6) may fail.

For example, let $\boldsymbol{\alpha}$ be $\forall\mathbf{y}(\mathbf{x}=\mathbf{y})$, where $\mathbf{x}$ and $\mathbf{y}$ are distinct. If we were to define $\boldsymbol{\alpha}(\mathbf{x}/\mathbf{t})$ as $\forall\mathbf{y}(\mathbf{t}=\mathbf{y})$ for arbitrary $\mathbf{t}$, then taking $\mathbf{t}$ as $\mathbf{y}$ itself we would get $\boldsymbol{\alpha}(\mathbf{x}/\mathbf{t}) = \forall\mathbf{y}(\mathbf{y}=\mathbf{y})$. Note that the new (second) occurrence of $\mathbf{y}$ got captured by a $\mathbf{y}$-quantifier. But then (6.6) would not always hold, because $\forall\mathbf{y}(\mathbf{y}=\mathbf{y})$ is satisfied by every valuation (it is logically true), whereas $\forall\mathbf{y}(\mathbf{x}=\mathbf{y})$ is satisfied just by valuations whose universe is a singleton.

Of course, this kind of complication, due to capturing, does not always arise. Instead of defining $\boldsymbol{\alpha}(\mathbf{x}/\mathbf{t})$ outright for all $\boldsymbol{\alpha}$, $\mathbf{x}$ and $\mathbf{t}$, we proceed in stages. First, we confine ourselves to cases in which capturing does not take place.

6.7. Definition

If no free occurrence of $\mathbf{x}$ in $\boldsymbol{\alpha}$ is within the scope of a $\mathbf{y}$-quantifier, where $\mathbf{y}$ is a variable that occurs in $\mathbf{t}$, then we say that $\mathbf{t}$ *is free [to be substituted] for* $\mathbf{x}$ *in* $\boldsymbol{\alpha}$; and in this case we define $\boldsymbol{\alpha}(\mathbf{x}/\mathbf{t})$ as the result obtained from $\boldsymbol{\alpha}$ when all free occurrences of $\mathbf{x}$ in $\boldsymbol{\alpha}$ are simultaneously replaced by $\mathbf{t}$. [For a more detailed version of this definition, proceeding by induction on $\deg\boldsymbol{\alpha}$, see B&M, p. 59f.]

It is now fairly easy to show that (6.6) holds in the special case where $\boldsymbol{\alpha}(\mathbf{x}/\mathbf{t})$ has so far been defined.

6.8. Theorem

(6.6) *holds whenever the term* $\mathbf{t}$ *is free for the variable* $\mathbf{x}$ *in the formula* $\boldsymbol{\alpha}$.

PROOF

DIY or see B&M, p. 60f. ■

6.9. Remark

There are two special cases where **t** is free for **x** in **α**. First, where **t** contains no variable other than **x**. Def. 6.7 therefore applies in this case. In particular, in the case where **t** is **x** itself, it is easy to see that **α(x/x)** is just **α**, as it ought to be. The second special case is where **t** contains no variable that occurs bound in **α**: in this case **α** does not contain any **y**-quantifier where **y** occurs in **t**.

In order to define **α(x/t)** in the remaining case – where **t** is not free for **x** in **α** – we must first modify the offending parts of **α** and make them harmless. The trouble is caused by free occurrences of **x** in **α** that fall within subformulas of **α** having the form **∀yβ**, where **y** is a variable that occurs in **t**. In order to make the substitution work, so that (6.6) is ensured, such subformulas of **α** must first be replaced by logically equivalent ones that use a harmless variable, say **z**, instead of **y**. This motivates the following

6.10. Definition

If **z** is a variable that does not occur free in **β** but is free for **y** in **β**, we say that the formula **∀z[β(y/z)]** arises from the formula **∀yβ** by *[correct] alphabetic change [of variable of quantification]*.

6.11. Remarks

(i) The reasons for requiring that **z** be free for **y** in **β** is that otherwise the substitution **β(y/z)** is not defined as yet. The reason for requiring that **z** has no free occurrences in **β** is that otherwise the formulas **∀z[β(y/z)]** and **∀yβ** may not be logically equivalent. For example, let **β** be **y=z**, where **y** and **z** are distinct variables. It is easy to see that **∀z(z=z)** and **∀y(y=z)** are not logically equivalent: the former is logically true, whereas the latter is satisfied by a valuation σ iff the universe of σ is a singleton.

(ii) If **z** does not occur at all in **β**, then **z** clearly fulfils the conditions in Def. 6.10.

(iii) It is not difficult to show that the operation of alphabetic change is reversible; in other words, if **∀z[β(y/z)]** arises from **∀yβ** by an alphabetic change, then the latter formula can be retrieved from the former by an alphabetic change (see B&M, p. 61).

6.12. Theorem

If $\forall\mathbf{z}[\boldsymbol{\beta}(\mathbf{y}/\mathbf{z})]$ *arises from* $\forall\mathbf{y}\boldsymbol{\beta}$ *by alphabetic change then these two formulas are logically equivalent.*

PROOF

DIY or see B&M, p. 61. ∎

6.13. Definition

(i) We say that a formula $\boldsymbol{\gamma}$ is obtained from a formula $\boldsymbol{\alpha}$ by an *alphabetic step* if $\boldsymbol{\alpha}$ has a subformula of the form $\forall\mathbf{y}\boldsymbol{\beta}$ and $\boldsymbol{\gamma}$ results from $\boldsymbol{\alpha}$ when one occurrence of $\forall\mathbf{y}\boldsymbol{\beta}$ is replaced by a formula $\forall\mathbf{z}[\boldsymbol{\beta}(\mathbf{y}/\mathbf{z})]$ that arises from it by alphabetic change.

(ii) We say that $\boldsymbol{\alpha}'$ is a *variant* of $\boldsymbol{\alpha}$, and write '$\boldsymbol{\alpha} \sim \boldsymbol{\alpha}'$', if $\boldsymbol{\alpha}'$ can be obtained from $\boldsymbol{\alpha}$ by a finite number of alphabetic steps.

6.14. Remarks

(i) The relation $\sim$ is easily seen to be an equivalence relation. It is reflexive: $\boldsymbol{\alpha} \sim \boldsymbol{\alpha}$ always holds because $\boldsymbol{\alpha}$ is obtained from itself by 0 alphabetic steps. It is symmetric: if $\boldsymbol{\alpha} \sim \boldsymbol{\alpha}'$ then also $\boldsymbol{\alpha}' \sim \boldsymbol{\alpha}$ because alphabetic changes, and hence also alphabetic steps, are reversible. Finally, it is clearly transitive: if $\boldsymbol{\alpha} \sim \boldsymbol{\alpha}'$ and $\boldsymbol{\alpha}' \sim \boldsymbol{\alpha}''$ then also $\boldsymbol{\alpha} \sim \boldsymbol{\alpha}''$.

(ii) By Thm. 6.12 and Rem. 4.13, if $\boldsymbol{\alpha} \sim \boldsymbol{\alpha}'$ then $\boldsymbol{\alpha} \equiv \boldsymbol{\alpha}'$.

We can now define the substitution $\boldsymbol{\alpha}(\mathbf{x}/\mathbf{t})$ in full generality.

6.15. Definition

Let a variable $\mathbf{x}$ and a term $\mathbf{t}$ be given. For any formula $\boldsymbol{\alpha}$, we select a formula $\boldsymbol{\alpha}'$ such that if $\mathbf{t}$ is free for $\mathbf{x}$ in $\boldsymbol{\alpha}$, then $\boldsymbol{\alpha}'$ is $\boldsymbol{\alpha}$ itself; but if $\mathbf{t}$ is not free for $\mathbf{x}$ in $\boldsymbol{\alpha}$, then $\boldsymbol{\alpha}'$ is a variant of $\boldsymbol{\alpha}$ in which $\mathbf{t}$ *is* free for $\mathbf{x}$. Thus $\boldsymbol{\alpha}'(\mathbf{x}/\mathbf{t})$ is already defined in Def. 6.7. We now define $\boldsymbol{\alpha}(\mathbf{x}/\mathbf{t})$ to be the same as $\boldsymbol{\alpha}'(\mathbf{x}/\mathbf{t})$. [For details see B&M, p. 63. If $\mathbf{t}$ is not free for $\mathbf{x}$ in $\boldsymbol{\alpha}$, then it does not really matter which variant of $\boldsymbol{\alpha}$ is selected to be $\boldsymbol{\alpha}'$, so long as $\mathbf{t}$ is free for $\mathbf{x}$ in $\boldsymbol{\alpha}'$. But a definition must be unambiguous, so a particular variant $\boldsymbol{\alpha}'$ must be selected. This is done by induction on $\deg\boldsymbol{\alpha}$. The gist of the choice is that each offending subformula $\forall\mathbf{y}\boldsymbol{\beta}$ of $\boldsymbol{\alpha}$ is replaced by $\forall\mathbf{z}[\boldsymbol{\beta}(\mathbf{y}/\mathbf{z})]$, where $\mathbf{z}$ is the first variable in the alphabetic list of $\mathcal{L}$-variables – that is, the $\mathbf{v}_i$ with the least i – such that this is a correct alphabetic change and such that $\mathbf{z}$ does not occur in $\mathbf{t}$.]

6.16. Problem

Show that (6.6) holds for *all* $\boldsymbol{\alpha}$, $\mathbf{x}$, $\mathbf{t}$ and σ.

§7. Hintikka sets

We shall introduce first-order Hintikka sets because, as in the propositional case, it is relatively easy to prove that such sets are satisfiable. It will follow that any set included in a Hintikka set is also satisfiable. This will come in handy later on, when we shall want to prove the appropriate completeness theorem.

7.1. Definition

A *[first-order] Hintikka* set *[in $\mathcal{L}$]* is a set $\boldsymbol{\Phi}$ of $\mathcal{L}$-formulas satisfying the following nine conditions:

(1) If $\boldsymbol{\alpha}$ is any atomic formula such that $\boldsymbol{\alpha} \in \boldsymbol{\Phi}$, then $\neg\boldsymbol{\alpha} \notin \boldsymbol{\Phi}$.
(2) If $\boldsymbol{\alpha}$ is any formula such that $\neg\neg\boldsymbol{\alpha} \in \boldsymbol{\Phi}$, then also $\boldsymbol{\alpha} \in \boldsymbol{\Phi}$.
(3) If $\boldsymbol{\alpha}$ and $\boldsymbol{\beta}$ are any formulas such that $\boldsymbol{\alpha}\to\boldsymbol{\beta} \in \boldsymbol{\Phi}$, then $\neg\boldsymbol{\alpha} \in \boldsymbol{\Phi}$ or $\boldsymbol{\beta} \in \boldsymbol{\Phi}$.
(4) If $\boldsymbol{\alpha}$ and $\boldsymbol{\beta}$ are any formulas such that $\neg(\boldsymbol{\alpha}\to\boldsymbol{\beta}) \in \boldsymbol{\Phi}$, then $\boldsymbol{\alpha} \in \boldsymbol{\Phi}$ and $\neg\boldsymbol{\beta} \in \boldsymbol{\Phi}$.
(5) If $\boldsymbol{\alpha}$ is any formula and $\mathbf{x}$ is any variable such that $\forall\mathbf{x}\boldsymbol{\alpha} \in \boldsymbol{\Phi}$, then $\boldsymbol{\alpha}(\mathbf{x}/\mathbf{t}) \in \boldsymbol{\Phi}$ for every $\mathcal{L}$-term $\mathbf{t}$.
(6) If $\boldsymbol{\alpha}$ is any formula and $\mathbf{x}$ is any variable such that $\neg\forall\mathbf{x}\boldsymbol{\alpha} \in \boldsymbol{\Phi}$, then $\neg\boldsymbol{\alpha}(\mathbf{x}/\mathbf{t}) \in \boldsymbol{\Phi}$ for some $\mathcal{L}$-term $\mathbf{t}$.
(7) If $\mathcal{L}$ is a language with equality, then $\mathbf{t}{=}\mathbf{t} \in \boldsymbol{\Phi}$ for every $\mathcal{L}$-term $\mathbf{t}$.
(8) If $n \geq 1$ and $\mathbf{s}_1, \mathbf{s}_2, \ldots, \mathbf{s}_n$ and $\mathbf{t}_1, \mathbf{t}_2, \ldots, \mathbf{t}_n$ are any $2n$ $\mathcal{L}$-terms such that for each $i = 1, 2, \ldots, n$ the equation $\mathbf{s}_i{=}\mathbf{t}_i$ is in $\boldsymbol{\Phi}$, then it follows that for every n-ary function symbol $\mathbf{f}$ of $\mathcal{L}$ the equation $\mathbf{f}\mathbf{s}_1\mathbf{s}_2 \ldots \mathbf{s}_n{=}\mathbf{f}\mathbf{t}_1\mathbf{t}_2 \ldots \mathbf{t}_n$ is also in $\boldsymbol{\Phi}$.
(9) If $n \geq 1$ and $\mathbf{s}_1, \mathbf{s}_2, \ldots, \mathbf{s}_n$ and $\mathbf{t}_1, \mathbf{t}_2, \ldots, \mathbf{t}_n$ are any $2n$ $\mathcal{L}$-terms such that for each $i = 1, 2, \ldots, n$ the equation $\mathbf{s}_i{=}\mathbf{t}_i$ is in $\boldsymbol{\Phi}$, and if $\mathbf{P}$ is any n-ary predicate symbol such that the atomic formula $\mathbf{P}\mathbf{s}_1\mathbf{s}_2 \ldots \mathbf{s}_n$ is in $\boldsymbol{\Phi}$, then the formula $\mathbf{P}\mathbf{t}_1\mathbf{t}_2 \ldots \mathbf{t}_n$ is also in $\boldsymbol{\Phi}$.

7.2. Remarks

(i) Conditions (8) and (9) of the definition are vacuous if $\mathscr{L}$ is a language without equality. The reason for excluding the case $n = 0$ in (8) is that for $n = 0$ this condition would have reduced to requiring that if $\mathbf{c}$ is any individual constant of $\mathscr{L}$ then $\mathbf{c{=}c} \in \mathbf{\Phi}$, which is already covered by condition (7).

(ii) Condition (9) applies in particular to the case where $n = 2$ and $\mathbf{P}$ is = itself. In this special case the condition says that if $\mathbf{s}_1$, $\mathbf{s}_2$, $\mathbf{t}_1$ and $\mathbf{t}_2$ are any four terms such that the three equations $\mathbf{s}_1{=}\mathbf{t}_1$, $\mathbf{s}_2{=}\mathbf{t}_2$ and $\mathbf{s}_1{=}\mathbf{s}_2$ are in $\mathbf{\Phi}$, then the equation $\mathbf{t}_1{=}\mathbf{t}_2$ is also in $\mathbf{\Phi}$. Fig. 2 can be used as a mnemonic for this statement. The four terms are represented by the four corners of the square; the three equations *assumed* to belong to $\mathbf{\Phi}$ are represented by the three solid sides, reading from top to bottom and from left to right; and the fourth equation, which is then *required* to belong to $\mathbf{\Phi}$, is represented by the dotted side, again reading from left to right.

For the rest of this section, we let $\mathbf{\Phi}$ be a fixed but arbitrary Hintikka set. We shall refer to the nine conditions of Def. 7.1 simply as '(1)', '(2)' and so on.

Our aim is to prove that $\mathbf{\Phi}$ is satisfiable. We shall define a particular valuation σ and show that $\sigma \models \mathbf{\Phi}$. In order to define σ, we must specify its various ingredients: first, we must specify its universe U; next, for each variable $\mathbf{x}$ we shall have to specify its value $\mathbf{x}^\sigma$, which must of course be a member of U; then, for each function symbol $\mathbf{f}$ we must specify the corresponding operation $\mathbf{f}^\sigma$ on U; finally, for each extra-logical predicate symbol $\mathbf{P}$ we have to specify the corresponding relation $\mathbf{P}^\sigma$ on U. (As for the logical predicate symbol =, if it is present in $\mathscr{L}$, we have no choice: $=^\sigma$ has to be the diagonal relation on U.)

Of all the ingredients of σ, the first – the universe U – turns out to

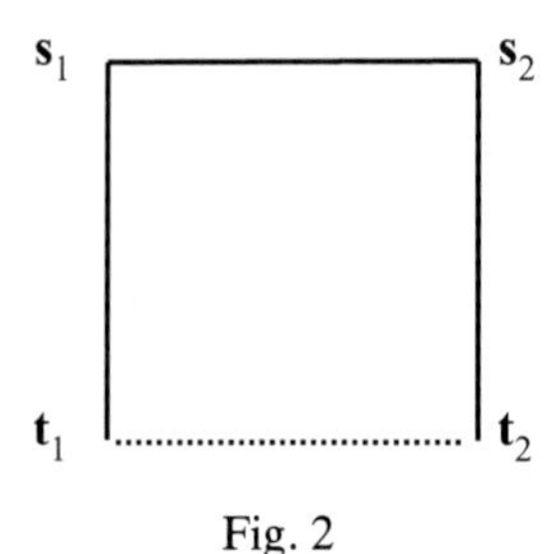

Fig. 2

require most work. Once U has been properly set up, the rest will follow quite smoothly. The nature of the members of U (that is, what 'stuff' they are made of) is clearly of no importance; what is vital is that for each term $\mathbf{t}$ there should be a member of U to serve as the value $\mathbf{t}^\sigma$. In general, the universe of a valuation may have members that do not serve as the value of any term under that valuation; but in the present case Occam's razor turns out to be useful. So we shall define an object $[\mathbf{t}]$ for each term $\mathbf{t}$ and – even before deciding what $[\mathbf{t}]$ is to be – we put

7.3. *Definition*

$U =_{\text{df}} \{[\mathbf{t}] : \mathbf{t} \text{ is an } \mathcal{L}\text{-term}\}.$

Our plan is to define σ in such a way that $\mathbf{t}^\sigma = [\mathbf{t}]$ for every term $\mathbf{t}$. As we have said, the nature of $[\mathbf{t}]$ is unimportant; but we must decide whether distinct terms are to have distinct values; in other words, if $\mathbf{s}$ and $\mathbf{t}$ are distinct, should $[\mathbf{s}]$ and $[\mathbf{t}]$ also be distinct? The simplest choice is to answer this question in the affirmative. The good news is that if $\mathcal{L}$ is without equality then this simplest choice actually works.

The bad news is that it does not work if $\mathcal{L}$ has equality. The snag is that in this case $\mathbf{\Phi}$ may contain equations $\mathbf{s}{=}\mathbf{t}$, where $\mathbf{s}$ and $\mathbf{t}$ are distinct terms. If σ is to satisfy $\mathbf{\Phi}$, it must in particular satisfy these equations, which (by the BSD F1) means that $\mathbf{s}^\sigma$ and $\mathbf{t}^\sigma$ must be the same. As we intend these values to be $[\mathbf{s}]$ and $[\mathbf{t}]$ respectively, we are forced to allow $[\mathbf{s}]$ and $[\mathbf{t}]$ to be equal whenever $\mathbf{s}{=}\mathbf{t} \in \mathbf{\Phi}$, even though $\mathbf{s}$ and $\mathbf{t}$ may be distinct. This motivates the following definition of the relation E between terms:

7.4. *Definition*

The relation E holds between two terms $\mathbf{s}$ and $\mathbf{t}$ – briefly, $\mathbf{s}E\mathbf{t}$ – if either $\mathcal{L}$ is without equality and $\mathbf{s}$ is the same as $\mathbf{t}$, or $\mathcal{L}$ has equality and the equation $\mathbf{s}{=}\mathbf{t}$ is in $\mathbf{\Phi}$.

7.5. *Lemma*

E is an equivalence relation: it is reflexive, symmetric and transitive.

PROOF

The case where $\mathcal{L}$ is without equality is trivial. Now suppose that $\mathcal{L}$ does have equality.

The reflexivity of E follows at once from (7).

To prove that E is symmetric, assume that $\mathbf{s}=\mathbf{t} \in \mathbf{\Phi}$. We must show that also $\mathbf{t}=\mathbf{s} \in \mathbf{\Phi}$. We shall make use of Rem. 7.2(ii). Choosing $\mathbf{s}_1$, $\mathbf{s}_2$, $\mathbf{t}_1$ and $\mathbf{t}_2$ as $\mathbf{s}$, $\mathbf{s}$, $\mathbf{t}$ and $\mathbf{s}$ respectively, we get the configuration shown in Fig. 3.

The equation $\mathbf{s}=\mathbf{t}$ (left side of the square) belongs to $\mathbf{\Phi}$ by assumption; and the equation $\mathbf{s}=\mathbf{s}$ (right and top sides) belongs to $\mathbf{\Phi}$ by (7). Hence by Rem. 7.2(ii) the equation $\mathbf{t}=\mathbf{s}$ (bottom side) must also be in $\mathbf{\Phi}$. So E is symmetric.

To prove that E is transitive, assume that $\mathbf{r}=\mathbf{s}$ and $\mathbf{s}=\mathbf{t}$ are in $\mathbf{\Phi}$. We must show that also $\mathbf{r}=\mathbf{t} \in \mathbf{\Phi}$. Again, we use Rem. 7.2(ii). This time we choose $\mathbf{s}_1$, $\mathbf{s}_2$, $\mathbf{t}_1$ and $\mathbf{t}_2$ as $\mathbf{r}$, $\mathbf{s}$, $\mathbf{r}$ and $\mathbf{t}$ respectively, and obtain the configuration shown in Fig. 4. The equation $\mathbf{r}=\mathbf{r}$ (left side of the square) is in $\mathbf{\Phi}$ by (7); and the equations $\mathbf{r}=\mathbf{s}$ and $\mathbf{s}=\mathbf{t}$ (top and right sides) are in $\mathbf{\Phi}$ by assumption. Hence also the equation $\mathbf{r}=\mathbf{t}$ (bottom side) is in $\mathbf{\Phi}$. So E is transitive. ■

7.6. Definition

For each term $\mathbf{t}$, we define $[\mathbf{t}]$ as the E-class of $\mathbf{t}$ (see Def. 2.3.4). Thus,

$$[\mathbf{t}] =_{\mathrm{df}} [\mathbf{t}]_E = \{\mathbf{s} : \mathbf{s}E\mathbf{t}\}.$$

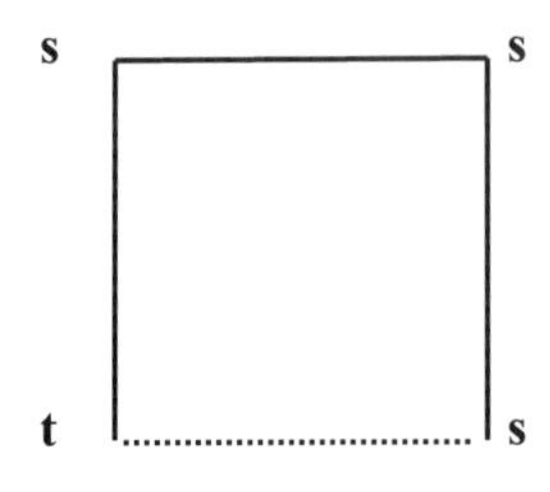

Fig. 3

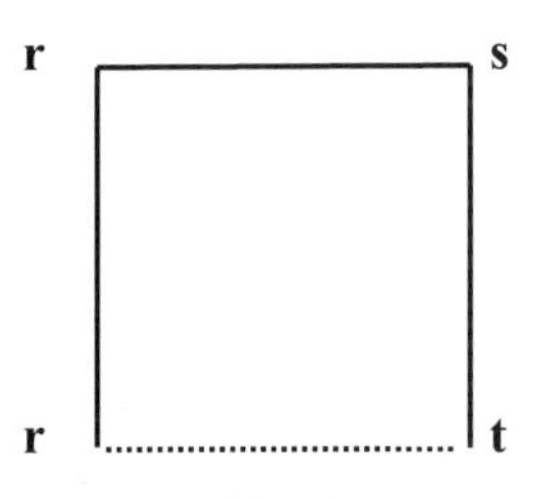

Fig. 4

7.7. Remarks

(i) If $\mathscr{L}$ is without equality, then $[\mathbf{t}]$ is simply $\{\mathbf{t}\}$, so that if $\mathbf{s}$ and $\mathbf{t}$ are distinct terms then $[\mathbf{s}]$ and $[\mathbf{t}]$ are also distinct. If $\mathscr{L}$ does have equality, then $[\mathbf{t}]$ is a class of terms that may have several – indeed even infinitely many – members.

(ii) Recall that by Thm. 2.3.5, $[\mathbf{s}] = [\mathbf{t}]$ iff $\mathbf{s}E\mathbf{t}$. Also, by Cor. 2.3.6, each term belongs to a unique E-class.

(iii) The class of all $\mathscr{L}$-strings is a set by Thm. 6.3.9. Hence by AS the class $\mathbf{T}$ of all terms is also a set. For each $\mathbf{t}$, $[\mathbf{t}]$ is a subset of $\mathbf{T}$ and so, by Def. 7.3, $U \subseteq \mathbf{PT}$. Thus U is a set by AP and AS.

Our intention was to have $\mathbf{t}^\sigma = [\mathbf{t}]$ for very term $\mathbf{t}$. For the particular case where $\mathbf{t}$ is a variable we are free to decree this as part of the specification of σ.

7.8. Definition

We put $\mathbf{x}^\sigma = [\mathbf{x}]$ for each variable $\mathbf{x}$.

Next, for each n-ary function symbol $\mathbf{f}$ we must define the n-ary operation on U that is to serve as $\mathbf{f}^\sigma$. To define $\mathbf{f}^\sigma$, we must specify, for each n-tuple of members of U, the member of U produced by applying $\mathbf{f}^\sigma$ to that n-tuple. Take n arbitrary members of U; by Def. 7.3 they are of the form $[\mathbf{t}_1], [\mathbf{t}_2], \ldots, [\mathbf{t}_n]$, where $\mathbf{t}_1, \mathbf{t}_2, \ldots, \mathbf{t}_n$ are terms. We must specify a member of U as $\mathbf{f}^\sigma([\mathbf{t}_1], [\mathbf{t}_2], \ldots, [\mathbf{t}_n])$. This individual (again by Def. 7.3) must have the form $[\mathbf{t}]$ where $\mathbf{t}$ is some term. How shall we choose this $\mathbf{t}$? Clearly, $\mathbf{t}$ must involve $\mathbf{f}$ and $\mathbf{t}_1, \mathbf{t}_2, \ldots, \mathbf{t}_n$. So an obvious choice is

7.9. Definition

If $\mathbf{f}$ is any n-ary function symbol and $\mathbf{t}_1, \mathbf{t}_2, \ldots, \mathbf{t}_n$ are any terms,

$$\mathbf{f}^\sigma([\mathbf{t}_1], [\mathbf{t}_2], \ldots, [\mathbf{t}_n]) =_{\mathrm{df}} [\mathbf{f}\mathbf{t}_1\mathbf{t}_2 \ldots \mathbf{t}_n].$$

7.10. Legitimation

If $n > 0$ – in which case, as stipulated in Sp. 1.1, $\mathscr{L}$ must have equality – then this definition needs to be legitimized. The point is that one and the same member of U may be represented in more than one way: $\mathbf{r}$ and $\mathbf{s}$ may be distinct terms such that the object $[\mathbf{r}]$ is the same as $[\mathbf{s}]$.

However, the definiendum $\mathbf{f}^\sigma([\mathbf{t}_1], [\mathbf{t}_2], \ldots, [\mathbf{t}_n])$ must depend only on the objects $[\mathbf{t}_1], [\mathbf{t}_2], \ldots, [\mathbf{t}_n]$ and not on the particular terms $\mathbf{t}_1, \mathbf{t}_2, \ldots, \mathbf{t}_n$ that happen to represent them. So we have to prove that the definiens $[\mathbf{ft}_1\mathbf{t}_2 \ldots \mathbf{t}_n]$ depends only on the objects $[\mathbf{t}_1], [\mathbf{t}_2], \ldots, [\mathbf{t}_n]$ rather than on the particular terms $\mathbf{t}_1, \mathbf{t}_2, \ldots, \mathbf{t}_n$ used to represent them. We must therefore show that if $[\mathbf{s}_i] = [\mathbf{t}_i]$ for $i = 1, 2, \ldots, n$, then also

$$[\mathbf{fs}_1\mathbf{s}_2 \ldots \mathbf{s}_n] = [\mathbf{ft}_1\mathbf{t}_2 \ldots \mathbf{t}_n].$$

This is easily done. Indeed, if $[\mathbf{s}_i] = [\mathbf{t}_i]$ for $i = 1, 2, \ldots, n$, then by Rem. 7.7(ii) for each i the equation $\mathbf{s}_i = \mathbf{t}_i$ is in $\mathbf{\Phi}$. So by (8) the equation $\mathbf{fs}_1\mathbf{s}_2 \ldots \mathbf{s}_n = \mathbf{ft}_1\mathbf{t}_2 \ldots \mathbf{t}_n$ is also in $\mathbf{\Phi}$ and $[\mathbf{fs}_1\mathbf{s}_2 \ldots \mathbf{s}_n] = [\mathbf{ft}_1\mathbf{t}_2 \ldots \mathbf{t}_n]$. ■

We have not completed our definition of σ: we still have to specify the relations $\mathbf{P}^\sigma$. But we are already in a position to prove

7.11. Lemma

$\mathbf{t}^\sigma = [\mathbf{t}]$ *for every term* $\mathbf{t}$.

PROOF

We proceed by induction on $\deg \mathbf{t}$. The case where $\mathbf{t}$ is a variable is covered by Def. 7.8. Now let $\mathbf{t}$ be $\mathbf{ft}_1\mathbf{t}_2 \ldots \mathbf{t}_n$. Then

$$\begin{aligned} \mathbf{t}^\sigma = (\mathbf{ft}_1\mathbf{t}_2 \ldots \mathbf{t}_n)^\sigma &= \mathbf{f}^\sigma(\mathbf{t}_1{}^\sigma, \mathbf{t}_2{}^\sigma, \ldots, \mathbf{t}_n{}^\sigma) && \text{by BSD F2,} \\ &= \mathbf{f}^\sigma([\mathbf{t}_1], [\mathbf{t}_2], \ldots, [\mathbf{t}_n]) && \text{by ind. hyp.,} \\ &= [\mathbf{ft}_1\mathbf{t}_2 \ldots \mathbf{t}_n] && \text{by Def. 7.9,} \\ &= [\mathbf{t}]. \end{aligned}$$

■

To complete the definition of σ, we have to define for each extralogical n-ary predicate symbol $\mathbf{P}$ an n-ary relation $\mathbf{P}^\sigma$ on U; that is, $\mathbf{P}^\sigma$ must be defined as a subset of U^n. To do this, we have to specify, for any n objects $[\mathbf{t}_1], [\mathbf{t}_2], \ldots, [\mathbf{t}_n]$, whether the n-tuple $\langle[\mathbf{t}_1], [\mathbf{t}_2], \ldots, [\mathbf{t}_n]\rangle$ is to belong to $\mathbf{P}^\sigma$. How are we to do this? Note that as we have just proved, $\langle[\mathbf{t}_1], [\mathbf{t}_2], \ldots, [\mathbf{t}_n]\rangle$ is $\langle\mathbf{t}_1{}^\sigma, \mathbf{t}_2{}^\sigma, \ldots, \mathbf{t}_n{}^\sigma\rangle$. Now, by the BSD F1, the atomic formula $\mathbf{Pt}_1\mathbf{t}_2 \ldots \mathbf{t}_n$ is going to be satisfied by σ iff $\langle\mathbf{t}_1{}^\sigma, \mathbf{t}_2{}^\sigma, \ldots, \mathbf{t}_n{}^\sigma\rangle \in \mathbf{P}^\sigma$. But remember what σ is for: it is supposed to satisfy $\mathbf{\Phi}$. Therefore, if $\mathbf{Pt}_1\mathbf{t}_2 \ldots \mathbf{t}_n$ is in $\mathbf{\Phi}$ we would like the n-tuple $\langle\mathbf{t}_1{}^\sigma, \mathbf{t}_2{}^\sigma, \ldots, \mathbf{t}_n{}^\sigma\rangle$ to be in $\mathbf{P}^\sigma$. This suggests

7.12. Definition

If $\mathbf{P}$ is any n-ary extralogical predicate symbol, then $\mathbf{P}^\sigma$ is defined to be the subset of U^n such that for any n terms $\mathbf{t}_1, \mathbf{t}_2, \ldots, \mathbf{t}_n$,

$$\langle[\mathbf{t}_1], [\mathbf{t}_2], \ldots, [\mathbf{t}_n]\rangle \in \mathbf{P}^\sigma \Leftrightarrow \mathbf{P}\mathbf{t}_1\mathbf{t}_2 \ldots \mathbf{t}_n \in \mathbf{\Phi}.$$

7.13. Legitimation

This definition too needs legitimation. We must make sure that whether or not $\langle[\mathbf{t}_1], [\mathbf{t}_2], \ldots, [\mathbf{t}_n]\rangle \in \mathbf{P}^\sigma$ holds depends on the objects $[\mathbf{t}_1], [\mathbf{t}_2], \ldots, [\mathbf{t}_n]$ rather than on the terms that happen to represent them. In other words, it must be proved that if $[\mathbf{s}_i] = [\mathbf{t}_i]$ for $i = 1, 2, \ldots, n$, then

$$\mathbf{P}\mathbf{s}_1\mathbf{s}_2 \ldots \mathbf{s}_n \in \mathbf{\Phi} \Leftrightarrow \mathbf{P}\mathbf{t}_1\mathbf{t}_2 \ldots \mathbf{t}_n \in \mathbf{\Phi}.$$

This is easy. DIY, using (9). ■

7.14. Remark

As mentioned before, if $\mathscr{L}$ has equality we have no choice as to the relation $=^\sigma$; we must put, for all terms $\mathbf{s}$ and $\mathbf{t}$,

$$\langle[\mathbf{s}], [\mathbf{t}]\rangle \in =^\sigma \Leftrightarrow [\mathbf{s}] = [\mathbf{t}].$$

But by Rem. 7.7(ii) this amounts to

$$\langle[\mathbf{s}], [\mathbf{t}]\rangle \in =^\sigma \Leftrightarrow \mathbf{s}=\mathbf{t} \in \mathbf{\Phi}.$$

This means that Def. 7.12 extends automatically also to the logical predicate symbol $=$.

Having completed the definition of σ, we can prove

7.15. Theorem

For any formula $\boldsymbol{\varphi}$,

$$\text{(a) } \boldsymbol{\varphi} \in \mathbf{\Phi} \Rightarrow \boldsymbol{\varphi}^\sigma = \top, \qquad \text{(b) } \neg\boldsymbol{\varphi} \in \mathbf{\Phi} \Rightarrow \boldsymbol{\varphi}^\sigma = \bot.$$

PROOF

We shall prove this double claim simultaneously by induction on $\deg \boldsymbol{\varphi}$. We distinguish four cases, corresponding to the clauses of Def. 1.7.

Case 1: $\boldsymbol{\varphi}$ is atomic; say $\boldsymbol{\varphi} = \mathbf{Pt}_1\mathbf{t}_2 \ldots \mathbf{t}_n$.

$$
\begin{aligned}
\text{(1a)}\quad \boldsymbol{\varphi} \in \boldsymbol{\Phi} &\Rightarrow \mathbf{Pt}_1\mathbf{t}_2 \ldots \mathbf{t}_n \in \boldsymbol{\Phi} \\
&\Rightarrow \langle [\mathbf{t}_1], [\mathbf{t}_2], \ldots, [\mathbf{t}_n] \rangle \in \mathbf{P}^\sigma \\
&\qquad\qquad \text{by Def. 7.12 and Rem. 7.14,} \\
&\Rightarrow \langle \mathbf{t}_1{}^\sigma, \mathbf{t}_2{}^\sigma, \ldots, \mathbf{t}_n{}^\sigma \rangle \in \mathbf{P}^\sigma \qquad \text{by Lemma 7.11,} \\
&\Rightarrow (\mathbf{Pt}_1\mathbf{t}_2 \ldots \mathbf{t}_n)^\sigma = \top \qquad \text{by BSD F1,} \\
&\Rightarrow \boldsymbol{\varphi}^\sigma = \top.
\end{aligned}
$$

$$
\begin{aligned}
\text{(1b)}\quad \neg\boldsymbol{\varphi} \in \boldsymbol{\Phi} &\Rightarrow \boldsymbol{\varphi} \notin \boldsymbol{\Phi} \qquad \text{by (1),} \\
&\Rightarrow \mathbf{Pt}_1\mathbf{t}_2 \ldots \mathbf{t}_n \notin \boldsymbol{\Phi} \\
&\Rightarrow \langle [\mathbf{t}_1], [\mathbf{t}_2], \ldots, [\mathbf{t}_n] \rangle \notin \mathbf{P}^\sigma \\
&\qquad\qquad \text{by Def. 7.12 and Rem. 7.14,} \\
&\Rightarrow \langle \mathbf{t}_1{}^\sigma, \mathbf{t}_2{}^\sigma, \ldots, \mathbf{t}_n{}^\sigma \rangle \notin \mathbf{P}^\sigma \qquad \text{by Lemma 7.11,} \\
&\Rightarrow (\mathbf{Pt}_1\mathbf{t}_2 \ldots \mathbf{t}_n)^\sigma = \bot \qquad \text{by BSD F1,} \\
&\Rightarrow \boldsymbol{\varphi}^\sigma = \bot.
\end{aligned}
$$

Case 2: $\boldsymbol{\varphi}$ is a negation formula. Similar to Case 2 in the proof of Thm. 7.10.3.

Case 3: $\boldsymbol{\varphi}$ is an implication formula. Similar to Case 3 in the proof of Thm. 7.10.3.

Case 4: $\boldsymbol{\varphi}$ is a universal formula; say $\boldsymbol{\varphi} = \forall\mathbf{x}\boldsymbol{\alpha}$.

$$
\begin{aligned}
\text{(4a)}\quad \boldsymbol{\varphi} \in \boldsymbol{\Phi} &\Rightarrow \forall\mathbf{x}\boldsymbol{\alpha} \in \boldsymbol{\Phi} \\
&\Rightarrow \boldsymbol{\alpha}(\mathbf{x}/\mathbf{t}) \in \boldsymbol{\Phi} \text{ for every term } \mathbf{t} \qquad \text{by (5),} \\
&\Rightarrow \boldsymbol{\alpha}(\mathbf{x}/\mathbf{t})^\sigma = \top \text{ for every term } \mathbf{t} \qquad \text{by ind. hyp.,} \\
&\Rightarrow \boldsymbol{\alpha}^{\sigma(\mathbf{x}/t)} = \top \text{ (where } t = \mathbf{t}^\sigma\text{) for every term } \mathbf{t} \\
&\qquad\qquad \text{by Prob. 6.16,} \\
&\Rightarrow \boldsymbol{\alpha}^{\sigma(\mathbf{x}/[\mathbf{t}])} = \top \text{ for every term } \mathbf{t} \qquad \text{by Lemma 7.11,} \\
&\Rightarrow \boldsymbol{\alpha}^{\sigma(\mathbf{x}/u)} = \top \text{ for every } u \in U \qquad \text{by Def. 7.3,} \\
&\Rightarrow (\forall\mathbf{x}\boldsymbol{\alpha})^\sigma = \top \qquad \text{by BSD F4,} \\
&\Rightarrow \boldsymbol{\varphi}^\sigma = \top.
\end{aligned}
$$

$$
\begin{aligned}
\text{(4b)}\quad \neg\boldsymbol{\varphi} \in \boldsymbol{\Phi} &\Rightarrow \neg\forall\mathbf{x}\boldsymbol{\alpha} \in \boldsymbol{\Phi} \\
&\Rightarrow \neg\boldsymbol{\alpha}(\mathbf{x}/\mathbf{t}) \in \boldsymbol{\Phi} \text{ for some term } \mathbf{t} \qquad \text{by (6),} \\
&\Rightarrow \boldsymbol{\alpha}(\mathbf{x}/\mathbf{t})^\sigma = \bot \text{ for some term } \mathbf{t} \qquad \text{by ind. hyp.,} \\
&\Rightarrow \boldsymbol{\alpha}^{\sigma(\mathbf{x}/t)} = \bot \text{ (where } t = \mathbf{t}^\sigma\text{) for some term } \mathbf{t} \\
&\qquad\qquad \text{by Prob. 6.16,} \\
&\Rightarrow \boldsymbol{\alpha}^{\sigma(\mathbf{x}/[\mathbf{t}])} = \bot \text{ for some term } \mathbf{t} \qquad \text{by Lemma 7.11,}
\end{aligned}
$$

$\Rightarrow \boldsymbol{\alpha}^{\sigma(\mathbf{x}/u)} = \bot$ for some $u \in U$ by Def. 7.3,
$\Rightarrow (\forall \mathbf{x}\boldsymbol{\alpha})^{\sigma} = \bot$ by BSD F4,
$\Rightarrow \boldsymbol{\varphi}^{\sigma} = \bot$. ■

We have thus shown that the valuation σ – specified by Defs. 7.3, 7.6, 7.8, 7.9 and 7.12 – satisfies the Hintikka set $\boldsymbol{\Phi}$. We shall now obtain an upper bound for the cardinality of the universe of σ.

7.16. Definition

The cardinality of the set of all primitive symbols of $\mathcal{L}$ is called the *cardinality of* $\mathcal{L}$ and denoted by '$\|\mathcal{L}\|$'.

7.17. Theorem

Given a Hintikka set $\boldsymbol{\Phi}$ in $\mathcal{L}$, we can define an $\mathcal{L}$-valuation σ such that the cardinality of the universe of σ is at most $\|\mathcal{L}\|$ and such that $\sigma \models \boldsymbol{\Phi}$.

PROOF

Take σ as the valuation specified above. By AC, there exists a choice function on the universe U of σ: a function that selects a single term in each E-class of terms. Since by Rem. 7.7(ii) distinct E-classes are disjoint, the choice function is an injection from U to the set of all $\mathcal{L}$-strings, whose cardinality, by Thm. 6.3.9, is exactly $\|\mathcal{L}\|$. ■

§8. Prenex formulas; parity

8.1 Definition

(i) A formula is said to be *prenex* if it is of the form

$$\mathbf{Q}_1\mathbf{x}_1\mathbf{Q}_2\mathbf{x}_2 \ldots \mathbf{Q}_k\mathbf{x}_k\boldsymbol{\beta},$$

where $k \geqslant 0$ and, for each i, $\mathbf{Q}_i$ is either $\forall$ or $\exists$, and $\boldsymbol{\beta}$ is quantifier-free (that is, contains no quantifiers). In this connection the string $\mathbf{Q}_1\mathbf{x}_1\mathbf{Q}_2\mathbf{x}_2 \ldots \mathbf{Q}\mathbf{x}_k$ is called the *prefix* and $\boldsymbol{\beta}$ the *matrix*. If moreover the variables $\mathbf{x}_1, \mathbf{x}_2, \ldots, \mathbf{x}_k$ in the prefix are distinct and all of them are free in the matrix $\boldsymbol{\beta}$, then the formula is said to be prenex *normal*.

(ii) A *prenex normal form for* a formula $\boldsymbol{\alpha}$ is a prenex normal formula logically equivalent to $\boldsymbol{\alpha}$.

8.2. Problem

(i) Let $\boldsymbol{\varphi}$ be a formula containing $n+1$ occurrences of $\forall$. Show how to find a formula of the form $\mathbf{Qx}\boldsymbol{\psi}$ – where $\mathbf{Q}$ is $\forall$ or $\exists$ and $\boldsymbol{\psi}$ contains only n occurrences of $\forall$ – which is logically equivalent to $\boldsymbol{\varphi}$. (Proceed by [strong] induction on $\deg\boldsymbol{\varphi}$. In the case where $\boldsymbol{\varphi}$ is $\boldsymbol{\alpha}\rightarrow\boldsymbol{\beta}$, we may assume, by the induction hypothesis, that $\boldsymbol{\varphi}$ is logically equivalent to a formula of the form $\mathbf{Qx}\boldsymbol{\gamma}\rightarrow\boldsymbol{\beta}$ or $\boldsymbol{\alpha}\rightarrow\mathbf{Qy}\boldsymbol{\delta}$, and by alphabetic change we can arrange that $\mathbf{x}$ is not free in $\boldsymbol{\beta}$ and $\mathbf{y}$ is not free in $\boldsymbol{\alpha}$. Then use Prob. 5.13(v)–(viii).)

(ii) Hence show how to obtain a prenex normal form for any given formula.

8.3. Definition

By induction on $\deg\boldsymbol{\alpha}$, we assign to each formula $\boldsymbol{\alpha}$ a *parity* $\operatorname{pr}\boldsymbol{\alpha}$, which is either 0 or 1, as follows:

(1) If $\boldsymbol{\alpha}$ is atomic, then $\operatorname{pr}\boldsymbol{\alpha} = 0$.
(2) If $\boldsymbol{\alpha} = \neg\boldsymbol{\beta}$, then $\operatorname{pr}\boldsymbol{\alpha} = 1 - \operatorname{pr}\boldsymbol{\beta}$.
(3) If $\boldsymbol{\alpha} = \boldsymbol{\beta}\rightarrow\boldsymbol{\gamma}$, then $\operatorname{pr}\boldsymbol{\alpha} = (1 - \operatorname{pr}\boldsymbol{\beta})\cdot\operatorname{pr}\boldsymbol{\gamma}$.
(4) If $\boldsymbol{\alpha} = \forall\mathbf{x}\boldsymbol{\beta}$, then $\operatorname{pr}\boldsymbol{\alpha} = \operatorname{pr}\boldsymbol{\beta}$.

We say that $\boldsymbol{\alpha}$ is *even* or *odd* according as $\operatorname{pr}\boldsymbol{\alpha}$ is 0 or 1.

8.4. Problem

(i) Show that the set of all even formulas is a Hintikka set, and hence is satisfiable.

(ii) Without using (i), define directly a valuation σ such that $\sigma \vDash \boldsymbol{\alpha}$ iff $\boldsymbol{\alpha}$ is even. (Take the universe of σ to be a singleton.)

§9. The first-order predicate calculus

We designate as *first-order axioms* all $\mathscr{L}$-formulas of the following eight groups:

9.1. Axiom group 1

All propositional axioms (7.6.3–7.6.7).

9.2. *Axiom group 2*

$\forall \mathbf{x}(\boldsymbol{\alpha}\to\boldsymbol{\beta})\to\forall \mathbf{x}\boldsymbol{\alpha}\to\forall \mathbf{x}\boldsymbol{\beta}$, for any formulas $\boldsymbol{\alpha}$ and $\boldsymbol{\beta}$ and any variable $\mathbf{x}$.

9.3. *Axiom group 3*

$\boldsymbol{\alpha}\to\forall \mathbf{x}\boldsymbol{\alpha}$, for any formula $\boldsymbol{\alpha}$ and any variable $\mathbf{x}$ that is not free in $\boldsymbol{\alpha}$.

9.4. *Axiom group 4*

$\forall \mathbf{x}\boldsymbol{\alpha}\to\boldsymbol{\alpha}(\mathbf{x}/\mathbf{t})$, for any formula $\boldsymbol{\alpha}$, variable $\mathbf{x}$ and term $\mathbf{t}$.

9.5. *Axiom group 5*

$\mathbf{t}=\mathbf{t}$, for any term $\mathbf{t}$.

9.6. *Axiom group 6*

$$\mathbf{s}_1=\mathbf{t}_1\to\mathbf{s}_2=\mathbf{t}_2\to\cdots\to\mathbf{s}_n=\mathbf{t}_n\to\mathbf{f}\mathbf{s}_1\mathbf{s}_2\ldots\mathbf{s}_n=\mathbf{f}\mathbf{t}_1\mathbf{t}_2\ldots\mathbf{t}_n,$$

for any $n\geqslant 1$, any $2n$ terms $\mathbf{s}_1, \mathbf{s}_2, \ldots, \mathbf{s}_n, \mathbf{t}_1, \mathbf{t}_2, \ldots, \mathbf{t}_n$ and any n-ary function symbol $\mathbf{f}$.

9.7. *Axiom group 7*

$$\mathbf{s}_1=\mathbf{t}_1\to\mathbf{s}_2=\mathbf{t}_2\to\cdots\to\mathbf{s}_n=\mathbf{t}_n\to\mathbf{P}\mathbf{s}_1\mathbf{s}_2\ldots\mathbf{s}_n\to\mathbf{P}\mathbf{t}_1\mathbf{t}_2\ldots\mathbf{t}_n,$$

for any $n\geqslant 1$, any $2n$ terms $\mathbf{s}_1, \mathbf{s}_2, \ldots, \mathbf{s}_n, \mathbf{t}_1, \mathbf{t}_2, \ldots, \mathbf{t}_n$ and any n-ary predicate symbol $\mathbf{P}$.

9.8. *Axiom group 8*

$\forall \mathbf{x}_1\forall \mathbf{x}_2\ldots\forall \mathbf{x}_k\boldsymbol{\alpha}$, for any $k\geqslant 1$, any variables $\mathbf{x}_1, \mathbf{x}_2, \ldots, \mathbf{x}_k$ (not necessarily distinct) and any $\mathscr{L}$-formula $\boldsymbol{\alpha}$ belonging to any of the preceding axiom groups.

9.9. *Remarks*

(i) Six of the eight groups of axioms are given by means of schemes; but the first and last groups are miscellanies. We shall refer to these eight groups of axioms briefly as 'Ax. 1', 'Ax. 2' and so on.

(ii) If $\mathcal{L}$ is without equality then Ax. 5, 6 and 7 are vacuous, because then there are no such $\mathcal{L}$-formulas.

(iii) In Ax. 7, **P** can be the equality symbol =. In this case $n = 2$ and we obtain the axiom scheme

$$\mathbf{s}_1\mathbf{=t}_1\rightarrow\mathbf{s}_2\mathbf{=t}_2\rightarrow\mathbf{s}_1\mathbf{=s}_2\rightarrow\mathbf{t}_1\mathbf{=t}_2.$$

Fig. 2 of Rem. 7.2(ii) can be used here too as a mnemonic, with the proviso that the equations are to be read off the square in the order: left side, right side, top, bottom.

9.10. Definition

(i) The *[classical] first-order predicate calculus [in $\mathcal{L}$]* (briefly, *Fopcal*) is the linear calculus based on the first-order axioms listed above, and on *modus ponens* as sole rule of inference.

(ii) *First-order deduction* is defined in the same way as *propositional deduction* (Def. 7.6.8), except that 'propositional axiom' is replaced by 'first-order axiom'.

(iii) We use '$\vdash$' to denote first-order deducibility – that is, deducibility in Fopcal – in the same way as '$\vdash_0$' was used to denote propositional deducibility.

(iv) All terminological and notational definitions and conventions laid down in §§6–8 and §12 of Ch. 7 in connection with $\vdash_0$ and Propcal are hereby adopted, *mutatis mutandis*, in connection with $\vdash$ and Fopcal.

9.11. Theorem

The Cut Rule, the Deduction Theorem, the Inconsistency Effect, reductio ad absurdum *and the Principle of Indirect Proof hold for Fopcal.* ■

9.12. Remark

In B&M a similar system of axioms is used, but Ax. 4 is subject to the proviso that $\mathbf{t}$ be free for $\mathbf{x}$ in $\boldsymbol{\alpha}$. The two versions of Fopcal are equivalent; the B&M version is more economical whereas the present one is a bit more user-friendly.

9.13. Warning

Versions of the classical Fopcal found in the literature fall into two groups. One group consists of *strong* versions that are equivalent to ours. The other group consists of *weak* versions that are equivalent to each other, but not to ours. To describe the relationship between the two groups, let us denote by '$\vdash^{\forall}$' the relation of deducibility in a weak version of Fopcal. The following four facts must be noted.

(i) Whenever $\mathbf{\Phi} \vdash \boldsymbol{\alpha}$ then also $\mathbf{\Phi} \vdash^{\forall} \boldsymbol{\alpha}$, but the converse does not always hold – it is in this sense that $\vdash$ is stronger than $\vdash^{\forall}$.

(ii) For any set $\mathbf{\Phi}$ of formulas, let $\mathbf{\Phi}^{\forall}$ be a set of sentences obtained from $\mathbf{\Phi}$ upon replacing each $\boldsymbol{\varphi} \in \mathbf{\Phi}$ by $\forall \mathbf{x}_1 \forall \mathbf{x}_2 \ldots \forall \mathbf{x}_k \boldsymbol{\varphi}$, where $\mathbf{x}_1, \mathbf{x}_2, \ldots, \mathbf{x}_k$ are the free variables of $\boldsymbol{\varphi}$. Then $\mathbf{\Phi} \vdash^{\forall} \boldsymbol{\alpha}$ iff $\mathbf{\Phi}^{\forall} \vdash \boldsymbol{\alpha}$.

(iii) While DT holds for $\vdash$ outright (see Thm. 9.11), only a restricted version of it, subject to certain conditions, holds for $\vdash^{\forall}$.

(iv) An unrestricted *rule of generalization* holds for $\vdash^{\forall}$: if $\mathbf{\Phi} \vdash^{\forall} \boldsymbol{\alpha}$ then also $\mathbf{\Phi} \vdash^{\forall} \forall \mathbf{x} \boldsymbol{\alpha}$, where $\mathbf{x}$ is any variable. For $\vdash$ only a restricted version of this rule holds, as we shall see.

9.14. Theorem (Semantic soundness of Fopcal)

If $\mathbf{\Phi} \vdash \boldsymbol{\alpha}$ *then also* $\mathbf{\Phi} \models \boldsymbol{\alpha}$. *In particular, if* $\vdash \boldsymbol{\alpha}$ *then also* $\models \boldsymbol{\alpha}$.

PROOF

Similar to the proof of the soundness of the propositional calculus (Thm. 7.6.12), except that now it needs to be verified that all first-order axioms are logically valid. This is straightforward; DIY. ■

9.15. Theorem

If $\mathbf{\Phi} \vdash_0 \boldsymbol{\alpha}$ *then also* $\mathbf{\Phi} \vdash \boldsymbol{\alpha}$. *In particular, if* $\vdash_0 \boldsymbol{\alpha}$ *then also* $\vdash \boldsymbol{\alpha}$. ■

9.16. Problem

Prove that $\vdash \boldsymbol{\alpha}(\mathbf{x}/\mathbf{t}) \rightarrow \exists \mathbf{x} \boldsymbol{\alpha}$.

9.17. Problem

Prove that $\vdash \exists \mathbf{x}(\mathbf{t}=\mathbf{x})$, provided $\mathbf{x}$ does not occur in $\mathbf{t}$. Point out where you use the assumption about $\mathbf{x}$ and $\mathbf{t}$.

§ 10. Rules of instantiation and generalization

10.1. Theorem (Rule of Universal Instantiation)

If $\mathbf{\Phi} \vdash \forall \mathbf{x}\alpha$ *then* $\mathbf{\Phi} \vdash \alpha(\mathbf{x}/\mathbf{t})$ *for any term* $\mathbf{t}$. ■

10.2. Remarks

(i) For brevity we shall refer to this rule as 'UI'.

(ii) Clearly, UI holds for any linear calculus with *modus ponens* as a rule of inference and all formulas of the form $\forall \mathbf{x}\alpha \rightarrow \alpha(\mathbf{x}/\mathbf{t})$ as theorems.

(iii) The only purpose of adopting Ax. 4 was to enable us to establish UI. Now that we have done so, Ax. 4 need not be invoked again. Indeed, it is easy to see that any calculus for which UI and DT hold has all formulas of the form $\forall \mathbf{x}\alpha \rightarrow \alpha(\mathbf{x}/\mathbf{t})$ as theorems.

(iv) Closely related to UI is the Rule of Existential Generalization (briefly, EG): *If* $\mathbf{\Phi} \vdash \alpha(\mathbf{x}/\mathbf{t})$ *for some term* $\mathbf{t}$, *then* $\mathbf{\Phi} \vdash \exists \mathbf{x}\alpha$. This rule follows at once from Prob. 9.16.

10.3. Definition

A variable is said to be *free* in a set (or a sequence) of formulas, if that variable is free in some formula belonging to the set (or the sequence).

10.4. Theorem

Given a deduction $\boldsymbol{D}$ *of a formula* α *from a set* $\mathbf{\Phi}$ *of hypotheses, if* $\mathbf{x}$ *is a variable that is not free in* $\mathbf{\Phi}$ *then we can construct a deduction* $\boldsymbol{D}'$ *of* $\forall \mathbf{x}\alpha$ *from* $\mathbf{\Phi}$ *such that* $\mathbf{x}$ *is not free in* $\boldsymbol{D}'$ *and every variable free in* $\boldsymbol{D}'$ *is free in* $\boldsymbol{D}$ *as well.*

PROOF

Let $\boldsymbol{D}$ be $\varphi_1, \varphi_2, \ldots, \varphi_n$; so $\varphi_n = \alpha$. We shall show by induction on k ($k = 1, 2, \ldots, n$) how to construct a deduction $\boldsymbol{D}_k$ of $\forall \mathbf{x}\varphi_k$ from $\mathbf{\Phi}$, such that $\mathbf{x}$ is not free in $\boldsymbol{D}_k$, and every variable free in $\boldsymbol{D}_k$ is free also in $\boldsymbol{D}$. Then we can take $\boldsymbol{D}_n$ as the required $\boldsymbol{D}'$.

Case 1: φ_k is an axiom of Fopcal. Then $\forall \mathbf{x}\varphi_k$ is likewise an axiom – Ax. 8 – and we can take $\boldsymbol{D}_k$ as this formula by itself.

Case 2: $\varphi_k \in \mathbf{\Phi}$. Then by assumption $\mathbf{x}$ is not free in φ_k, and we can take $\boldsymbol{D}_k$ to be

$$\boldsymbol{\varphi}_k, \qquad \text{(hyp.)}$$
$$\boldsymbol{\varphi}_k \rightarrow \forall \mathbf{x} \boldsymbol{\varphi}_k, \qquad \text{(Ax. 3)}$$
$$\forall \mathbf{x} \boldsymbol{\varphi}. \qquad \text{(m.p.)}$$

Case 3: $\boldsymbol{\varphi}_k$ is obtained by *modus ponens* from two earlier formulas in $\boldsymbol{D}$. Then there are i, $j < k$ such that $\boldsymbol{\varphi}_j = \boldsymbol{\varphi}_i \rightarrow \boldsymbol{\varphi}_k$. By the induction hypothesis, we already possess deductions with the required properties, $\boldsymbol{D}_i$ and $\boldsymbol{D}_j$ of $\forall \mathbf{x} \boldsymbol{\varphi}_i$ and $\forall \mathbf{x}(\boldsymbol{\varphi}_i \rightarrow \boldsymbol{\varphi}_k)$ respectively. It is now enough to show that from these two formulas the formula $\forall \mathbf{x} \boldsymbol{\varphi}_k$ can be deduced by means of a deduction in which $\mathbf{x}$ is not free and whose free variables are all included among those of $\boldsymbol{D}$. Here is such a deduction:

$$\forall \mathbf{x} \boldsymbol{\varphi}_i, \qquad \text{(hyp.)}$$
$$\forall \mathbf{x}(\boldsymbol{\varphi}_i \rightarrow \boldsymbol{\varphi}_k), \qquad \text{(hyp.)}$$
$$\forall \mathbf{x}(\boldsymbol{\varphi}_i \rightarrow \boldsymbol{\varphi}_k) \rightarrow \forall \mathbf{x} \boldsymbol{\varphi}_i \rightarrow \forall \mathbf{x} \boldsymbol{\varphi}_k, \qquad \text{(Ax. 2)}$$
$$\forall \boldsymbol{\varphi}_i \rightarrow \forall \mathbf{x} \boldsymbol{\varphi}_k, \qquad \text{(m.p.)}$$
$$\forall \mathbf{x} \boldsymbol{\varphi}_k. \qquad \text{(m.p.)} \quad \blacksquare$$

10.5. *Corollary (Rule of Universal Generalization on a Variable)*

If $\boldsymbol{\Phi} \vdash \boldsymbol{\alpha}$ *and* $\mathbf{x}$ *is not free in* $\boldsymbol{\Phi}$ *then* $\boldsymbol{\Phi} \vdash \forall \mathbf{x} \boldsymbol{\alpha}$. ■

10.6. *Remarks*

(i) We shall refer to this rule briefly as 'UGV'.

(ii) The only purpose of adopting Ax. 2, Ax. 3 and Ax. 8 was to enable us to prove Thm. 10.4. Now that this has been done these axioms need not be invoked again.

(iii) It is obvious that if $\vdash^*$ is the relation of deducibility in any calculus for which UGV holds, then from $\vdash^* \boldsymbol{\alpha}$ it follows that also $\vdash^* \forall \mathbf{x} \boldsymbol{\alpha}$ for any variable $\mathbf{x}$ (cf. Ax. 8). If in addition DT also holds for $\vdash^*$, then $\vdash^* \boldsymbol{\alpha} \rightarrow \forall \mathbf{x} \boldsymbol{\alpha}$ for any formula $\boldsymbol{\alpha}$ and any variable $\mathbf{x}$ that is not free in $\boldsymbol{\alpha}$ (cf. Ax. 3). See also Prob. 10.7 below.

(iv) Thm. 10.4 can be strengthened: it is enough to require that $\mathbf{x}$ is not free in any formula of $\boldsymbol{\Phi}$ used as a hypothesis in the given deduction (although it may be free in formulas of $\boldsymbol{\Phi}$ that are not so used). To see this, let $\boldsymbol{\Phi}_0$ be the set of those members of $\boldsymbol{\Phi}$ that are used in the given deduction $\boldsymbol{D}$, and apply the theorem to $\boldsymbol{\Phi}_0$. Similarly, in Cor. 10.5 it is enough to require that $\mathbf{x}$ is not free in members of $\mathbf{\Phi}$ used as hypotheses in some particular deduction of $\boldsymbol{\alpha}$ from $\mathbf{\Phi}$. Similar remarks apply also to other results in the present section.

(v) On the other hand, the proviso that $\mathbf{x}$ must not be free in the hypotheses used to deduce $\boldsymbol{\alpha}$ is essential. For example, let $\boldsymbol{\alpha}$ be $\mathbf{x}\neq\mathbf{y}$, where $\mathbf{x}$ and $\mathbf{y}$ are distinct variables. If not for the proviso in Cor. 10.5, we would have $\mathbf{x}\neq\mathbf{y} \vdash \forall\mathbf{x}(\mathbf{x}\neq\mathbf{y})$ and hence, by Thm. 9.14, also $\mathbf{x}\neq\mathbf{y} \models \forall\mathbf{x}(\mathbf{x}\neq\mathbf{y})$. But this is absurd, as $\mathbf{x}\neq\mathbf{y}$ is clearly satisfied by any valuation that assigns $\mathbf{x}$ and $\mathbf{y}$ distinct values, whereas $\forall\mathbf{x}(\mathbf{x}\neq\mathbf{y})$ is satisfied by no valuation.

10.7. Problem

Let $\vdash^*$ be the relation of deducibility in a calculus with *modus ponens* as a rule of inference and for which Cut, DT, UI and UGV hold. Show that $\vdash^*\forall\mathbf{x}(\boldsymbol{\alpha}\rightarrow\boldsymbol{\beta})\rightarrow\forall\mathbf{x}\boldsymbol{\alpha}\rightarrow\forall\mathbf{x}\boldsymbol{\beta}$ for any formulas $\boldsymbol{\alpha}$ and $\boldsymbol{\beta}$ and any variable $\mathbf{x}$.

10.8. Definition

For any formula $\boldsymbol{\alpha}$ and variable $\mathbf{x}$, we put

$$\exists!\mathbf{x}\boldsymbol{\alpha} =_{\text{df}} \exists\mathbf{y}\forall\mathbf{x}(\boldsymbol{\alpha}\leftrightarrow\mathbf{x}=\mathbf{y}),$$

where $\mathbf{y}$ is the first variable in alphabetic order that differs from $\mathbf{x}$ and is not free in $\boldsymbol{\alpha}$.

10.9. Problem

(i) Verify that $\sigma \models \exists!\mathbf{x}\boldsymbol{\alpha}$ iff $\sigma(\mathbf{x}/u) \models \boldsymbol{\alpha}$ for exactly one individual u in the universe U of σ.

(ii) Prove that $\vdash\exists!\mathbf{x}(\mathbf{t}=\mathbf{x})$, provided $\mathbf{x}$ does not occur in $\mathbf{t}$.

10.10. Theorem (Rule of Universal Generalization on a Constant)

If $\mathbf{\Phi} \vdash \boldsymbol{\alpha}(\mathbf{x}/\mathbf{c})$, *where* $\mathbf{c}$ *is a constant that occurs neither in* $\mathbf{\Phi}$ *nor in* $\boldsymbol{\alpha}$, *then also* $\mathbf{\Phi} \vdash \forall\mathbf{x}\boldsymbol{\alpha}$.

PROOF

Let D be a deduction $\boldsymbol{\varphi}_1, \boldsymbol{\varphi}_2, \ldots, \boldsymbol{\varphi}_n$ of $\boldsymbol{\alpha}(\mathbf{x}/\mathbf{c})$ from $\mathbf{\Phi}$. Thus $\boldsymbol{\varphi}_n = \boldsymbol{\alpha}(\mathbf{x}/\mathbf{c})$.

Now let $\mathbf{y}$ be a new variable, in the sense that it is distinct from $\mathbf{x}$

and does not occur at all (either free or bound) in the deduction $\boldsymbol{D}$. Let $\boldsymbol{D}'$ be the sequence $\boldsymbol{\varphi}_1', \boldsymbol{\varphi}_2', \ldots, \boldsymbol{\varphi}_n'$ of formulas obtained from $\boldsymbol{D}$ upon replacing $\mathbf{c}$ everywhere by $\mathbf{y}$. We claim that $\boldsymbol{D}'$ is a deduction of $\boldsymbol{\alpha}(\mathbf{x}/\mathbf{y})$ from $\boldsymbol{\Phi}$.

Indeed, for any k (where $1 \leq k \leq n$) three cases are possible. First, $\boldsymbol{\varphi}_k$ may be an axiom. In this case it is easy to verify that $\boldsymbol{\varphi}_k'$ is also an axiom. Second, $\boldsymbol{\varphi}_k$ may be a hypothesis, a member of $\boldsymbol{\Phi}$. In this case $\boldsymbol{\varphi}_k'$ is $\boldsymbol{\varphi}_k$ itself, because $\mathbf{c}$ does not occur in $\boldsymbol{\Phi}$. Finally, $\boldsymbol{\varphi}_k$ may have been obtained by *modus ponens* from two earlier formulas in $\boldsymbol{D}$, $\boldsymbol{\varphi}_i$ and $\boldsymbol{\varphi}_j$. In this case it is obvious that $\boldsymbol{\varphi}_k'$ is obtained by *modus ponens* from $\boldsymbol{\varphi}_i'$ and $\boldsymbol{\varphi}_j'$. Thus $\boldsymbol{D}'$ is a deduction of $\boldsymbol{\alpha}(\mathbf{x}/\mathbf{c})'$ from $\boldsymbol{\Phi}$.

We still have to show that $\boldsymbol{\alpha}(\mathbf{x}/\mathbf{c})'$ is in fact $\boldsymbol{\alpha}(\mathbf{x}/\mathbf{y})$. To see this, recall that $\mathbf{c}$ does not occur in $\boldsymbol{\alpha}$. Thus the occurrences of $\mathbf{c}$ in $\boldsymbol{\alpha}(\mathbf{x}/\mathbf{c})$ are just those that replace the free occurrences of $\mathbf{x}$ in $\boldsymbol{\alpha}$; there are no other occurrences of $\mathbf{c}$ in $\boldsymbol{\alpha}(\mathbf{x}/\mathbf{c})$. Now, $\boldsymbol{\alpha}(\mathbf{x}/\mathbf{c})'$ was obtained from $\boldsymbol{\alpha}(\mathbf{x}/\mathbf{c})$ upon replacing these occurrences of $\mathbf{c}$ by the new variable $\mathbf{y}$. Thus $\boldsymbol{\alpha}(\mathbf{x}/\mathbf{c})'$ can be obtained directly from $\boldsymbol{\alpha}$ upon replacing all free occurrences of $\mathbf{x}$ in $\boldsymbol{\alpha}$ by $\mathbf{y}$. But $\boldsymbol{\alpha}(\mathbf{x}/\mathbf{y})$ is obtained from $\boldsymbol{\alpha}$ in precisely the same way, because $\mathbf{y}$ is a new variable, not occurring in $\boldsymbol{\alpha}$, so that the substitution of $\mathbf{y}$ for $\mathbf{x}$ in $\boldsymbol{\alpha}$ does not involve any alphabetic changes.

We have now established that $\boldsymbol{D}'$ is indeed a deduction of $\boldsymbol{\alpha}(\mathbf{x}/\mathbf{y})$ from $\boldsymbol{\Phi}$. Moreover, note that $\mathbf{y}$ does not occur in those members of $\boldsymbol{\Phi}$ that are used as hypotheses in $\boldsymbol{D}'$: the only occurrences of $\mathbf{y}$ in $\boldsymbol{D}'$ are those that have replaced occurrences of $\mathbf{c}$, but $\mathbf{c}$ does not occur in $\boldsymbol{\Phi}$. Therefore by UGV we have $\boldsymbol{\Phi} \vdash \forall \mathbf{y}[\boldsymbol{\alpha}(\mathbf{x}/\mathbf{y})]$.

By UI we have $\forall \mathbf{y}[\boldsymbol{\alpha}(\mathbf{x}/\mathbf{y})] \vdash \boldsymbol{\alpha}(\mathbf{x}/\mathbf{y})(\mathbf{y}/\mathbf{x})$. But it is easy to see that $\boldsymbol{\alpha}(\mathbf{x}/\mathbf{y})(\mathbf{y}/\mathbf{x})$ is in fact $\boldsymbol{\alpha}$ itself; hence we have got $\forall \mathbf{y}[\boldsymbol{\alpha}(\mathbf{x}/\mathbf{y})] \vdash \boldsymbol{\alpha}$. Now, $\mathbf{x}$ is clearly not free in $\forall \mathbf{y}[\boldsymbol{\alpha}(\mathbf{x}/\mathbf{y})]$, so we can use UGV again and obtain $\forall \mathbf{y}[\boldsymbol{\alpha}(\mathbf{x}/\mathbf{y})] \vdash \forall \mathbf{x}\boldsymbol{\alpha}$.

By Cut we finally have $\boldsymbol{\Phi} \vdash \forall \mathbf{x}\boldsymbol{\alpha}$, as required. ■

10.11. Remark

We shall refer to this rule briefly as 'UGC'.

§11. Consistency

As decreed in Def. 9.10(iv), a set $\boldsymbol{\Phi}$ of $\mathcal{L}$-formulas is *[first-order] inconsistent* (briefly, $\boldsymbol{\Phi}\vdash$) if both members of a contradictory pair can be deduced from $\boldsymbol{\Phi}$ in Fopcal. Otherwise, $\boldsymbol{\Phi}$ is *[first-order] consistent*.

We have already noted (Thm. 9.11) that IE, *reductio* and PIP hold for Fopcal. The other results of § 8 of Ch. 7 also have counterparts in Fopcal. In particular, the following two results are proved similarly to Thm. 7.8.4 and Cor. 7.8.5.

11.1. Theorem

If $\mathbf{\Phi} \vdash$ *then* $\mathbf{\Phi} \models$. ■

11.2. Corollary (Consistency of Fopcal)

It is impossible that both $\vdash \boldsymbol{\alpha}$ and $\vdash \neg\boldsymbol{\alpha}$. ■

11.3. Remark

This proof of the consistency of Fopcal uses semantic notions which, generally speaking, require a relatively powerful set-theoretic ambient theory (see Rem. 4.14). On the other hand, since deductions are finite objects, proof-theoretic notions such as deducibility and consistency are quite elementary. It is therefore natural to ask whether the consistency of Fopcal can be proved in an elementary way, without appealing to semantics. Such a proof is outlined in the following problem.

11.4. Problem

(i) Show that if $\mathbf{\Phi} \vdash \boldsymbol{\alpha}$ and $\mathbf{\Phi}$ is a set of even formulas (see Def. 8.3) then $\boldsymbol{\alpha}$ is even as well. (Verify that all the axioms of Fopcal are even formulas and that *modus ponens* yields an even conclusion from even premisses.)

(ii) Hence prove the consistency of Fopcal.

We shall now prove a few results that have no counterpart in the propositional calculus. These results, which will be needed later, are concerned with a consistent set $\mathbf{\Phi}$ of formulas that contains formulas of the form $\neg\forall\mathbf{x}\boldsymbol{\alpha}$. We add to $\mathbf{\Phi}$ 'witnessing' formulas $\neg\boldsymbol{\alpha}(\mathbf{x}/\mathbf{c})$, where the 'witness' $\mathbf{c}$ is a fresh constant, that does not occur in $\mathbf{\Phi}$. We prove that the resulting set is consistent. First we consider the case where just one witnessing formula is added; then a finite number; and then an arbitrary set of such formulas.

11.5. Lemma

If $\mathbf{\Phi}$ is consistent and $\neg\forall\mathbf{x}\boldsymbol{\alpha} \in \mathbf{\Phi}$, and $\mathbf{c}$ is a constant that does not occur in $\mathbf{\Phi}$, then $\mathbf{\Phi} \cup \{\neg\boldsymbol{\alpha}(\mathbf{x}/\mathbf{c})\}$ is also consistent.

PROOF

If $\mathbf{\Phi}, \neg\boldsymbol{\alpha}($[illegible]$)\vdash$, then by PIP $\mathbf{\Phi} \vdash \boldsymbol{\alpha}(\mathbf{x}/\mathbf{c})$. As $\mathbf{c}$ does not occur in $\mathbf{\Phi}$ and as we a[illegible]uming that $\neg\forall\mathbf{x}\boldsymbol{\alpha} \in \mathbf{\Phi}$, $\mathbf{c}$ cannot occur in $\boldsymbol{\alpha}$ either. Therefore by UGC $\mathbf{\Phi} \vdash \forall\mathbf{x}\boldsymbol{\alpha}$. But this is impossible, since $\neg\forall\mathbf{x}\boldsymbol{\alpha} \in \mathbf{\Phi}$ and $\mathbf{\Phi}$ was assumed to be consistent. ■

11.6. Problem

Prove the Rule of Existential Instantiation with a Constant (EIC): *If $\mathbf{\Phi}$ is consistent and $\exists\mathbf{x}\boldsymbol{\alpha} \in \mathbf{\Phi}$, and $\mathbf{c}$ is a constant that does not occur in $\mathbf{\Phi}$, then $\mathbf{\Phi} \cup \{\boldsymbol{\alpha}(\mathbf{x}/\mathbf{c})\}$ is also consistent.*

11.7. Lemma

Let $\mathbf{\Phi}$ be consistent; for each $i = 1, 2, \ldots, k$, let $\neg\forall\mathbf{x}_i\boldsymbol{\alpha}_i \in \mathbf{\Phi}$, and let $\mathbf{c}_i$ be distinct constants that do not occur in $\mathbf{\Phi}$.

Then $\mathbf{\Phi} \cup \{\neg\boldsymbol{\alpha}_i(\mathbf{x}_i/\mathbf{c}_i) : i = 1, 2, \ldots, k\}$ is also consistent.

PROOF

DIY by [weak] induction on k, using Lemma 11.5. ■

11.8. Lemma

Let $\mathbf{\Phi}$ be consistent; let $\mathbf{\Phi}'$ be obtained from $\mathbf{\Phi}$ by adding, for every formula of the form $\neg\forall\mathbf{x}\boldsymbol{\alpha}$ in $\mathbf{\Phi}$, a 'witnessing' formula $\neg\boldsymbol{\alpha}(\mathbf{x}/\mathbf{c})$, where $\mathbf{c}$ does not occur in $\mathbf{\Phi}$ and where distinct constants $\mathbf{c}$ are used for distinct formulas of the form $\neg\forall\mathbf{x}\boldsymbol{\alpha}$. Then $\mathbf{\Phi}'$ is consistent as well.

PROOF

It is enough to prove that every finite subset of $\mathbf{\Phi}'$ is consistent. (Cf. Prob. 7.8.3(i): a similar result clearly holds for Fopcal.) However, a finite subset of $\mathbf{\Phi}'$ contains only a finite number of the new witnessing formulas, and is therefore included in a set of the form $\mathbf{\Phi} \cup \{\neg\boldsymbol{\alpha}_i(\mathbf{x}_i/\mathbf{c}_i) : i = 1, 2, \ldots, k\}$, which is consistent by Lemma 11.7. ■

In the sequel we shall need to consider, in addition to a given first-order language $\mathcal{L}$, languages obtained from it by adding new individual constants, which will be used in connection with Lemma 11.8. We shall need to be sure that a consistent set of $\mathcal{L}$-formulas remains consistent within such an extended language.

This is not entirely obvious. Suppose $\mathcal{L}^+$ is obtained from $\mathcal{L}$ by adding a set $\mathbf{C}$ of new constants. Let $\mathbf{\Phi}$ be a set of $\mathcal{L}$-formulas that is consistent within $\mathcal{L}$. This means that there exists an $\mathcal{L}$-formula $\boldsymbol{\alpha}$ that is not deducible from $\mathbf{\Phi}$ within $\mathcal{L}$. But in $\mathcal{L}^+$ there are formulas that do not belong to $\mathcal{L}$ and in particular there are more axioms – additional members of the eight axiom groups – containing new constants. Can the formula $\boldsymbol{\alpha}$ become deducible from $\mathbf{\Phi}$ within $\mathcal{L}^+$ by using these additional axioms?

We shall now show that this is in fact impossible.

11.9. Theorem

Let $\mathbf{\Phi}$ be a set of $\mathcal{L}$-formulas that is consistent within $\mathcal{L}$. Let $\mathcal{L}^+$ be obtained from $\mathcal{L}$ by adding a set $\mathbf{C}$ of new individual constants. Then $\mathbf{\Phi}$ is consistent within $\mathcal{L}^+$ as well.

PROOF

By assumption, there is an $\mathcal{L}$-formula $\boldsymbol{\alpha}$ not deducible from $\mathbf{\Phi}$ in $\mathcal{L}$. It is enough to show that this remains the case also in $\mathcal{L}^+$.

Suppose that $\boldsymbol{D}$ is a deduction of $\boldsymbol{\alpha}$ from $\mathbf{\Phi}$ within $\mathcal{L}^+$. Since $\boldsymbol{D}$ is a *finite* sequence of $\mathcal{L}^+$-formulas, it can contain only a finite number of new constants, say $\mathbf{c}_1, \mathbf{c}_2, \ldots, \mathbf{c}_m$. Now choose m distinct variables $\mathbf{y}_1, \mathbf{y}_2, \ldots, \mathbf{y}_m$ that do not occur (free or bound) in $\boldsymbol{D}$ and let $\boldsymbol{D}'$ be obtained from $\boldsymbol{D}$ upon replacing $\mathbf{c}_1, \mathbf{c}_2, \ldots, \mathbf{c}_m$ throughout by $\mathbf{y}_1, \mathbf{y}_2, \ldots, \mathbf{y}_m$ respectively.

An argument similar to that used in the proof of Thm. 10.10 shows that $\boldsymbol{D}'$ is a deduction from $\mathbf{\Phi}$. Indeed, when $\boldsymbol{D}$ was transformed into $\boldsymbol{D}'$ any axiom used in $\boldsymbol{D}$ was transformed into an axiom; any hypothesis remained unchanged (since $\mathbf{\Phi}$ is a set of $\mathcal{L}$-formulas, it contains no new constants); and any application of *modus ponens* in $\boldsymbol{D}$ was transformed into an application of *modus ponens*. Now, $\boldsymbol{D}'$ is a deduction within $\mathcal{L}$, because the new constants that were present have been supplanted by variables. The last member of $\boldsymbol{D}'$ is still $\boldsymbol{\alpha}$, which has remained unchanged as it does not contain any new constants. So now we have a deduction of $\boldsymbol{\alpha}$ from $\mathbf{\Phi}$ within $\mathcal{L}$ – contrary to our original assumption. ■

§12. Maximal consistency

By Def. 9.10(iv), a set of $\mathcal{L}$-formulas is *maximal [first-order] consistent [in $\mathcal{L}$]* if it is consistent but not included in any other consistent set of $\mathcal{L}$-formulas. As usual, we omit the qualifications 'first-order' and 'in $\mathcal{L}$' when there is no risk of confusion. The following two theorems are proved in exactly the same way as their propositional counterparts.

12.1. Theorem

If $\mathbf{\Phi}$ is a maximal consistent set and $\mathbf{\Phi} \vdash \boldsymbol{\alpha}$, then $\boldsymbol{\alpha} \in \mathbf{\Phi}$. ■

12.2. Theorem

A consistent set $\mathbf{\Phi}$ is maximal consistent iff for every formula $\boldsymbol{\alpha}$ either $\boldsymbol{\alpha} \in \mathbf{\Phi}$ or $\neg\boldsymbol{\alpha} \in \mathbf{\Phi}$. ■

12.3. Remark

From Thm. 12.2 it follows that if $\mathcal{L}$ is extended to a richer language $\mathcal{L}^+$, by adding new extralogical symbols (for example, new constants) then a set $\mathbf{\Phi}$ of $\mathcal{L}$-formulas that is maximal consistent in $\mathcal{L}$ will no longer be so in $\mathcal{L}^+$. Indeed, if $\boldsymbol{\alpha}$ is an $\mathcal{L}^+$-formula containing a new symbol (one that does not belong to $\mathcal{L}$) then $\boldsymbol{\alpha}$ is not an $\mathcal{L}$-formula, so neither $\boldsymbol{\alpha}$ nor $\neg\boldsymbol{\alpha}$ can belong to $\mathbf{\Phi}$. Of course, by Thm. 11.9 $\mathbf{\Phi}$ is still consistent in $\mathcal{L}^+$.

The following result is proved similarly to Thm. 7.12.6(i).

12.4. Theorem

For any valuation σ, the set $\{\boldsymbol{\varphi} : \boldsymbol{\varphi}^\sigma = \top\}$ is maximal consistent. ■

The counterpart of Thm. 7.12.6(ii) is also true: every maximal [first-order] consistent set has the form $\{\boldsymbol{\varphi} : \boldsymbol{\varphi}^\sigma = \top\}$ for a unique valuation σ. But in order to prove this we must first show that every maximal consistent set is satisfiable. In propositional logic we were able to show that every maximal [propositionally] consistent set is a [propositional] Hintikka set, and hence satisfiable. Here matters are not so simple.

12.5. Theorem

If $\mathbf{\Phi}$ *is maximal consistent, it fulfils conditions* (1)–(5) *and* (7)–(9) *of* Def. 7.1.

PROOF

Conditions (1)–(4) are verified as in the proof of Thm. 7.12.5. Conditions (5) and (7)–(9) are verified by invoking UI and Ax. 5–Ax. 7 respectively and using Thm. 12.1. ∎

The following problem provides a counter-example showing that a maximal consistent set need not fulfil the missing condition (6) of Def. 7.1, and hence need not be a Hintikka set.

12.6. Problem

Let $\mathscr{L}$ be a first-order language with equality but without any extra-logical symbols. Let σ be the $\mathscr{L}$-valuation whose universe is $U = \{u, v\}$, where u and v are distinct, and such that $\mathbf{x}^\sigma = u$ for every variable $\mathbf{x}$. Let $\mathbf{\Phi} = \{\boldsymbol{\varphi} : \boldsymbol{\varphi}^\sigma = \top\}$, so that by Thm. 12.4 $\mathbf{\Phi}$ is maximal consistent. Let $\boldsymbol{\alpha}$ be the formula $\mathbf{x}=\mathbf{y}$, where $\mathbf{x}$ and $\mathbf{y}$ are distinct variables.

Show that $\neg\forall\mathbf{x}\boldsymbol{\alpha} \in \mathbf{\Phi}$ but there is no $\mathscr{L}$-term $\mathbf{t}$ such that $\neg\boldsymbol{\alpha}(\mathbf{x}/\mathbf{t}) \in \mathbf{\Phi}$. (Note that the only terms of $\mathscr{L}$ are the variables.)

§ 13. Completeness

13.1. Preview

As in propositional logic, the [strong] completeness of Fopcal will follow immediately once we show that any given consistent set $\mathbf{\Phi}$ of $\mathscr{L}$-formulas is satisfiable. Also, exactly as in propositional logic, it is easy to see that the set of all consistent sets of $\mathscr{L}$-formulas is of finite character (cf. proof of Thm. 7.13.1); hence, by the Tukey–Teichmüller Lemma (Thm. 5.2.8), any consistent $\mathbf{\Phi}$ is included in some $\mathbf{\Psi}$ that is maximal consistent within $\mathscr{L}$. However, since a maximal consistent set may not be a Hintikka set, we have no *direct* way of showing that $\mathbf{\Psi}$ is satisfiable.

It is clear from Thm. 12.5 and Prob. 12.6 that the only reason that may prevent $\mathbf{\Psi}$ from being a Hintikka set is the absence in it of witnessing formulas. To overcome this obstacle, we use Lemma 11.8, and add to $\mathbf{\Psi}$ enough witnessing formulas, using constants as witnesses. However, in order to make sure that these witness constants do not occur in $\mathbf{\Psi}$ (as Lemma 11.8 requires) we extend $\mathscr{L}$ to a richer language $\mathscr{L}_1$ by adding an adequate supply of new constants. By Thm. 11.9 $\mathbf{\Psi}$ is still consistent in $\mathscr{L}_1$, so we may apply Lemma 11.8 there. Let $\mathbf{\Phi}_1$ be the set so obtained. Unfortunately, in $\mathscr{L}_1$ $\mathbf{\Psi}$ is no longer *maximal* consistent (see Rem. 12.3), nor does the addition of new witnessing formulas produce a maximal consistent set: all we can say about $\mathbf{\Phi}_1$ is that it is consistent. It seems as though we are back where we started.

Not despairing, we extend $\mathbf{\Phi}_1$ to a maximal consistent set $\mathbf{\Psi}_1$ within $\mathscr{L}_1$. Then we extend $\mathscr{L}_1$ to a richer language $\mathscr{L}_2$ by adding yet more new constants, and get $\mathbf{\Phi}_2$ from $\mathbf{\Psi}_1$ in the same way as we got $\mathbf{\Phi}_1$ from $\mathbf{\Psi}$.

The good news is that by iterating this procedure *ad infinitum* we obtain *in the limit* a set that is not only maximal consistent but also a Hintikka set, and includes our original set $\mathbf{\Phi}$.

Throughout this section we shall be working within set theory (that is, assume it as an ambient metatheory). In particular, as explained in Rem. 6.1.8, we shall identify the natural numbers with the finite ordinals (a.k.a. finite cardinals).

13.2. Definition

A set $\mathbf{\Phi}$ of $\mathscr{L}$-formulas is a *Henkin set in* $\mathscr{L}$ if $\mathbf{\Phi}$ is maximal consistent in $\mathscr{L}$ and, for any formula $\boldsymbol{\alpha}$ and variable $\mathbf{x}$, if $\neg\forall\mathbf{x}\boldsymbol{\alpha} \in \mathbf{\Phi}$ then $\neg\boldsymbol{\alpha}(\mathbf{x}/\mathbf{t}) \in \mathbf{\Phi}$ for some term $\mathbf{t}$.

13.3. Remark

From Thm. 12.5 and Def. 7.1 it follows at once that a Henkin set in $\mathscr{L}$ is also a Hintikka set in $\mathscr{L}$. Hence by Thm. 7.17 such a set is satisfied by some valuation whose universe has cardinality not greater than $\|\mathscr{L}\|$.

From now until the end of the proof of Thm. 13.8 we let $\mathbf{\Phi}$ be a fixed but arbitrary consistent set of $\mathscr{L}$-formulas.

By [weak] induction on n we define for each natural number n a first-order language $\mathscr{L}_n$, a set $\mathbf{\Phi}_n$ of $\mathscr{L}_n$-formulas, and a set $\mathbf{\Psi}_n$ of $\mathscr{L}_n$-formulas that is maximal consistent in $\mathscr{L}_n$.

13.4. Definition

Basis. We put $\mathcal{L}_0 = \mathcal{L}$ and $\mathbf{\Phi}_0 = \mathbf{\Phi}$. As $\mathbf{\Psi}_0$ we choose some set of formulas that is maximal consistent in $\mathcal{L}_0$ and includes $\mathbf{\Phi}_0$. (The existence of such $\mathbf{\Psi}_0$ is ensured by the Tukey–Teichmüller Lemma.)

Induction step. Assume as induction hypothesis that $\mathcal{L}_n$, $\mathbf{\Phi}_n$ and $\mathbf{\Psi}_n$ have been defined, and that $\mathbf{\Psi}_n$ is a set of $\mathcal{L}_n$-formulas that is maximal consistent within $\mathcal{L}_n$.

For each $\mathcal{L}_n$-formula $\boldsymbol{\varphi}$, let $\mathbf{c}_{\boldsymbol{\varphi}}$ be a new constant (not present in $\mathcal{L}_n$) such that if $\boldsymbol{\varphi}$ and $\boldsymbol{\psi}$ are distinct formulas then $\mathbf{c}_{\boldsymbol{\varphi}}$ and $\mathbf{c}_{\boldsymbol{\psi}}$ are distinct constants. Let $\mathbf{C}_n$ be the set of all these new constants:

$$\mathbf{C}_n = \{\mathbf{c}_{\boldsymbol{\varphi}} : \boldsymbol{\varphi} \text{ is an } \mathcal{L}_n\text{-formula}\}.$$

We define $\mathcal{L}_{n+1}$ as the language obtained by adding the set of constants $\mathbf{C}_n$ to $\mathcal{L}_n$.

Since $\mathbf{\Psi}_n$ is maximal consistent in $\mathcal{L}_n$, it follows from Thm. 11.9 that it is still consistent (albeit not maximally so) in the richer language $\mathcal{L}_{n+1}$. We define $\mathbf{\Phi}_{n+1}$ to be the set of formulas obtained from $\mathbf{\Psi}_n$ as follows: for each formula $\boldsymbol{\varphi} \in \mathbf{\Psi}_n$ of the form $\neg\forall\mathbf{x}\boldsymbol{\alpha}$, add to $\mathbf{\Psi}_n$ the formula $\neg\boldsymbol{\alpha}(\mathbf{x}/\mathbf{c}_{\boldsymbol{\varphi}})$, where $\mathbf{c}_{\boldsymbol{\varphi}}$ is the new constant in $\mathbf{C}_n$ corresponding to this particular formula $\boldsymbol{\varphi}$. Clearly, $\mathbf{\Phi}_{n+1}$ is a set of $\mathcal{L}_{n+1}$-formulas. And since $\mathbf{\Psi}_n$ is a set of $\mathcal{L}_n$-formulas, none of these new constants occur in it, so by Lemma 11.8 $\mathbf{\Phi}_{n+1}$ is consistent.

Finally, we choose as $\mathbf{\Psi}_{n+1}$ some set of formulas that is maximal consistent in $\mathcal{L}_{n+1}$ and includes $\mathbf{\Phi}_{n+1}$. (The existence of such a set is again ensured by the Tukey–Teichmüller Lemma.)

This concludes our inductive definition.

13.5. Remark

From Def. 13.4 it is evident that the $\mathbf{\Phi}_n$ and $\mathbf{\Psi}_n$ form a *chain* of sets:

$$\mathbf{\Phi} = \mathbf{\Phi}_0 \subseteq \mathbf{\Psi}_0 \subseteq \mathbf{\Phi}_1 \subseteq \mathbf{\Psi}_1 \ldots \subseteq \mathbf{\Phi}_n \subseteq \mathbf{\Psi}_n \subseteq \mathbf{\Phi}_{n+1} \subseteq \mathbf{\Psi}_{n+1} \subseteq \ldots$$

13.6. Definition

We define $\mathcal{L}_\omega$ as the union of all the languages $\mathcal{L}_n$; and $\mathbf{\Psi}_\omega$ as the union of all the sets $\mathbf{\Psi}_n$ for $n = 0, 1, 2, \ldots$.

Thus $\mathcal{L}_\omega$ is obtained from $\mathcal{L}$ by adding to the latter the union of all the sets $\mathbf{C}_n$, for $n = 0, 1, 2, \ldots$; and an $\mathcal{L}_\omega$-formula $\boldsymbol{\alpha}$ belongs to $\mathbf{\Psi}_\omega$ iff it belongs to $\mathbf{\Psi}_n$ for some n.

13.7. Remark

From Rem. 13.5 it follows that an $\mathscr{L}_\omega$-formula $\boldsymbol{\alpha}$ belongs to $\boldsymbol{\Psi}_\omega$ iff there is some n such that $\boldsymbol{\alpha} \in \boldsymbol{\Psi}_k$ for all $k \geqslant n$.

13.8. Theorem

$\boldsymbol{\Psi}_\omega$ is a Henkin set in $\mathscr{L}_\omega$.

PROOF

First, we show that $\boldsymbol{\Psi}_\omega$ is consistent. For the same reason as in propositional logic (cf. Prob. 7.8.3), it is enough to show that every *finite* subset of $\boldsymbol{\Psi}_\omega$ is consistent. So let $\boldsymbol{\alpha}_1, \boldsymbol{\alpha}_2, \ldots, \boldsymbol{\alpha}_m$ be members of $\boldsymbol{\Psi}_\omega$; we shall show that $\{\boldsymbol{\alpha}_1, \boldsymbol{\alpha}_2, \ldots, \boldsymbol{\alpha}_m\}$ is consistent.

Since $\boldsymbol{\alpha}_1 \in \boldsymbol{\Psi}_\omega$, it follows (see Rem. 13.7) that there is a number n_1 such that $\boldsymbol{\alpha}_1 \in \boldsymbol{\Psi}_k$ for all $k \geqslant n_1$. Similarly, there is a number n_2 such that $\boldsymbol{\alpha}_2 \in \boldsymbol{\Psi}_k$ for all $k \geqslant n_2$. And so on for each of the $\boldsymbol{\alpha}_j$, where $j = 1, 2, \ldots, m$. Now let k be any number greater than the m numbers $n_1, n_2, \ldots, n_m$. Then clearly $\boldsymbol{\alpha}_j \in \boldsymbol{\Psi}_k$ for $j = 1, 2, \ldots, m$. It follows that $\{\boldsymbol{\alpha}_1, \boldsymbol{\alpha}_2, \ldots, \boldsymbol{\alpha}_m\} \subseteq \boldsymbol{\Psi}_k$. But by Def. 13.4 $\boldsymbol{\Psi}_k$ is maximal consistent in $\mathscr{L}_k$, hence consistent. So its subset $\{\boldsymbol{\alpha}_1, \boldsymbol{\alpha}_2, \ldots, \boldsymbol{\alpha}_m\}$ is certainly consistent, as claimed.

By Thm. 12.2, in order to show that $\boldsymbol{\Psi}_\omega$ is *maximal* consistent in $\mathscr{L}_\omega$ it is enough to show that for any $\mathscr{L}_\omega$-formula $\boldsymbol{\alpha}$, either $\boldsymbol{\alpha}$ or $\neg\boldsymbol{\alpha}$ is in $\boldsymbol{\Psi}_\omega$. So let $\boldsymbol{\alpha}$ be any $\mathscr{L}_\omega$-formula. Now, $\boldsymbol{\alpha}$ can only contain a *finite* number of the new constants (those not in the original language $\mathscr{L}$); say these constants are $\mathbf{c}_1, \mathbf{c}_2, \ldots, \mathbf{c}_m$. An argument entirely similar to the one used in the preceding paragraph shows that if k is a sufficiently big number then all these m constants are present in $\mathscr{L}_k$. Thus $\boldsymbol{\alpha}$ is in fact an $\mathscr{L}_k$-formula for some k. But by Def. 13.4 $\boldsymbol{\Psi}_k$ is maximal consistent in $\mathscr{L}_k$, so $\boldsymbol{\alpha}$ or $\neg\boldsymbol{\alpha}$ must belong to $\boldsymbol{\Psi}_k$ and hence also to $\boldsymbol{\Psi}_\omega$, which includes $\boldsymbol{\Psi}_k$.

Having proved that $\boldsymbol{\Psi}_\omega$ is maximal consistent in $\mathscr{L}_\omega$, we need only show that it fulfils the additional condition: given that $\neg\forall\mathbf{x}\boldsymbol{\alpha} \in \boldsymbol{\Psi}_\omega$ we have to show that $\neg\boldsymbol{\alpha}(\mathbf{x}/\mathbf{t}) \in \boldsymbol{\Psi}_\omega$ for some term $\mathbf{t}$. However, if $\neg\forall\mathbf{x}\boldsymbol{\alpha} \in \boldsymbol{\Psi}_\omega$ then by Def. 13.6 $\neg\forall\mathbf{x}\boldsymbol{\alpha} \in \boldsymbol{\Psi}_n$ for some n. Therefore by Def. 13.4 a formula $\neg\boldsymbol{\alpha}(\mathbf{x}/\mathbf{c})$ – where $\mathbf{c}$ is a suitably chosen new constant belonging to $\mathbf{C}_n$ – was one of the formulas added to $\boldsymbol{\Psi}_n$ to obtain $\boldsymbol{\Phi}_{n+1}$. Thus $\neg\boldsymbol{\alpha}(\mathbf{x}/\mathbf{c}) \in \boldsymbol{\Phi}_{n+1} \subseteq \boldsymbol{\Psi}_{n+1} \subseteq \boldsymbol{\Psi}_\omega$. ■

13.9. Theorem

If $\mathbf{\Phi}$ is a consistent set of $\mathcal{L}$-formulas then $\mathbf{\Phi}$ is satisfied by some $\mathcal{L}$-valuation whose universe has cardinality not greater than $\|\mathcal{L}\|$.

PROOF

We have specified in Defs. 13.4 and 13.6 how to extend the language $\mathcal{L}$ to a language $\mathcal{L}_\omega$ by adding new constants, and how to define a set $\mathbf{\Psi}_\omega$ of $\mathcal{L}_\omega$-formulas such that $\mathbf{\Phi} \subseteq \mathbf{\Psi}_\omega$; and we have shown in Thm. 13.8 that $\mathbf{\Psi}_\omega$ is a Henkin set in $\mathcal{L}_\omega$.

By Rem. 13.3, $\mathbf{\Psi}_\omega$ – and hence also its subset $\mathbf{\Phi}$ – is satisfied by some $\mathcal{L}_\omega$-valuation, say σ_ω, as obtained in §7, whose universe has cardinality not greater than $\|\mathcal{L}_\omega\|$.

Let σ be the $\mathcal{L}$-valuation that agrees with σ_ω on all the variables, as well as on all the extralogical symbols of $\mathcal{L}$. (The only difference between σ_ω and σ is that the former assigns interpretations to the new constants, which are not in $\mathcal{L}$, while σ ignores them.) Then clearly σ is an $\mathcal{L}$-valuation that satisfies $\mathbf{\Phi}$.

The universe of σ is the same as that of σ_ω; so we shall complete the proof by showing that $\|\mathcal{L}_\omega\| = \|\mathcal{L}\|$. For brevity, we put $\lambda = \|\mathcal{L}\|$. Of course, λ is an infinite cardinal, because the set of variables is infinite; in fact, its cardinality is $\aleph_0$.

The set of all $\mathcal{L}$-formulas is included in the set of all $\mathcal{L}$-strings, hence by Thm. 6.3.9 the cardinality of the former set is $\leqslant\lambda$. (In fact, it is quite easy to show that its cardinality is *exactly* λ, but we shall not need this.) Recall that $\mathcal{L}_0$ is $\mathcal{L}$ itself; so by Def. 13.4 $\mathbf{C}_0$ is equipollent to the set of $\mathcal{L}$-formulas, hence $|\mathbf{C}_0| \leqslant \lambda$. By Def. 13.4 and Thm. 6.3.6 we have $\|\mathcal{L}_1\| = \lambda$. The same argument shows, by induction on n, that $\|\mathcal{L}_n\| = \lambda$ and $|\mathbf{C}_n| \leqslant \lambda$ for all n.

It now follows that $|\bigcup\{\mathbf{C}_n : n < \omega\}| \leqslant \aleph_0 \cdot \lambda$, which by Thm. 6.3.5 is exactly λ. Using Thm. 6.3.6 as before, we see that $\|\mathcal{L}_\omega\| = \lambda$. ■

We can now prove

13.10. Theorem (Strong semantic completeness of Fopcal)

For any set $\mathbf{\Phi}$ of formulas and any formula $\boldsymbol{\alpha}$, if *$\mathbf{\Phi} \models \boldsymbol{\alpha}$ then $\mathbf{\Phi} \vdash \boldsymbol{\alpha}$.*

PROOF

Similar to that of Thm. 7.13.2. ■

13.11. Remarks

(i) Conjoining Thms. 9.14 and 13.10 we have

$$\mathbf{\Phi} \models \boldsymbol{\alpha} \Leftrightarrow \mathbf{\Phi} \vdash \boldsymbol{\alpha}.$$

Similarly, from Thms. 11.1 and 13.9 we get

$$\mathbf{\Phi} \models \Leftrightarrow \mathbf{\Phi} \vdash.$$

(ii) As pointed out in Rem. 4.14, the notions of logical consequence and (un)satisfiability are essentially set-theoretic and thus presuppose a fairly strong ambient theory. In contrast, as pointed out in Rem. 11.3, the notions of deducibility and (in)consistency in Fopcal are relatively elementary and do not require an ambient theory that treats infinite pluralities as objects. It is therefore highly remarkable that logical consequence and unsatisfiability turn out to be equivalent to deducibility and inconsistency, respectively. Of course, the *proof* of this equivalence required rather powerful set theory.

(iii) Note however that if the primitive symbols of $\mathcal{L}$ are given by explicit enumeration, the proof can be made more elementary: in Def. 13.4, instead of invoking the TT Lemma we can obtain the maximal consistent sets $\mathbf{\Psi}_n$ as outlined in Rem. 7.13.3(i).

We conclude this chapter with two very important results.

13.12. Theorem (Compactness theorem for first-order logic)

If $\mathbf{\Phi}$ is a set of formulas such that every finite subset of $\mathbf{\Phi}$ is satisfiable, then so is $\mathbf{\Phi}$ itself.

PROOF

Similar to that of Thm. 7.13.4. ■

13.13. Theorem (Löwenheim–Skolem)

Let $\mathbf{\Phi}$ be a satisfiable set of $\mathcal{L}$-formulas. Then there exists a valuation σ such that $\sigma \models \mathbf{\Phi}$ and such that the universe of σ has cardinality not greater than $\|\mathcal{L}\|$.

PROOF

By Thm. 11.1, $\mathbf{\Phi}$ is consistent. Now apply Thm. 13.9. ■

9

Facts from recursion theory

§ 1. Preliminaries

1.1. Preview

In this chapter we put formal languages on one side and present some concepts and results from recursion theory that will be needed in the sequel.

Recursion theory was created in the 1930s by logicians (Alonzo Church, Kurt Gödel, Stephen Kleene, Emil Post, Alan Turing and others) mainly for the sake of its applications to logic. But the theory itself belongs to the abstract part of computing science. It is concerned with *computability* – roughly speaking, the property of being mechanically computable in principle (ignoring practical limitations of time and memory storage space).

Our exposition will be neither rigorous nor self-contained. For some of the key concepts, we shall provide intuitive explanations rather than precise definitions. Instead of proving all theorems rigorously, we shall in most cases present intuitive arguments. One major result – the MRDP Theorem – will be stated without proof.

For a rigorous coverage of all this material, see Ch. 6 of B&M. Alternative presentations of recursion theory can be found in books wholly devoted to this subject, as well as in books that combine it with logic. A classic of the first kind is

> Hartly Rogers, *Theory of recursive functions and effective computability.*

A fairly recent example of the second kind of book is

> Daniel E. Cohen, *Computability and logic.*

1.2. Conventions

(i) In this chapter, by *n-ary relation* we mean n-ary relation on the set N of natural numbers – that is, a subset of N^n. In particular, a *property* is a subset of N. By *relation* we mean n-ary relation for some $n \geqslant 1$.

(ii) By *n-ary function* we mean an n-ary operation on N (see Defs. 8.3.2 and 8.3.4). In particular, a 0-ary function is just a natural number. By *function* we mean n-ary function for some $n \geqslant 0$.

(iii) We use small italic letters – especially 'a', 'b', 'c', 'x', 'y' and 'z', with or without subscripts – as informal variables ranging over natural numbers; that is, the values of these variables are always assumed to be natural numbers.

(iv) We use small German letters as informal variables ranging over n-tuples of natural numbers. For the i-th component of such an n-tuple we use the corresponding italic letter with subscript 'i'. For example, $\mathfrak{a} = \langle a_1, a_2, \ldots, a_n \rangle$ and $\mathfrak{x} = \langle x_1, x_2, \ldots, x_n \rangle$.

(v) If P is an n-ary relation, we often write '$P\mathfrak{a}$' instead of '$\mathfrak{a} \in P$'.

1.3. Definition

(i) We define *propositional* (a.k.a. *Boolean*) operations on relations as follows. If P is an n-ary relation, then its *negation* $\neg P$ is defined by stipulating, for all $\mathfrak{x} \in N^n$:

$$\neg P\mathfrak{x} \Leftrightarrow P\mathfrak{x} \text{ does not hold.}$$

If P and Q are n-ary relations, we define their *disjunction* $P \vee Q$ by stipulating, for all $\mathfrak{x} \in N^n$:

$$(P \vee Q)\mathfrak{x} \Leftrightarrow P\mathfrak{x} \text{ or } Q\mathfrak{x}.$$

Other propositional operations, such as *conjunction* and *implication*, can be defined in the obvious way, either directly or from negation and disjunction. We shall usually write, e.g., '$P\mathfrak{x} \vee Q\mathfrak{x}$' instead of '$(P \vee Q)\mathfrak{x}$'.

(ii) If Q is an $(n + 1)$-ary relation, we can obtain an n-ary relation P by stipulating, for all $\mathfrak{x} \in N^n$:

$$P\mathfrak{x} \Leftrightarrow Q(\mathfrak{x}, y) \text{ holds for some } y.$$

We shall write, more briefly, $P\mathfrak{x} \Leftrightarrow \exists y Q(\mathfrak{x}, y)$, and say that P is obtained from Q by *existential quantification*.

The operation of *universal quantification* is defined in the obvious way, directly or in terms of negation and existential quantification.

(iii) The propositional operations as well as the two quantifications are called *logical* operations.

1.4. Warning

Take care not to confuse '¬', '∀', etc. with their bold-face counterparts, '**¬**', '**∀**', etc. The former denote operations on relations; the latter denote symbols in a formal language (which we are not studying in this chapter). The typographical similarity between the two sets of symbols is an intended pun and a mnemonic device, as will become clearer in the next chapter.

§2. Computers

We shall define the central concepts of recursion theory in terms of the notion of *computer*. The computers we have in mind are like real-life programmable digital computers, but idealized in one crucial respect (see Assumption 2.6 below). To help clarify this notion, we state in informal intuitive terms the most essential assumptions we will make about computers and the way they operate.

2.1. Assumption

A computer is a *digital* calculating machine: its states differ from each other in a discrete manner. (This rules out analogue calculating devices such as the slide-rule, whose states [are supposed to] vary continuously.)

2.2. Assumption

A computer is a *deterministic mechanism*: it operates by rigidly and deterministically following instructions stored in it in advance. (This rules out resort to chance or random devices.)

2.3. Assumption

A computer operates in a serial discrete step-wise manner.

2.4. Assumption

A computer has a *memory* capable of storing finitely many [representations of] natural numbers – which may be part of the input or the output or an intermediate stage of a computation – and instructions. (Without loss of generality, we may assume that instructions are coded by natural numbers, as is in fact the case in present-day programmable computers; so the content of the memory is always a finite sequence of numbers.)

2.5. Assumption

A computer operates according to a *program*, a finite list of instructions, stored in it in advance (see Assumptions 2.2 and 2.4). Each instruction requires the computer to execute a simple step such as to erase a number stored in a specified location in the memory, or increase by 1 the number stored in a specified location, or print out as *output* the number stored in a specified location, or simply to stop. After each step, the next instruction to be obeyed is determined by the content of the memory (including the program itself).

2.6. Assumption

The computer's memory has an *unlimited* storage capacity: it is able to store an arbitrarily long finite sequence of natural numbers, each of which can be arbitrarily large. (Thus, although the amount of information stored in the memory is always finite, we assume that this amount has no upper bound.)

2.7. Remarks

(i) Assumptions 2.1–2.5 are perfectly realistic: they are in fact satisfied by many existing machines, from giant super-computers down to modest programmable pocket calculators. Assumption 2.6, in contrast, is a far-reaching idealization: a real-life machine can only store a limited amount of information. While the storage capacity of many real machines can be enhanced by adding on peripheral devices such as magnetic tapes or disks, this cannot be done without limit.

(ii) In connection with Assumption 2.5 it is interesting to note that the repertory of commands that a computer is able to obey (that is, the range of elementary steps it is able to perform) need not be at all impressive: in this respect the powers of a modest programmable pocket calculator are more than adequate. Real-life computing machines vary enormously in memory size and speed of operation. But if we assume that restrictions of memory size are removed, then the only significant difference is that of speed. Provided it had access to unlimited storage capacity, a machine with fairly rudimentary powers could simulate (if only at much reduced speed) the operation of any computer that has so far been constructed or described.

(iii) Several computers can be combined to form a more complex system, which can itself be regarded as a computer.

§ 3. Recursiveness

3.1. Definition

Let P be an n-ary relation. By a *decide-P machine* we mean a computer with an input port and an output port, which is programmed so that if any n-tuple $x \in N^n$ is fed into the input port then after a finite number of steps the computer prints out an output – say 1 for *yes* and 0 for *no* – indicating whether Px holds or not.

A relation P is *recursive* (or *computable*) if a decide-P machine can be constructed (that is, if a computer can be programmed to act as a decide-P machine).

3.2. Remarks

(i) Naturally, the length of the computation, the number of steps required by the machine to produce an output, will in general depend on the input n-tuple x. We impose no bound on the length of the computation but merely require it to be finite. Thus we ignore real-life limitations of time: in practice a computation that may take a million years is useless.

(ii) To be precise we should have said that the inputs fed into the computer are not n-tuples of numbers (which are abstract entities) but *representations* of such n-tuples. Similarly what the computer prints out is not a number, 0 or 1, but a *representation* of a number. Similar – quite harmless – lapses will be committed throughout this chapter.

(iii) Any relation you are likely to think of, off-hand, is certain to be recursive – unless you are already familiar with some of the tricks of recursion theory or are exceptionally ingenious. (We shall meet examples of non-recursive relations in the next chapter.)

(iv) Nevertheless, set-theoretically speaking, the overwhelming majority of relations are non-recursive. (Here is an outline of a proof. Working within ZF set theory, we identify N with the set of finite cardinals. Using Thm. 6.3.7 and Cantor's Thm. 3.6.8, it is easy to show that for each $n \geqslant 1$ the set of all n-ary relations has cardinality $>\aleph_0$. On the other hand, a computer program is a finite string of instructions, each of which is a finite string of symbols in some programming language with a countable set of primitive symbols. Hence by Thm. 6.3.9 the set of all programs is countable. If follows that the set of all recursive relations must also be countable.)

3.3 Definition

Let P be an n-ary relation. By an *enumerate-P machine* we mean a computer with an output port and programmed so that it prints out, one by one, all the n-tuples $\mathfrak{x} \in N^n$ for which $P\mathfrak{x}$ holds, and no others.

A relation P is said to be *recursively enumerable* – briefly, *r.e.* – if an enumerate-P can be constructed (that is, if a computer can be programmed to act as an enumerate-P machine).

3.4. Remarks

(i) If P is infinite (that is, holds for infinitely many n-tuples) then an enumerate-P machine, once switched on, will never stop unless it is switched off. We impose no bound on the number of computation steps the machine may make between printing out two successive n-tuples; we only require it to be finite.

(ii) An r.e. relation is sometimes said to be *semi-recursive*. The reason for this will soon become clear.

3.5. Lemma

The n-ary relation N^n (the set of all n-tuples of natural numbers) is r.e.

PROOF

All n-tuples can be arranged in some systematic order. For example, we may order them according to the following two rules:

1. If the maximal component of $\mathfrak{a}$ is smaller than that of $\mathfrak{b}$, then $\mathfrak{a}$ will precede $\mathfrak{b}$.
2. All n-tuples with the same maximal component will be ordered lexicographically.

(The *maximal component* of an n-tuple $\mathfrak{x}$ is the greatest among the numbers $x_1, x_2, \ldots, x_n$. *Lexicographic* order is the order in which words are listed in a dictionary. Here we regard an n-tuple $\mathfrak{x}$ as a 'word' with x_1 as its first letter, x_2 as its second, and so on.) As an illustration, take $n = 2$. The pairs of natural numbers will be ordered as follows (cf. proof of Thm. 6.3.2):

$$\langle 0, 0\rangle,$$
$$\langle 0, 1\rangle, \langle 1, 0\rangle, \langle 1, 1\rangle,$$
$$\langle 0, 2\rangle, \langle 1, 2\rangle, \langle 2, 0\rangle, \langle 2, 1\rangle, \langle 2, 2\rangle,$$
$$\langle 0, 3\rangle, \langle 1, 3\rangle, \langle 2, 3\rangle, \langle 3, 0\rangle, \langle 3, 1\rangle, \langle 3, 2\rangle, \langle 3, 3\rangle, \ldots.$$

Clearly, this procedure can be mechanized: a computer can be programmed to spew out all n-tuples of natural numbers in this order. ■

3.6. Theorem

Let P be an n-ary relation. Then P is recursive iff both P and $\neg P$ are r.e.

PROOF

($\Rightarrow$). Suppose P is recursive. Then we can construct a decide-P machine $\mathfrak{D}$. As we have just seen, we can also construct an enumerate-N^n machine $\mathfrak{E}$. We set $\mathfrak{E}$ to work, and compile a final output by modifying the output of $\mathfrak{E}$ as follows. We feed a copy of each n-tuple $\mathfrak{a}$ that $\mathfrak{E}$ prints out into $\mathfrak{D}$. If the latter says that $P\mathfrak{a}$ holds, $\mathfrak{a}$ is left in the final output; but if $\mathfrak{D}$ says that $P\mathfrak{a}$ does not hold, then $\mathfrak{a}$ is eliminated from the final output. This procedure can be mechanized, yielding an enumerate-P machine. An enumerate-$\neg P$ machine can be constructed in a similar way.

($\Leftarrow$). Now suppose both P and $\neg P$ are r.e. Then we have at our disposal both an enumerate-P machine and an enumerate-$\neg P$

machine. These can be used to construct a (rather inefficient but quite legitimate) decide-P machine, as follows. We set both enumerating machines to work. Given any n-tuple $\mathfrak{a} \in N^n$, the outputs of both machines are monitored, until $\mathfrak{a}$ emerges from one of them (this is bound to happen sooner or later!) and then it is noted from which of our two machines $\mathfrak{a}$ has emerged. (All this monitoring and noting can of course be done automatically.) If $\mathfrak{a}$ has come out of the enumerate-P machine, then $P\mathfrak{a}$ holds; whereas if $\mathfrak{a}$ has come out of the other machine, $P\mathfrak{a}$ does not hold. ■

3.7. Remarks

(i) Note that in the second half of this proof we needed *both* enumerating machines. If we only had an enumerate-P machine, and we tried to use it for testing whether $P\mathfrak{a}$ holds, then if the answer happened to be negative we would never find that out.

(ii) By Thm. 3.6, every recursive relation is r.e. We shall see in the next chapter that the converse of this is false.

3.8. Theorem

If P is obtained from Q by existential quantification and Q is r.e., then P is r.e. as well.

PROOF

Suppose $P\mathfrak{x} = \exists y Q(\mathfrak{x}, y)$. Since Q is r.e., we can construct an enumerate-Q machine. Set this machine to work, and let its output be modified as follows. Whenever an $(n + 1)$-tuple $\langle \mathfrak{a}, b \rangle$ pops out, the last component b is erased, leaving the n-tuple $\mathfrak{a}$. (This modification can of course be done automatically.) It is easy to see that we now have an enumerate-P machine. ■

3.9. Definition

Let f be an n-ary function. By a *compute-f machine* we mean a computer with an input port and an output port, and programmed so that if any $\mathfrak{x} \in N^n$ is fed into the input port, then after a finite number of steps the computer prints out as output the value $f\mathfrak{x}$.

We say that f is a *recursive* (or *computable*) function if a compute-f machine can be constructed.

Recall that the *graph* of an n-ary function f is the $(n + 1)$-ary relation P such that

$$P(\mathfrak{x}, y) \Leftrightarrow (f\mathfrak{x} = y)$$

for all $\mathfrak{x} \in N^n$ and all $y \in N$. (As a matter of fact, if $n \geqslant 1$ then from Convention 1.2(ii), Def. 8.3.2 and Prob. 2.3.3 it follows that the graph of f is f itself; but this is not important just now.)

3.10. Theorem

For any function f, the following three conditions are equivalent:

(i) *f is a recursive function (in the sense of Def. 3.9);*
(ii) *the graph of f is recursive (in the sense of Def. 3.1);*
(iii) *the graph of f is r.e.*

PROOF

Let f be an n-ary function, and let P be its graph.
(i) $\Rightarrow$ (ii). Assuming that f is recursive, we can construct a compute-f machine $\mathfrak{C}$. We can employ $\mathfrak{C}$ to find out, for any $(n + 1)$-tuple $\langle \mathfrak{a}, b \rangle \in N^n$, whether $P(\mathfrak{a}, b)$ holds or not, as follows.

Given any $(n + 1)$-tuple $\langle \mathfrak{a}, b \rangle$, we split it into the n-tuple $\mathfrak{a}$ and the number b. We make a record of the latter, and feed the former into $\mathfrak{C}$. When $\mathfrak{C}$ prints out the value $f\mathfrak{a}$, we compare it with our record of b and see whether they are equal. $P(\mathfrak{a}, b)$ holds iff $f\mathfrak{a} = b$.

The procedure described in the previous paragraph can obviously be automated, yielding a decide-P machine.
(ii) $\Rightarrow$ (iii) is immediate from Thm. 3.6.
(iii) $\Rightarrow$ (i). Assuming that P is r.e., we can construct an enumerate-P machine $\mathfrak{E}$. We can use $\mathfrak{E}$ in the following way to calculate $f\mathfrak{a}$ for any $\mathfrak{a} \in N^n$.

Upon receiving $\mathfrak{a}$, we set $\mathfrak{E}$ to work and monitor its output, checking each $(n + 1)$-tuple as it is printed out, to see whether it is of the form $\langle \mathfrak{a}, b \rangle$, having $\mathfrak{a}$ as its first n components. Sooner or later, such an $(n + 1)$-tuple is bound to turn up. When it does, we know that its last component, b, is the value $f\mathfrak{a}$.

The procedure described in the previous paragraph can obviously be automated, yielding a compute-f machine. (No prizes for efficiency, but it is perfectly legitimate.) ■

3.11. Remarks

(i) Recursion theory studies functions of a more general kind: an n-ary function is allowed to have any subset of N^n as its domain (instead of the whole of N^n, as we insist here). The definition of a *compute-f machine* must then be modified by stipulating that the machine prints out the correct value $f\mathfrak{a}$ for any input $\mathfrak{a} \in \text{dom}f$; but for an input $\mathfrak{a} \notin \text{dom}f$ it goes on computing for ever, without producing any output. For these more general functions it is not difficult to show that conditions (i) and (iii) of Thm. 3.10 are still equivalent to each other; but they do not imply condition (ii).

(ii) The first rigorous description of a computer satisfying Assumptions 2.1–2.6, devised expressly for the purpose of explicating the intuitive notion of computability, was published by Turing in 1936. Since then many alternative machines satisfying Assumptions 2.1–2.6 have been invented. (For a description of Turing machines see the books by Rogers and D. E. Cohen cited in § 1; the latter contains also descriptions of several other alternatives.) In each case it was easy to prove that the operation of the alternative machine can be simulated by a Turing machine; the converse also holds, provided the alternative machine satisfies some modest requirements.

This and other evidence lends overwhelming support to the claim – known as *Church's Thesis* – that any function that is mechanically computable in the intuitive sense is computable by a Turing machine (or, for that matter, by one of its equivalent alternatives). Church's Thesis is equivalent to the claim that any relation that is mechanically decidable (or enumerable) in the intuitive sense can be decided (or enumerated, respectively) by a Turing machine.

(iii) Although a recursive or r.e. relation may well be infinite in extension, and a recursive function is necessarily infinite in extension, each such entity is completely determined by a computer program, which is a finite object. For this reason, recursion theory does not on the whole require powerful set-theoretic presuppositions. Even without such presuppositions it is possible to treat recursive and r.e. relations and recursive functions as objects: if need be, programs can play this role vicariously, standing in for the more abstract entities they characterize.

§4. Closure results

4.1. Theorem

The class of recursive relations is closed under all propositional operations.

PROOF

Let P and Q be n-ary recursive relations. Thus we can construct a decide-P machine $\mathfrak{D}_P$ and a decide-Q machine $\mathfrak{D}_Q$. Then $\mathfrak{D}_P$ can be turned into a decide-$\neg P$ machine, simply be reversing its outputs. Therefore $\neg P$ is recursive.

To construct a decide-$(P \vee Q)$ machine, let $\mathfrak{D}_P$ and $\mathfrak{D}_Q$ operate alongside each other. Given any n-tuple $\mathfrak{a} \in N^n$, a copy of it is fed into each of these two machines. Their two outputs are channelled into a collating unit. This unit checks the two outputs, and if at least one of them is 'yes' it gives out a final output 'yes'; but if both $\mathfrak{D}_P$ and $\mathfrak{D}_Q$ say 'no', then the collating unit gives out a final output 'no'. We have now got a decide-$(P \vee Q)$ machine, showing that $P \vee Q$ is recursive. The other Boolean operations can be reduced to negation and disjunction.

■

4.2. Remark

According to Assumption 2.3, a computer is supposed to operate in a serial manner. This seems to be violated by the decide-$(P \vee Q)$ machine just described, which has $\mathfrak{D}_P$ and $\mathfrak{D}_Q$ as two components working in parallel. The apparent difficulty can be resolved by assuming that the two components operate alternately, as in bipedal walking: each one pausing while the other performs a step.

4.3. Theorem

The class of r.e. relations is closed under disjunction, conjunction and existential quantification.

PROOF

Let P and Q be n-ary r.e. relations. So, we can construct an enumerate-P machine $\mathfrak{E}_P$ and an enumerate-Q machine $\mathfrak{E}_Q$. We set these two machines to operate alongside each other (see Rem. 4.2).

To get an enumerate-$(P \vee Q)$ machine, we channel the outputs of $\mathfrak{E}_P$ and $\mathfrak{E}_Q$ into a collating unit that combines these two outputs into a single list. The combined list is the output of an enumerate-$(P \vee Q)$ machine. Hence $P \vee Q$ is r.e.

To get an enumerate-$(P \wedge Q)$ machine, we need, in addition to a collating unit, two waiting lists or *buffers* in which information can be accumulated – one each for P and Q. Initially both buffers are empty. The collating unit examines in turn each fresh n-tuple that pops out of $\mathfrak{E}_P$ or $\mathfrak{E}_Q$. The two buffers as well as a *final list* are compiled according to the following rules. Each time a fresh n-tuple $\mathfrak{a}$ comes out of $\mathfrak{E}_P$, the collating unit checks whether an identical n-tuple is already stored in the Q-buffer. If $\mathfrak{a}$ is found to be in the Q-buffer, then it is put onto the final list; but if $\mathfrak{a}$ is not in the Q-buffer then the collating unit adds it to the P-buffer. Similarly, each time a fresh n-tuple $\mathfrak{b}$ comes out of $\mathfrak{E}_Q$, the collating unit checks whether $\mathfrak{b}$ is stored in the P-buffer. If $\mathfrak{b}$ is found to be stored there, then it is put onto the final list; otherwise, it is added to the Q-buffer. It is easy to see that the final list is the output of an enumerate-$(P \wedge Q)$ machine, showing that $P \wedge Q$ is r.e.

As for closure under existential quantification – this has already been proved (see Thm. 3.8). ■

Next, we show that the class of r.e. relations is closed under the operation of *adding a redundant variable*.

4.4. Theorem

Let P be an n-ary relation. Let Q be the $(n + 1)$-ary relation such that, for all $\mathfrak{x} \in N^n$ and all $y \in N$,

$$Q(\mathfrak{x}, y) \Leftrightarrow P\mathfrak{x}.$$

If P is r.e., then Q is r.e. as well.

PROOF

By hypothesis we can construct an enumerate-P machine $\mathfrak{E}_P$. Also, by Lemma 3.5 we can construct an enumerate-$N^{(n+1)}$ machine $\mathfrak{E}$.

To get an enumeration of Q, we set both $\mathfrak{E}_P$ and $\mathfrak{E}$ to work. As in the proof of the $\wedge$ part of Thm. 4.3, we compile a final list as well as two buffers, one each for P and $N^{(n+1)}$. When an n-tuple $\mathfrak{a}$ pops out of $\mathfrak{E}_P$, it is added to the P buffer; and every $(n + 1)$-tuple of the form $\langle \mathfrak{a}, b \rangle$ that is already stored in the $N^{(n+1)}$ buffer is added to the final list.

When any $(n + 1)$-tuple $\langle \mathfrak{a}, b \rangle$ pops out of $\mathfrak{E}$, we check whether $\mathfrak{a}$ is present in the P buffer; if it is, $\langle \mathfrak{a}, b \rangle$ goes on the final list; if not, it goes to the $N^{(n+1)}$ buffer. ■

4.5. Remarks

(i) Results similar to Thm. 4.4 hold also for the class of recursive relations and the class of recursive functions; but they are too obvious to be stated as theorems.

(ii) Using these facts, we can deal with disjunctions and conjunctions of r.e. or recursive relations that are not of the same n-arity. For example, if P and Q are binary, we can form a quaternary relation R by stipulating that for all w, x, y and z,

$$R(w, x, y, z) \Leftrightarrow P(w, x) \wedge Q(y, z).$$

By adding y and z to P and w and x to Q as redundant variables, we can see that if P and Q are r.e. (or recursive) then so is R.

For the final theorem of this section, we let $f_1, f_2, \ldots, f_k$ be n-ary functions. Let g be a k-ary function and let the function h be obtained by composing g with $f_1, f_2, \ldots, f_k$; in other words, for all $\mathfrak{x} \in N^n$,

$$h\mathfrak{x} = g(f_1\mathfrak{x}, f_2\mathfrak{x}, \ldots, f_k\mathfrak{x}).$$

Let P be a k-ary relation and let the relation Q be obtained by composing P with $f_1, f_2, \ldots, f_k$; in other words, for all $\mathfrak{x} \in N^n$,

$$Q\mathfrak{x} \Leftrightarrow P(f_1\mathfrak{x}, f_2\mathfrak{x}, \ldots, f_k\mathfrak{x}).$$

4.6. Theorem

Let $f_1, f_2, \ldots, f_k$ be recursive functions.

(i) *If g is a recursive function as well, then so is h.*
(ii) *If P is a recursive relation, then so is Q.*
(iii) *If P is r.e., then so is Q.*

PROOF

(i) By hypothesis, we can construct machines $\mathfrak{F}_1, \mathfrak{F}_2, \ldots, \mathfrak{F}_k$ that compute $f_1, f_2, \ldots, f_k$ respectively; also, we can construct a compute-g machine, $\mathfrak{G}$. To compute h, we proceed as follows.

Given any n-tuple $\mathfrak{a} \in N^n$, copies of it are fed into the input ports of $\mathfrak{F}_1, \mathfrak{F}_2, \ldots, \mathfrak{F}_k$. When these k machines have produced their outputs, $b_1, b_2, \ldots, b_k$, they are put together as a k-tuple $\langle b_1, b_2, \ldots, b_k \rangle$, which is fed into the input port of $\mathfrak{G}$. The output produced by the latter is the required value $h\mathfrak{a}$.

This procedure can be mechanized, yielding a compute-h machine.

The proof of (ii) is similar. To prove (iii), we note that

$$Q\mathfrak{x} \Leftrightarrow \exists y_1 \exists y_2 \ldots \exists y_k[(f_1\mathfrak{x} = y_1) \wedge (f_2\mathfrak{x} = y_2) \wedge \ldots, \wedge (f_k\mathfrak{x} = y_k) \wedge P(y_1, y_2, \ldots, y_k)].$$

By Thm. 3.10, the graphs of $f_1, f_2, \ldots, f_k$ are r.e., and P is r.e. by hypothesis. Hence Q is r.e. by Thm. 4.3. and Rem. 4.5(ii). ∎

§5. The MRDP Theorem

5.1. Preview

In 1970, Yuri Matiyasevič – building upon work done during the preceding two decades by Julia Robinson, Martin Davis and Hilary Putnam – completed the proof of a remarkable theorem that characterizes r.e. relations in extremely elementary terms. We refer to this result by the acronym 'MRDP', for the four names just mentioned.

In view of Thms. 3.6 and 3.10, the MRDP Thm. also provides elementary characterizations of the other two central concepts of recursion theory: recursive relations and recursive functions. These characterizations simplify the application of recursion theory to logic.

We shall present the MRDP Thm. without proof, which is too long to be included here.

5.2. Definition

(i) An n-ary function f is a *monomial* if for some natural number a (called the *coefficient*) and natural numbers $k_1, k_2, \ldots, k_n$ (called the *exponents*) the equality

$$f\mathfrak{x} = a{x_1}^{k_1}{x_2}^{k_2} \ldots {x_n}^{k_n}$$

holds for all $\mathfrak{x} \in N^n$.

(ii) An n-ary function f is a *polynomial* if it is a sum of monomials; that is, for some monomials $f_1, f_2, \ldots, f_m$ the equality

$$f\mathfrak{x} = f_1\mathfrak{x} + f_2\mathfrak{x} + \cdots + f_m\mathfrak{x}$$

holds for all $\mathfrak{x} \in N^n$.

5.3. *Definition*

(i) An n-ary relation P is *elementary* if there are n-ary polynomials f and g such that, for all $x \in N^n$,

$$Px \Leftrightarrow (fx = gx).$$

(ii) An n-ary relation P is said to be *diophantine* if it can be obtained by a finite number of existential quantifications from an elementary relation; in other words, there are $(n + m)$-ary polynomials f and g such that, for all $x \in N^n$,

$$Px \Leftrightarrow \exists y_1 \exists y_2 \ldots \exists y_m [f(x, y_1, y_2, \ldots, y_m) = g(x, y_1, y_2, \ldots, y_m)].$$

(Here m may be 0, so every elementary relation is *a fortiori* diophantine.)

5.4. *Theorem (MRDP)*

A relation is r.e. iff it is diophantine. ■

5.5. *Remarks*

(i) The $\Leftarrow$ part of the theorem is simple to prove. First, let P be an n-ary elementary relation, and let f and g be polynomials satisfying the condition of Def. 5.3(i). For any given $x \in N^n$ we can calculate the values fx and gx – this involves a finite number of additions and multiplications of natural numbers. Then the two values can be compared to see whether Px holds or not. This procedure can clearly be mechanized, yielding a decide-P machine. Thus every elementary relation is recursive, and hence r.e. by Thm. 3.6. Now, by Def. 5.3(ii), any diophantine relation is obtainable from an elementary relation by a finite number of existential quantifications; so it is r.e. by Thm. 3.8.

(ii) The $\Rightarrow$ part of the MRDP Thm. is far harder to prove. The original proof (including Robinson's early results and her joint work with Davis and Putnam) is reproduced in B&M, pp. 284–311. A shorter and more direct version of the proof is presented in pp. 111–123 of Cohen's book cited in § 1.

(iii) The proof of the MRDP Thm. is *effective*: it provides us with a method whereby from a given description (program) of an enumerate-P machine it is possible in principle (granted enough time and patience) to obtain polynomials f and g in terms of which P can be presented as prescribed in Def. 5.3 (ii). Conversely, given such a presentation, it is easy to construct a program under which a computer will operate as an enumerate-P machine.

10
Limitative results

§ 1. Preliminaries

1.1. Preview

The main results in this chapter reveal the inherent limitations of formalism and the formalist approach to mathematics. For the sake of simplicity we confine ourselves to a very basic part of mathematics: *elementary arithmetic* (a.k.a. *elementary number theory*), whose subject-matter is the elementary structure of natural numbers (see Ex. 8.3.6). However, these results can be generalized without much difficulty to richer and more elaborate mathematical contexts.

1.2. Convention

We shall often write 'number' as short for 'natural number'. Unless stated otherwise, we shall follow the notation and terminology of Ch. 9 (see Conv. 9.1.2). Also, we use 'k', 'm', 'n' and 'p' as informal variables ranging over numbers.

1.3. Specification

From now on, unless stated otherwise, our formal object language $\mathscr{L}$ will be the *first-order language of arithmetic*; namely, the first-order language with equality $=$, whose extralogical symbols are:

(i) One individual constant, $\mathbf{0}$;
(ii) One unary function symbol, $\mathbf{s}$;
(iii) Two binary function symbols, $+$ and $\times$.

1.4. Remarks

(i) Note that $\mathscr{L}$ has no extralogical predicate symbols, so its only atomic formulas are equations.

(ii) Since '$\mathbf{s}$' is now used as a syntactic constant denoting the unary function-symbol of $\mathcal{L}$, we cannot use it any longer as a syntactic variable ranging over $\mathcal{L}$-terms. For this purpose we shall use '$\mathbf{q}$', '$\mathbf{r}$' and '$\mathbf{t}$', with or without subscripts.

(iii) The terms of $\mathcal{L}$ evidently fall into the following five mutually exclusive categories:
(1) Terms of the form $\mathbf{x}$, consisting of a single occurrence of a variable;
(2) The single term $\mathbf{0}$;
(3) Terms of the form $\mathbf{st}$, where $\mathbf{t}$ is any term;
(4) Terms of the form $\mathbf{+rt}$, where $\mathbf{r}$ and $\mathbf{t}$ are any terms;
(5) Terms of the form $\mathbf{\times rt}$, where $\mathbf{r}$ and $\mathbf{t}$ are any terms.
Terms of the last three categories will be referred to as '$\mathbf{s}$-*terms*', '$\mathbf{+}$-*terms*' and '$\mathbf{\times}$-*terms*' respectively.

1.5. Definition

In addition to Def. 8.2.2, which remains in force here – and for similar reasons – we put, for any terms $\mathbf{r}$ and $\mathbf{t}$:

(i) $(\mathbf{r+t}) =_{\text{df}} \mathbf{+rt}$,
(ii) $(\mathbf{r \times t}) =_{\text{df}} \mathbf{\times rt}$.

In using this metalinguistic notation, brackets are required. To prevent proliferation of brackets, which would impair legibility, we omit brackets subject to three simple conventions. First, the Greek cross '$\mathbf{+}$' is deemed to separate more strongly than the St Andrew cross '$\mathbf{\times}$'. Second, of any two occurrences of '$\mathbf{+}$' (or of '$\mathbf{\times}$') enclosed within the same pairs of brackets, the one further to the left is deemed to separate more strongly. Third, we do not omit any pair of brackets whose left member comes immediately after an occurrence of '$\mathbf{s}$'; hence, when restoring brackets, no new left bracket should be placed immediately after an '$\mathbf{s}$'. For example,

$$\begin{aligned}&\mathbf{s0+ss0\times s0\times sss0+0} = \mathbf{s0+ss0\times(s0\times sss0)+0}\\ &= \mathbf{s0+[ss0\times(s0\times sss0)]+0}\\ &= \mathbf{s0+\{[ss0\times(s0\times sss0)]+0\}} = \mathbf{\{s0+\{[ss0\times(s0\times sss0)]+0\}\}}.\end{aligned}$$

1.6. Definition

Proceeding by induction, we define, for each natural number k, an $\mathcal{L}$-term $\mathbf{s}_k$, called the *k-th $\mathcal{L}$-numeral*:

$$\mathbf{s}_0 = \mathbf{0}, \qquad \mathbf{s}_{k+1} = \mathbf{ss}_k.$$

Thus $\mathbf{s}_k$ is the $\mathcal{L}$-term consisting of a single occurrence of $\mathbf{0}$ preceded by k occurrences of $\mathbf{s}$.

1.7. Recapitulation

Applying Def. 8.4.2 to our present language $\mathcal{L}$, we see that an $\mathcal{L}$-interpretation (a.k.a. $\mathcal{L}$-structure) $\mathfrak{U}$ is completely determined by the following ingredients.

(i) A non-empty set U – the domain of $\mathfrak{U}$.
(ii) An individual $\mathbf{0}^{\mathfrak{U}} \in U$ – the individual denoted by $\mathbf{0}$ under the interpretation $\mathfrak{U}$.
(iii) A unary operation $\mathbf{s}^{\mathfrak{U}}$ on U – the operation that interprets $\mathbf{s}$ under $\mathfrak{U}$.
(iv) Two binary operations $\mathbf{+}^{\mathfrak{U}}$ and $\mathbf{\times}^{\mathfrak{U}}$ on U – the operations that interpret $\mathbf{+}$ and $\mathbf{\times}$ respectively under $\mathfrak{U}$.

Apart from the conditions we have just specified, these ingredients of an $\mathcal{L}$-interpretation can be quite arbitrary. Thus U can be a set of any cardinality whatsoever, so long as it is non-empty; the nature of the individuals (members of U) is immaterial; and $\mathbf{0}^{\mathfrak{U}}$ can be any member of U. Similarly, $\mathbf{s}^{\mathfrak{U}}$ can be an arbitrary unary operation on U; and $\mathbf{+}^{\mathfrak{U}}$ and $\mathbf{\times}^{\mathfrak{U}}$ can be arbitrary binary operations on U.

However, of the huge variety of possible $\mathcal{L}$-interpretations we single out one, for which the language $\mathcal{L}$ was designed in the first place.

1.8. Definition

The *intended* or *standard* $\mathcal{L}$-interpretation $\mathfrak{N}$ is characterized as follows:

(i) $\mathfrak{N}$ has as its domain the set N of natural numbers.
(ii) $\mathbf{0}^{\mathfrak{N}} = 0$ (the number zero).
(iii) $\mathbf{s}^{\mathfrak{N}} = s$, the *successor* function (that is, $sx = x + 1$ for each number x).
(iv) $\mathbf{+}^{\mathfrak{N}} = +$ and $\mathbf{\times}^{\mathfrak{N}} = \times$ (the operations of natural-number addition and multiplication, respectively).

1.9. Definition

(i) If $\mathbf{t}$ is a closed $\mathscr{L}$-term, we call $\mathbf{t}^{\mathfrak{N}}$ the *numerical value* of $\mathbf{t}$ (cf. Def. 8.5.6).
(ii) We say that an $\mathscr{L}$-sentence $\boldsymbol{\varphi}$ is *true* or *false* according as $\mathfrak{N} \models \boldsymbol{\varphi}$ or $\mathfrak{N} \not\models \boldsymbol{\varphi}$ (cf. Def. 8.5.10).

1.10. Remarks

(i) We have chosen the syntactic constants '**0**', '**s**', '+' and '×' advisedly, so as to serve a mnemonic purpose: each of these symbols graphically suggests the standard interpretation of the $\mathscr{L}$-symbol that it denotes. This punning mnemonic role of the four syntactic constants is made manifest in clauses (ii), (iii) and (iv) of Def. 1.8. For example, '**0**' has been chosen as the name (in our metalanguage) for the individual constant of $\mathscr{L}$. The shape (if any!) of the latter constant is left unspecified, but *under the standard interpretation of* $\mathscr{L}$ it is treated as a *name of* the number zero, that number which is conventionally denoted by the numeral '0'. Since '**0**' was chosen for its present role precisely because it looks like '0', we have a mnemonically useful pun: $\mathbf{0}^{\mathfrak{N}} = 0$.

A similar mnemonic purpose is served by the choice of '=' as the syntactic constant denoting the equality symbol of $\mathscr{L}$, except that in this case the pun is not confined to the standard interpretation. Indeed, by Def. 8.4.2(iii), under *any* $\mathscr{L}$-interpretation $\mathfrak{U}$ the equality symbol of $\mathscr{L}$ is interpreted as denoting the identity relation on the domain U of $\mathfrak{U}$. As a result, we have (as part of clause F1 of the BSD) the mnemonically useful pun:

$$(\mathbf{r}{=}\mathbf{t})^{\sigma} = \top \text{ iff } \mathbf{r}^{\sigma} = \mathbf{t}^{\sigma},$$

for any $\mathscr{L}$-valuation σ and any $\mathscr{L}$-terms $\mathbf{r}$ and $\mathbf{t}$.

(ii) A practical advantage of the choice of '**0**', '**s**', '+' and '×' is that when we refer to an $\mathscr{L}$-term by means of this metalinguistic notation, it is often quite easy to work out by inspection the value of that term under any valuation based on $\mathfrak{N}$. (This value must be a number, because the domain of $\mathfrak{N}$ is the set N of numbers.)

For example, consider the term $\mathbf{x}{\times}\mathbf{x} + \mathbf{ss0}{\times}\mathbf{x}{\times}\mathbf{y} + \mathbf{y}{\times}\mathbf{y}$, where

$\mathbf{x}$ and $\mathbf{y}$ are variables. If σ is a valuation based on $\mathfrak{N}$, it is easy to see that $(\mathbf{x}\times\mathbf{x} + \mathbf{ss0}\times\mathbf{x}\times\mathbf{y} + \mathbf{y}\times\mathbf{y})^\sigma = x^2 + 2xy + y^2$, where x and y are the numbers $\mathbf{x}^\sigma$ and $\mathbf{y}^\sigma$ respectively.

In particular, if $\mathbf{t}$ is a closed term, it is a simple matter to work out the numerical value $\mathbf{t}^{\mathfrak{N}}$ of $\mathbf{t}$.

Similarly, when we refer to an $\mathcal{L}$-formula by means of our metalinguistic notation, it is often quite easy to work out by inspection the truth value of that formula under any valuation based on $\mathfrak{N}$. In particular, if $\boldsymbol{\varphi}$ is an $\mathcal{L}$-sentence it may be quite easy to work out by inspection whether $\mathfrak{N} \models \boldsymbol{\varphi}$ – that is, whether $\boldsymbol{\varphi}$ is true. For example, it is not difficult to verify that

$$\mathfrak{N} \models \forall\mathbf{x}\forall\mathbf{y}[(\mathbf{x}+\mathbf{y})\times(\mathbf{x}+\mathbf{y})=\mathbf{x}\times\mathbf{x} + \mathbf{ss0}\times\mathbf{x}\times\mathbf{y} + \mathbf{y}\times\mathbf{y}].$$

1.11. Warning

Beware, however, of being deceived by this suggestive notation: Rem. 1.10(ii) works for the standard interpretation, but not necessarily for other interpretations. Thus, for example, you must not assume that $\mathbf{0}$ always denotes the number 0. Rather, under an arbitrary $\mathcal{L}$-interpretation $\mathfrak{U}$, the object $\mathbf{0}^{\mathfrak{U}}$ denoted by $\mathbf{0}$ need not be a number at all, let alone the number 0; in fact, it can be any object whatsoever.

Or, to take another simple example, you must not assume that the sentence $\mathbf{0+0=0}$ is true under an arbitrary $\mathcal{L}$-interpretation. Of course, this sentence is easily seen to be true in the sense of Def. 1.9(ii). It is clearly satisfied in the standard structure $\mathfrak{N}$. But it is not *logically* true: If σ is a valuation based on an arbitrary interpretation $\mathfrak{U}$, then we find (using the BSD) that $(\mathbf{0+0=0})^\sigma = \top$ iff $f(a, a) = a$, where $f = +^{\mathfrak{U}}$ and $a = \mathbf{0}^{\mathfrak{U}}$ (that is, f and a are the binary operation and individual named by $+$ and $\mathbf{0}$ respectively under $\mathfrak{U}$). It is quite possible that $f(a, a) \neq a$, in which case $\mathfrak{U} \not\models \mathbf{0+0=0}$.

1.12. Problem

Show that $\mathbf{s}_k{}^{\mathfrak{N}} = k$ (see Def. 1.6).

1.13. Problem

Let $\mathbf{x}$, $\mathbf{y}$ and $\mathbf{z}$ be distinct variables. Let σ be a valuation based on $\mathfrak{N}$ and let x and y be the numbers $\mathbf{x}^\sigma$ and $\mathbf{y}^\sigma$ respectively. For each of the following five formulas state a condition involving x and y, which is necessary and sufficient in order that σ satisfy the formula in question.

(i) $\exists \mathbf{z}(\mathbf{x}+\mathbf{z}=\mathbf{y})$,
(ii) $\exists \mathbf{z}(\mathbf{x}+\mathbf{s}\mathbf{z}=\mathbf{y})$,
(iii) $\forall \mathbf{y}(\mathbf{x}\neq \mathbf{s}\mathbf{y})$,
(iv) $\exists \mathbf{y}(\mathbf{x}=\mathbf{s}_2\times \mathbf{y})$,
(v) $\exists \mathbf{z}(\mathbf{x}=\mathbf{y}\times \mathbf{z})$.

§2. Theories

2.1. Definition

For any number n, we let $\mathbf{\Phi}_n$ be the set of all $\mathscr{L}$-formulas whose free variables are among $\mathbf{v}_1, \mathbf{v}_2, \ldots, \mathbf{v}_n$, the first n variables of $\mathscr{L}$ in alphabetic order (cf. Spec. 8.1.1(i)). In particular, $\mathbf{\Phi}_0$ is the set of all $\mathscr{L}$-sentences.

2.2. Remark

If $\varphi \in \mathbf{\Phi}_n$, it does not follow that all the variables $\mathbf{v}_1, \mathbf{v}_2, \ldots, \mathbf{v}_n$ must be free in φ; but only that no *other* variables are free in φ. Hence $\mathbf{\Phi}_n \subseteq \mathbf{\Phi}_{n+1}$ for all n.

2.3. Definition

(i) If $\mathbf{\Gamma}$ is any set of sentences (that is, $\mathbf{\Gamma} \subseteq \mathbf{\Phi}_0$) we put

$$\mathbf{Dc}\mathbf{\Gamma} =_{\mathrm{df}} \{\varphi \in \mathbf{\Phi}_0 : \mathbf{\Gamma} \vdash \varphi\}.$$

$\mathbf{Dc}\mathbf{\Gamma}$ is called the *deductive closure* of $\mathbf{\Gamma}$.

(ii) We put $\mathbf{\Lambda} =_{\mathrm{df}} \mathbf{Dc}\varnothing$.

2.4. Remarks

By definition, $\mathbf{Dc}\mathbf{\Gamma}$ is the set of all sentences that can be deduced from $\mathbf{\Gamma}$ in Fopcal. However, by the soundness and completeness of Fopcal (Thms. 8.9.14 and 8.13.10), $\mathbf{Dc}\mathbf{\Gamma}$ is also the set of all sentences that are logical consequences of $\mathbf{\Gamma}$; in particular $\mathbf{\Lambda}$ is the set of all logically true sentences (cf. Def. 8.4.10). '$\mathbf{\Lambda}$' is mnemonic for 'logic'.

2.5. Definition

An $\mathscr{L}$*-theory* is a set $\mathbf{\Sigma} \subseteq \mathbf{\Phi}_0$ such that $\mathbf{\Sigma} = \mathbf{Dc\Sigma}$; in other words, it is a set of $\mathscr{L}$-sentences closed (or saturated) under deducibility of $\mathscr{L}$-sentences.

2.6. Problem

If $\mathbf{\Gamma}$ is any set of sentences, show that $\mathbf{Dc\Gamma}$ is a theory that includes $\mathbf{\Gamma}$ itself. Moreover, $\mathbf{Dc\Gamma}$ is the *smallest* such theory: if $\mathbf{\Sigma}$ is any theory that includes $\mathbf{\Gamma}$, then $\mathbf{Dc\Gamma} \subseteq \mathbf{\Sigma}$.

2.7. Definition

If $\mathbf{\Sigma}$ is a theory, then a *postulate set for* $\mathbf{\Sigma}$ is any set $\mathbf{\Gamma}$ of sentences such that $\mathbf{\Sigma} = \mathbf{Dc\Gamma}$.

2.8. Remark

The ideas we have just introduced may be applied in two mutually converse ways. In some cases we start with a given set $\mathbf{\Gamma}$ of sentences as postulates, and wish to investigate the resulting theory $\mathbf{Dc\Gamma}$. In other cases we start with a given theory $\mathbf{\Sigma}$ and wish to find a set of postulates for it that has some desirable property. (Of course, by Defs. 2.5 and 2.7 every theory is a postulate set for itself; but the point is to find a simpler set.)

2.9. Examples

(i) Consider $\mathbf{\Lambda} = \mathbf{Dc}\varnothing$. By Prob. 2.6, $\mathbf{\Lambda}$ is a theory; moreover, it is the *smallest* theory, in the sense that it is included in every theory.

(ii) The set $\mathbf{\Phi}_0$ of *all* sentences is evidently a theory. Moreover, it is the *largest* theory, in the sense that it includes every theory. Clearly, $\mathbf{\Phi}_0$ is inconsistent. Moreover, it is the *only* inconsistent theory. Indeed, if $\mathbf{\Sigma}$ is an inconsistent theory, then for every sentence φ we have $\mathbf{\Sigma} \vdash \varphi$ by IE, hence $\varphi \in \mathbf{\Sigma}$ because $\mathbf{\Sigma}$ is a theory. So $\mathbf{\Sigma}$ must be $\mathbf{\Phi}_0$.

2.10. Definition

For any $\mathscr{L}$-structure $\mathfrak{U}$ we put

$$\mathbf{Th}\mathfrak{U} =_{\mathrm{df}} \{\boldsymbol{\varphi} \in \boldsymbol{\Phi}_0 : \mathfrak{U} \vDash \boldsymbol{\varphi}\}.$$

$\mathbf{Th}\mathfrak{U}$ is called *the theory of* $\mathfrak{U}$; it is the set of all sentences that hold in $\mathfrak{U}$.

2.11. Remark

It is easy to see that $\mathbf{Th}\mathfrak{U}$ is indeed a theory in the sense of Def. 2.5: if $\boldsymbol{\psi}$ is a sentence such that $\mathbf{Th}\mathfrak{U} \vdash \boldsymbol{\psi}$ then, by the soundness of Fopcal, $\mathfrak{U} \vDash \boldsymbol{\psi}$; therefore $\boldsymbol{\psi} \in \mathbf{Th}\mathfrak{U}$.

2.12. Definition

A theory $\boldsymbol{\Sigma}$ is *complete* if it is consistent, and for any sentence $\boldsymbol{\varphi}$ either $\boldsymbol{\varphi} \in \boldsymbol{\Sigma}$ or $\neg\boldsymbol{\varphi} \in \boldsymbol{\Sigma}$.

2.13. Problem

(i) Show that a consistent theory $\boldsymbol{\Sigma}$ is complete iff it is maximal among consistent theories, that is, it is not included in any other consistent theory.
(ii) Show that, for any $\mathscr{L}$-structure $\mathfrak{U}$, $\mathbf{Th}\mathfrak{U}$ is a complete theory.
(iii) Show that any consistent theory is included in a complete theory.
(iv) Show that any complete theory is of the form $\mathbf{Th}\mathfrak{U}$ for some $\mathfrak{U}$.

2.14. Definition

(i) We put

$$\boldsymbol{\Omega} =_{\mathrm{df}} \mathbf{Th}\mathfrak{N}.$$

The theory $\boldsymbol{\Omega}$, consisting of all true sentences (in the sense of Def. 1.9(ii)) is called *complete first-order arithmetic*.
(ii) A set of sentences – and, in particular, a theory – is said to be *sound* if it is included in $\boldsymbol{\Omega}$; in other words, if all the sentences belonging to it are true.

2.15. Remarks

(i) By Prob. 2.13(ii), $\boldsymbol{\Omega}$ is indeed a complete theory. By Def. 2.14, $\boldsymbol{\Omega}$ is a sound theory. In fact, $\boldsymbol{\Omega}$ is the *only* complete sound theory. Indeed, if $\mathbf{\Sigma}$ is sound, then $\mathbf{\Sigma} \subseteq \mathbf{\Omega}$; but if $\mathbf{\Sigma}$ is also a complete theory then by Prob. 2.13(i) it cannot be included in *another* consistent theory, so $\mathbf{\Sigma}$ must coincide with $\mathbf{\Omega}$.

(ii) $\mathbf{\Omega}$ can be regarded as *the whole truth about* $\mathfrak{N}$ *in* $\mathcal{L}$, in the sense that it consists of all $\mathcal{L}$-sentences that are true in $\mathfrak{N}$. But is it really the whole truth about $\mathfrak{N}$? We shall address this question in the next section.

§ 3. Skolem's Theorem

3.1. Preview

In this section we show that $\mathfrak{N}$ cannot be uniquely characterized in $\mathcal{L}$: even $\mathbf{\Omega}$ – the whole truth about $\mathfrak{N}$ in $\mathcal{L}$ – is not sufficient to single out $\mathfrak{N}$ because $\mathbf{\Omega}$ has, apart fromm $\mathfrak{N}$ itself, other models that are not isomorphic to $\mathfrak{N}$.

3.2. Convention

We shall often wish to consider the standard structure $\mathfrak{N}$ alongside some $\mathcal{L}$-structure, which may or may not be the standard one. In such cases it will be convenient to denote the latter structure by '$^*\mathfrak{N}$'. Whenever we use this notation, we shall take it for granted that

(i) $^*N$ is the domain of $^*\mathfrak{N}$,
(ii) ^{*0} is $\mathbf{0}^{*\mathfrak{N}}$ (the designated individual of $^*N$),
(iii) $^*s$ is $\mathbf{s}^{*\mathfrak{N}}$ (the basic unary operation of $^*\mathfrak{N}$),
(iv) $^*+$ and $^*\times$ are $+^{*\mathfrak{N}}$ and $\times^{*\mathfrak{N}}$ respectively (the basic binary operations of $^*\mathfrak{N}$).

The prefix '*' is pronounced as 'pseudo'.

3.3. Remark

The purpose of this convention is to stress both the similarities and dissimilarities (if any) between $\mathfrak{N}$ and $^*\mathfrak{N}$.

3.4. Definition

(i) An *embedding* of the structure $\mathfrak{N}$ in the structure $^*\mathfrak{N}$ is an injection from N to $^*N$ (that is, a 1-1 mapping from N into $^*N$)

such that

$$(*) \qquad \begin{gathered} f0 = {}^*0, \qquad f(m+1) = {}^*s(fm), \\ f(m+n) = fm \;{}^*\!+\; fn, \qquad f(mn) = fm \;{}^*\!\times\; fn, \end{gathered}$$

for all numbers m and n.

(ii) If, in addition, f is a surjection from N to ${}^*N$ (that is, f maps N *onto* ${}^*N$) then f is called an *isomorphism* between $\mathfrak{N}$ and ${}^*\mathfrak{N}$, and the two structures are said to be *isomorphic* to each other.

3.5. Remarks

(i) If f is an isomorphism between $\mathfrak{N}$ and ${}^*\mathfrak{N}$, then ${}^*\mathfrak{N}$ is an exact replica of $\mathfrak{N}$: each number n has a unique counterpart fn and each individual of ${}^*\mathfrak{N}$ is the counterpart of a unique number; and, moreover, by $(*)$ the basic operations on numbers are exactly mimicked by the corresponding basic operations on their counterparts. The two structures are structurally indistinguishable.

For this reason we shall from now on refer not just to $\mathfrak{N}$ itself but also to any $\mathscr{L}$-structure isomorphic to it as *the standard structure*.

(ii) If f is merely an embedding of $\mathfrak{N}$ in ${}^*\mathfrak{N}$, then this means that ${}^*\mathfrak{N}$ has a substructure isomorphic to $\mathfrak{N}$.

3.6. Problem

Let f be an embedding of $\mathfrak{N}$ in ${}^*\mathfrak{N}$. For any valuation σ based on $\mathfrak{N}$, we define $f\sigma$ as the valuation based on ${}^*\mathfrak{N}$ such that, for each variable $\mathbf{y}$,

$$\mathbf{y}^{f\sigma} = f(\mathbf{y}^{\sigma}).$$

(i) Show that $\mathbf{t}^{f\sigma} = f(\mathbf{t}^{\sigma})$ for any term $\mathbf{t}$. Hence, in particular, if $\mathbf{t}$ is a closed term it follows that $\mathbf{t}^{*\mathfrak{N}} = f(\mathbf{t}^{\mathfrak{N}})$. (Use induction on $\deg \mathbf{t}$, distinguishing the five cases mentioned in Rem. 1.4(iii). Note that the fact that f is injective need not be used in the proof.)

(ii) Show that $f[\sigma(\mathbf{x}/n)] = (f\sigma)(\mathbf{x}/fn)$, where $\mathbf{x}$ is any variable and n is any number.

(iii) Show that if f is an isomorphism between $\mathfrak{N}$ and ${}^*\mathfrak{N}$ then $\boldsymbol{\alpha}^{f\sigma} = \boldsymbol{\alpha}^{\sigma}$ for any formula $\boldsymbol{\alpha}$. In particular, ${}^*\mathfrak{N} \vDash \boldsymbol{\varphi}$ iff $\mathfrak{N} \vDash \boldsymbol{\varphi}$ for any sentence $\boldsymbol{\varphi}$.

3.7. Remark

by Def. 2.14, $\mathfrak{N} \models \boldsymbol{\varphi}$ iff $\boldsymbol{\varphi} \in \boldsymbol{\Omega}$; thus $\mathfrak{N}$ is a model for $\boldsymbol{\Omega}$ (see Def. 8.5.10). From Prob. 3.6(iii) it follows that any structure $^*\mathfrak{N}$ isomorphic to $\mathfrak{N}$ is likewise a model for $\boldsymbol{\Omega}$. This is hardly surprising, since such $^*\mathfrak{N}$ is a carbon copy of $\mathfrak{N}$. The surprising fact, which will be proved next, is that not all models for $\boldsymbol{\Omega}$ are standard.

3.8. Theorem (Skolem, 1934)

There exists a nonstandard model for $\boldsymbol{\Omega}$ – that is, a model for $\boldsymbol{\Omega}$ that is not isomorphic to $\mathfrak{N}$. Moreover, there is such a model whose domain is denumerable.

PROOF

Choose any variable $\mathbf{x}$, and for each number n let $\boldsymbol{\varphi}_n$ be the formula $\mathbf{x} \neq \mathbf{s}_n$. Now consider the following set of formulas:

$$\boldsymbol{\Phi} = \boldsymbol{\Omega} \cup \{\boldsymbol{\varphi}_n : n \in N\}.$$

We claim that $\boldsymbol{\Phi}$ is satisfiable. By the Compactness Thm. 8.13.12, this claim will be proved if we show that every *finite* subset of $\boldsymbol{\Phi}$ is satisfiable.

So let $\boldsymbol{\Phi}'$ be any finite subset of $\boldsymbol{\Phi}$. Clearly, $\boldsymbol{\Phi}'$ can only contain a finite number of formulas $\boldsymbol{\varphi}_n$; hence $\boldsymbol{\Phi}'$ is included in the set $\boldsymbol{\Omega} \cup \{\boldsymbol{\varphi}_n : n < p\}$, provided p is sufficiently large. So in order to show that $\boldsymbol{\Phi}'$ is satisfiable, we need only show that $\boldsymbol{\Omega} \cup \{\boldsymbol{\varphi}_n : n < p\}$ is satisfiable. However, the latter set is satisfied by any valuation σ based on $\mathfrak{N}$, provided $\mathbf{x}^\sigma \geqslant p$. Indeed, since σ is based on $\mathfrak{N}$, it satisfies $\boldsymbol{\Omega}$. Furthermore, $\mathbf{s}_n{}^\sigma = n$ (see Prob. 1.12); hence if $\mathbf{x}^\sigma \geqslant p$ then σ also satisfies the formulas $\boldsymbol{\varphi}_n$ – that is, $\mathbf{x} \neq \mathbf{s}_n$ – for every $n < p$.

We have thus proved our claim that $\boldsymbol{\Phi}$ is satisfiable. Let τ be a valuation that satisfies $\boldsymbol{\Phi}$ and let $^*\mathfrak{N}$ be its underlying structure. $^*\mathfrak{N}$ is a model for $\boldsymbol{\Omega}$, because τ satisfies $\boldsymbol{\Phi}$, which includes $\boldsymbol{\Omega}$.

As the language $\mathscr{L}$ is denumerable, it follows from the Löwenheim–Skolem Thm. 8.13.13 that we may take the domain $^*N$ of $^*\mathfrak{N}$ to be countable (that is, finite or denumerable). However, $^*N$ cannot be finite, because $\boldsymbol{\Omega}$ contains the sentences $\mathbf{s}_m \neq \mathbf{s}_n$ for all pairs of distinct numbers m and n, and therefore all these sentences must be satisfied in $^*\mathfrak{N}$, which can only happen if $^*N$ is infinite. Thus $^*N$ is denumerable.

It remains to show that $^*\mathfrak{N}$ is nonstandard; in other words, that it is not isomorphic to $\mathfrak{N}$. Suppose f is an embedding of $\mathfrak{N}$ in $^*\mathfrak{N}$. We shall prove that f cannot be surjective (that is, cannot map N *onto* $^*N$). Indeed, for each number n our valuation τ satisfies the formula $\boldsymbol{\varphi}_n$, that is, $\mathbf{x} \neq \mathbf{s}_n$. Hence (by the BSD) we must have

$$\mathbf{x}^\tau \neq \mathbf{s}_n{}^\tau \text{ for every number } n.$$

However, by Probs. 3.6(i) and 1.12 we have

$$\mathbf{s}_n{}^\tau = \mathbf{s}_n{}^{*\mathfrak{N}} = f(\mathbf{s}_n{}^{\mathfrak{N}}) = fn.$$

Thus $\mathbf{x}^\tau$ – which must belong to $^*N$, the universe of τ – cannot be fn for any number n. This shows that f is not surjective. ■

3.9. Problem

Let $^*\mathfrak{N}$ be any model for $\mathbf{\Omega}$. Let f be the mapping from N to $^*N$ defined by:

$$fn = \mathbf{s}_n{}^{*\mathfrak{N}} \text{ for all } n.$$

(i) Show that f is injective. (If $m \neq n$ then $\mathbf{s}_m \neq \mathbf{s}_n$ is in $\mathbf{\Omega}$ and so must hold in $^*\mathfrak{N}$.) Prove:
(ii) f is an embedding of $\mathfrak{N}$ in $^*\mathfrak{N}$.
(iii) f is the *only* embedding of $\mathfrak{N}$ in $^*\mathfrak{N}$. (Use Prob. 3.6(i).)
(iv) Hence $^*\mathfrak{N}$ is a standard model of $\mathbf{\Omega}$ iff $^*N = \{\mathbf{s}_n{}^{*\mathfrak{N}} : n \in N\}$.

3.10. Remark

Skolem's Theorem means that the whole truth about $\mathfrak{N}$ cannot be expressed in $\mathscr{L}$. As we have noted, $\mathbf{\Omega}$ is all that can be said in $\mathscr{L}$ about $\mathfrak{N}$; but $\mathbf{\Omega}$ fails to pin $\mathfrak{N}$ down uniquely (even up to isomorphism). At first sight it may seem that is perhaps due to some accidental defect of $\mathscr{L}$. Can $\mathscr{L}$ perhaps be enriched (and $\mathfrak{N}$ correspondingly elaborated) so that in the richer formal language the correspondingly more elaborate structure of natural numbers may be characterized uniquely up to isomorphism? For a discussion of this question, and a pessimistic answer, see B&M, pp. 320–324. We shall return to this issue in the Appendix.

§4. Representability

4.1. Preview

This section is devoted to defining new concepts rather than to proving major results. We shall introduce two ways in which a relation on N may be formally expressed or *represented* in a theory $\mathbf{\Sigma}$.

4.2. Reminder

We recall some of the conventions introduced in Ch. 9. Lower-case German letters '$\mathfrak{a}$', '$\mathfrak{b}$', '$\mathfrak{x}$' and '$\mathfrak{y}$' are used as informal variables ranging over the set N^n of all n-tuples of numbers. Where a German letter is used for an n-tuple, the corresponding italic letter is used for the components of that n-tuple. Thus, for example, $\mathfrak{a} = \langle a_1, a_2, \ldots, a_n \rangle$ and $\mathfrak{x} = \langle x_1, x_2, \ldots, x_n \rangle$.

Note that the number of components of a tuple denoted by a German letter is always assumed to be n (rather than k or m etc.).

Recall that by *relation* we mean relation on N. If P is an n-ary relation, we usually write, for example, '$P\mathfrak{a}$' as short for '$\mathfrak{a} \in P$'.

4.3. Remark

The symbols '$\mathfrak{a}$' and '$\mathfrak{x}$' do not refer to, or have anything to do with, the formal language $\mathscr{L}$; they are ordinary mathematical symbols used as variables in our own language.

4.4. Definition (abbreviated notation for substitution)

For any terms $\mathbf{r}, \mathbf{t}_1, \mathbf{t}_2, \ldots, \mathbf{t}_n$ and any formula $\boldsymbol{\alpha}$ we put

(i) $\mathbf{r}(\mathbf{t}_1, \mathbf{t}_2, \ldots, \mathbf{t}_n) =_{\text{df}} \mathbf{r}(\mathbf{v}_1/\mathbf{t}_1, \mathbf{v}_2/\mathbf{t}_2, \ldots, \mathbf{v}_n/\mathbf{t}_n)$,
(ii) $\boldsymbol{\alpha}(\mathbf{t}_1, \mathbf{t}_2, \ldots, \mathbf{t}_n) =_{\text{df}} \boldsymbol{\alpha}(\mathbf{v}_1/\mathbf{t}_1, \mathbf{v}_2/\mathbf{t}_2, \ldots, \mathbf{v}_n/\mathbf{t}_n)$.

4.5. Remarks

(i) Here the terms $\mathbf{t}_1, \mathbf{t}_2, \ldots, \mathbf{t}_n$ are substituted simultaneously for all free occurrences of $\mathbf{v}_1, \mathbf{v}_2, \ldots, \mathbf{v}_n$ respectively – the first n variables in alphabetic order (cf. Spec. 8.1.1(i)). So, for example, '$\boldsymbol{\alpha}(\mathbf{t})$' is short for '$\boldsymbol{\alpha}(\mathbf{v}_1/\mathbf{t})$'. If $\mathbf{t}$ is to be substituted for a variable $\mathbf{x}$ other than $\mathbf{v}_1$, we cannot use the abbreviated notation but have to write '$\boldsymbol{\alpha}(\mathbf{x}/\mathbf{t})$' in full.

(ii) When substituting several terms in a formula, as in Def. 4.4(ii), alphabetic changes of bound variables may be necessary in order to prevent capture. Also, it is important that the terms are substituted *simultaneously* rather than successively. (For a detailed precise treatment of the technicalities involved in simultaneous substitution, see B&M, pp. 65–67.) However, in many cases when the abbreviated notation is used below, the terms that are substituted will be closed terms; so no changes of bound variables will be required. In such cases it is also unimportant whether the substitution is made simultaneously or successively.

Next, for the case where the terms to be substituted for the variables $\mathbf{v}_1, \mathbf{v}_2, \ldots, \mathbf{v}_n$ are numerals, we introduce a further useful abbreviation which slightly stretches the use of lower-case German letters.

4.6. Definition

For any term $\mathbf{r}$, any formula $\boldsymbol{\alpha}$ and any $\mathfrak{a} \in N^n$, we put

(i) $\mathbf{r}(\mathbf{s}_\mathfrak{a}) =_{\text{df}} \mathbf{r}(\mathbf{s}_{a_1}, \mathbf{s}_{a_2}, \ldots, \mathbf{s}_{a_n})$,
(ii) $\boldsymbol{\alpha}(\mathbf{s}_\mathfrak{a}) =_{\text{df}} \boldsymbol{\alpha}(\mathbf{s}_{a_1}, \mathbf{s}_{a_2}, \ldots, \mathbf{s}_{a_n})$.

Thus, $\boldsymbol{\alpha}(\mathbf{s}_\mathfrak{a})$ is obtained from $\boldsymbol{\alpha}$ by substituting the a_i-th numeral for all free occurrences of $\mathbf{v}_i$, where $i = 1, 2, \ldots, n$.

If $\boldsymbol{\alpha} \in \boldsymbol{\Phi}_n$, then – for any $\mathfrak{a} \in N^n$ – $\boldsymbol{\alpha}(\mathbf{s}_\mathfrak{a})$ is a sentence. If $\boldsymbol{\Sigma}$ is a theory, it makes sense to enquire whether the sentence $\boldsymbol{\alpha}(\mathbf{s}_\mathfrak{a})$ belongs to $\boldsymbol{\Sigma}$; similarly, we may enquire whether its negation, the sentence $\neg\boldsymbol{\alpha}(\mathbf{s}_\mathfrak{a})$, belongs to that theory. This gives rise to the following important definition.

4.7. Definition

Let P be any n-ary relation and let $\boldsymbol{\Sigma}$ be a theory.

(i) A formula $\boldsymbol{\alpha} \in \boldsymbol{\Phi}_n$ *represents P weakly in* $\boldsymbol{\Sigma}$ if, for all $\mathfrak{x} \in N^n$,

$$P\mathfrak{x} \Leftrightarrow \boldsymbol{\alpha}(\mathbf{s}_\mathfrak{x}) \in \boldsymbol{\Sigma}.$$

P is weakly representable in $\boldsymbol{\Sigma}$ if it is weakly represented in $\boldsymbol{\Sigma}$ by some $\boldsymbol{\alpha} \in \boldsymbol{\Phi}_n$.

(ii) A formula $\boldsymbol{\alpha} \in \boldsymbol{\Phi}_n$ *represents P strongly in* $\boldsymbol{\Sigma}$ if, for all $\mathfrak{x} \in N^n$,

$$P\mathfrak{x} \Rightarrow \boldsymbol{\alpha}(\mathbf{s}_\mathfrak{x}) \in \boldsymbol{\Sigma}, \qquad \neg P\mathfrak{x} \Rightarrow \neg\boldsymbol{\alpha}(\mathbf{s}_\mathfrak{x}) \in \boldsymbol{\Sigma}.$$

P is strongly representable in $\boldsymbol{\Sigma}$ if it is strongly represented in $\boldsymbol{\Sigma}$ by some $\boldsymbol{\alpha} \in \boldsymbol{\Phi}_n$.

4.8. Remarks

(i) Recall that $\neg$ is the (informal) negation operation on relations; thus $\neg Px$ holds iff Px does not.

(ii) Use of the adverbs 'weakly' and 'strongly' is justified because, *for a consistent theory*, weak representation follows from strong representation: if $\boldsymbol{\alpha}$ represents P strongly in $\boldsymbol{\Sigma}$ and Px does not hold, then $\neg\boldsymbol{\alpha}(\mathbf{s}) \in \boldsymbol{\Sigma}$; and – provided $\boldsymbol{\Sigma}$ is consistent – it follows that $\boldsymbol{\alpha}(\mathbf{s}_x) \notin \boldsymbol{\Sigma}$. Thus $\boldsymbol{\alpha}$ also represents P weakly in $\boldsymbol{\Sigma}$.

If $\boldsymbol{\Sigma}$ is the inconsistent theory, the above argument fails. Weak and strong representability in this theory are, however, trivial notions. (See Prob. 4.9.)

(iii) For any $\boldsymbol{\alpha} \in \boldsymbol{\Phi}_n$ and any theory $\boldsymbol{\Sigma}$, there is always a unique n-ary relation P that is weakly represented by $\boldsymbol{\alpha}$ in $\boldsymbol{\Sigma}$, because Def. 4.7(i) determines such P uniquely.

On the other hand, $\boldsymbol{\alpha}$ may not represent any relation strongly in $\boldsymbol{\Sigma}$, because for some x it may happen that neither $\boldsymbol{\alpha}(\mathbf{s}_x) \in \boldsymbol{\Sigma}$ nor $\neg\boldsymbol{\alpha}(\mathbf{s}_x) \in \boldsymbol{\Sigma}$.

(iv) However, if $\boldsymbol{\Sigma}$ is a complete theory (cf. Def. 2.12) then $\neg\boldsymbol{\alpha}(\mathbf{s}_x) \in \boldsymbol{\Sigma}$ iff $\boldsymbol{\alpha}(\mathbf{s}_x) \notin \boldsymbol{\Sigma}$; so in this case strong representation is equivalent to weak representation. In other words, in a complete theory any $\boldsymbol{\alpha} \in \boldsymbol{\Phi}_n$ represents a unique n-ary relation both weakly and strongly. In connection with a complete theory we shall therefore omit these qualifications and say simply that a given formula *represents* the relation.

4.9. Problem

Let $\boldsymbol{\alpha} \in \boldsymbol{\Phi}_n$, where $n > 0$. Determine the n-ary relations that $\boldsymbol{\alpha}$ represents weakly/strongly in the inconsistent theory.

§ 5. Arithmeticity

5.1. Preview

In this section we investigate an important class of relations: those representable in complete first-order arithmetic, $\boldsymbol{\Omega}$. In view of Rem. 4.8(iv), in the present context we need not distinguish between weak and strong representation, so we say simply that a given formula *represents* a relation in $\boldsymbol{\Omega}$.

5.2. Definition

A relation is *arithmetical* if it is representable in $\mathbf{\Omega}$.

5.3. Remark

Thus by Def. 4.7, an n-ary relation P is arithmetical iff there is a representing formula $\boldsymbol{\alpha} \in \mathbf{\Phi}_n$ such that

$$(*) \qquad Px \Leftrightarrow \boldsymbol{\alpha}(\mathbf{s}_x) \in \mathbf{\Omega}$$

for all $x \in N^n$.

And since by Def. 2.14(i) $\mathbf{\Omega} = \mathbf{Th}\mathfrak{N}$, condition $(*)$ is tantamount to:

$$(**) \qquad Px \Leftrightarrow \mathfrak{N} \models \boldsymbol{\alpha}(\mathbf{s}_x).$$

5.4. Definition

Let $\boldsymbol{\alpha} \in \mathbf{\Phi}_n$ and $\mathfrak{a} \in N^n$. If $\boldsymbol{\alpha}$ is satisfied by some valuation σ based on $\mathfrak{N}$ such that $\mathbf{v}_i{}^\sigma = a_i$ for $i = 1, 2, \ldots, n$, we write:

$$`\mathfrak{N} \models \boldsymbol{\alpha}[\mathfrak{a}]'.$$

5.5. Remarks

(i) If $\mathfrak{N} \models \boldsymbol{\alpha}[\mathfrak{a}]$, then by Thm. 8.5.8 $\boldsymbol{\alpha}$ is satisfied by *every* valuation σ based on $\mathfrak{N}$ such that $\mathbf{v}_i{}^\sigma = a_i$ for $i = 1, 2, \ldots, n$.

(ii) Def. 5.4 is a contextual definition: it defines the whole expression '$\mathfrak{N} \models \boldsymbol{\alpha}[\mathfrak{a}]$' as a package. The part '$\boldsymbol{\alpha}[\mathfrak{a}]$' of this package has no meaning on its own: it does not denote anything whatsoever. In particular, '$\boldsymbol{\alpha}[\mathfrak{a}]$' must not be confused with '$\boldsymbol{\alpha}(\mathbf{s}_\mathfrak{a})$', which *does* have meaning on its own: it denotes the $\mathscr{L}$-sentence obtained from $\boldsymbol{\alpha}$ by substituting the n-tuple of numerals $\mathbf{s}_\mathfrak{a}$ for the first n variables of $\mathscr{L}$. However –

5.6. Lemma

Let $\boldsymbol{\alpha} \in \mathbf{\Phi}_n$ and $\mathfrak{a} \in N^n$. Then $\mathfrak{N} \models \boldsymbol{\alpha}(\mathbf{s}_\mathfrak{a})$ iff $\mathfrak{N} \models \boldsymbol{\alpha}[\mathfrak{a}]$.

PROOF

We consider in detail the case $n = 1$. In this case $\boldsymbol{\alpha}$ has no free variable other than $\mathbf{v}_1$, and we must show, for any number a, that $\mathfrak{N} \models \boldsymbol{\alpha}(\mathbf{s}_a)$ *iff*

$\mathfrak{N} \models \boldsymbol{\alpha}[a]$. Here goes:

$$\begin{aligned} \mathfrak{N} \models \boldsymbol{\alpha}(\mathbf{s}_a) &\Leftrightarrow \boldsymbol{\alpha}(\mathbf{s}_a)^\sigma = \top \quad \text{for some valuation } \sigma \text{ based on } \mathfrak{N} \\ &\qquad\qquad\qquad\qquad\qquad \text{by Def. 8.5.10,} \\ &\Leftrightarrow \boldsymbol{\alpha}^{\sigma(\mathbf{v}_1/a)} = \top \qquad \text{by Prob. 8.6.16, since } {\mathbf{s}_a}^\sigma = a, \\ &\Leftrightarrow \mathfrak{N} \models \boldsymbol{\alpha}[a] \qquad\qquad\qquad \text{by Def. 5.4.} \end{aligned}$$

The general case, for arbitrary n, is treated similarly. Of course, it utilizes the generalization of Eqn. 8.6.6 to the case of simultaneous substitution of n terms. (See B&M, p. 65.) ■

5.7. Remark

From this lemma it now follows that conditions (*) and (**) of Rem. 5.3 are equivalent to

$$(***) \qquad\qquad Px \Leftrightarrow \mathfrak{N} \models \boldsymbol{\alpha}[x].$$

5.8. Examples

Because condition (***) refers to the *standard* interpretation, it is always straightforward to work out the n-ary relation represented in $\boldsymbol{\Omega}$ by a given $\boldsymbol{\alpha} \in \boldsymbol{\Phi}_n$. All that we need to do is to 'deformalize' $\boldsymbol{\alpha}$ by 'translating' it from $\mathcal{L}$ into the metalanguage (see Rem. 1.10).

(i) Consider the formula $\mathbf{v}_1+\mathbf{v}_3=\mathbf{v}_2$. It belongs to $\boldsymbol{\Phi}_3$ and hence represents in $\boldsymbol{\Omega}$ a ternary relation P. Moreover, P is evidently the relation determined by

$$P(x_1, x_2, x_3) \Leftrightarrow x_1 + x_3 = x_2.$$

Equivalently, $P = \{\langle x_1, x_2, x_3\rangle \in N^3 : x_1 + x_3 = x_2\}$.

Note that our formula also belongs to $\boldsymbol{\Phi}_4$ (as well as to $\boldsymbol{\Phi}_n$ for any $n \geqslant 3$). So it represents in $\boldsymbol{\Omega}$ a quaternary relation Q, which is given by

$$Q(x_1, x_2, x_3, x_4) \Leftrightarrow x_1 + x_3 = x_2,$$

or

$$Q = \{\langle x_1, x_2, x_3, x_4\rangle \in N : x_1 + x_3 = x_2\}.$$

Of course, Q does not depend on its fourth argument; but it is nevertheless a quaternary relation!

(ii) Next, consider the formula $\exists\mathbf{v}_3(\mathbf{v}_1+\mathbf{v}_3=\mathbf{v}_2)$. It belongs to $\boldsymbol{\Phi}_2$ and therefore represents in $\mathbf{\Omega}$ a binary relation R. By direct 'deformalization' we see at once that R is given by

$$R(x_1, x_2) \Leftrightarrow \exists x_3(x_1 + x_3 = x_2).$$

It does not require much knowledge of arithmetic to realize that R is the relation $\leqslant$; more explicitly:

$$R(x_1, x_2) \Leftrightarrow x_1 \leqslant x_2 \text{ or } R = \{\langle x_1, x_2\rangle \in N^2 : x_1 \leqslant x_2\}.$$

This example should look familiar; it is of course Prob. 1.13(i) in a slightly different guise.

(iii) Now consider the formula $\forall \mathbf{v}_2(\mathbf{v}_1 \neq \mathbf{s}\mathbf{v}_2)$. This belongs to $\mathbf{\Phi}_1$ and therefore represents in $\mathbf{\Omega}$ a property S. By direct 'deformalization' we see:

$$Sx_1 \Leftrightarrow \forall x_2(x_1 \neq x_2 + 1),$$

and, using a tiny bit of knowledge of arithmetic, we realize that $Sx_1 \Leftrightarrow x_1 = 0$, so that $S = \{0\}$. Of course, S is also represented in $\mathbf{\Omega}$ by other formulas, for example $\mathbf{v}_1\mathbf{=0}$.

5.9. Lemma

If the equation $\mathbf{r=t}$ *belongs to* $\mathbf{\Phi}_n$ *then it represents in* $\mathbf{\Omega}$ *an elementary n-ary relation. Conversely, every elementary relation is represented in* $\mathbf{\Omega}$ *by an equation.*

PROOF

First, suppose that $\mathbf{r=t}$ belongs to $\mathbf{\Phi}_n$. This simply means that every variable occurring in $\mathbf{r}$ or $\mathbf{t}$ is among $\mathbf{v}_1, \mathbf{v}_2, \ldots, \mathbf{v}_n$. In addition to variables, $\mathbf{r}$ and $\mathbf{t}$ may contain occurrences of $\mathbf{0}$, $\mathbf{s}$, $\mathbf{+}$ and $\mathbf{\times}$.

Let P be the n-ary relation represented by this equation in $\mathbf{\Omega}$. To determine P we use the process of 'deformalization' illustrated in Ex. 5.8. We get, for all $x \in N^n$:

$$(*) \qquad Px \Leftrightarrow fx = gx,$$

where fx and gx are obtained from $\mathbf{r}$ and $\mathbf{t}$ respectively in the obvious way: each $\mathbf{v}_i$ is 'translated' as 'x_i', $\mathbf{0}$ is 'translated' as '0', and so on. Thus fx and gx are given by expressions (in our metalanguage) made up of variables 'x_1', 'x_2', . . . , 'x_n' numerals '0' and '1' (the latter comes from translating the symbol $\mathbf{s}$ of $\mathscr{L}$) and operation symbols '+' and '×'. Simplifying these expressions by the rules of elementary algebra, we see that fx and gx are polynomials; hence P is elementary (cf. Defs. 9.5.2 and 9.5.3).

Conversely, suppose that P is an n-ary elementary relation. Then P satisfies an equivalence of the form (*), where fx and gx are polynomials. To obtain an $\mathcal{L}$-formula that represents P in $\mathbf{\Omega}$, all we have to do is to *formalize* the equation $fx = gx$ – translating it in the obvious way into $\mathcal{L}$. We get an equation $\mathbf{r}=\mathbf{t}$ that represents P in $\mathbf{\Omega}$. ■

5.10. Warning

Not every formula that represents in $\mathbf{\Omega}$ an elementary relation is an equation. What we have shown is that among the (infinitely many) formulas representing in $\mathbf{\Omega}$ a given elementary relation there must be an equation.

5.11. Theorem

The following two conditions are equivalent:

(i) *P is an arithmetical relation;*
(ii) *P can be obtained from elementary relations by a finite number of applications of logical operations.*

PROOF

(i) ⇒ (ii). Let P be an n-ary arithmetical relation. Then P is represented in $\mathbf{\Omega}$ by some formula $\boldsymbol{\alpha} \in \mathbf{\Phi}_n$. We shall show by induction on $\deg \boldsymbol{\alpha}$ that (ii) holds.

Case 1: $\boldsymbol{\alpha}$ is an equation. Then by Lemma 5.9 P is itself elementary, so (ii) clearly holds.

Case 2: $\boldsymbol{\alpha} = \neg\boldsymbol{\beta}$. Let Q be the n-ary relation represented in $\mathbf{\Omega}$ by $\boldsymbol{\beta}$. Then it is easy to see that $P = \neg Q$. By the induction hypothesis, Q is obtainable from elementary relations by a finite number of applications of logical operations. Since P is obtained from Q by an application of $\neg$, it is clear that (ii) holds.

Case 3: $\boldsymbol{\alpha} = \boldsymbol{\beta}\to\boldsymbol{\gamma}$. Let Q and R be the n-ary relations represented in $\mathbf{\Omega}$ by $\boldsymbol{\beta}$ and $\boldsymbol{\gamma}$ respectively. Then it is easy to see that $P = Q \to R = \neg Q \vee R$. By the induction hypothesis, both Q and R are obtainable from elementary relations by a finite number of applications of logical operations. Hence the same holds for P.

Case 4: $\boldsymbol{\alpha} = \forall \mathbf{y} \boldsymbol{\beta}$. Without loss of generality, we may assume that $\mathbf{y}$ is $\mathbf{v}_{n+1}$ (otherwise, by appropriate alphabetic changes, we can obtain from $\boldsymbol{\alpha}$ a variant $\forall \mathbf{v}_{n+1} \boldsymbol{\beta}'$, which is logically equivalent to $\boldsymbol{\alpha}$, has the same degree as $\boldsymbol{\alpha}$ and, like $\boldsymbol{\alpha}$, represents P in $\boldsymbol{\Omega}$). Therefore $\boldsymbol{\beta} \in \boldsymbol{\Phi}_{n+1}$, so $\boldsymbol{\beta}$ represents in $\boldsymbol{\Omega}$ an $(n+1)$-ary relation Q. Then clearly P is obtained from Q by (informal) universal quantification: $Px \Leftrightarrow \forall y Q(x, y)$. By the induction hypothesis, Q is obtainable from elementary relations by a finite number of applications of logical operations. Hence the same holds for P.

(ii) $\Rightarrow$ (i). Assume (ii). Then P is obtainable from elementary relations by a finite number, say k, of applications of the three logical operations: negation, implication and universal quantification. (The other logical operations can be reduced to these.) We proceed by induction on k.

Case 1: P itself is elementary. Then P is arithmetical by Lemma 5.9.

Case 2: $P = \neg Q$, where Q is obtainable from elementary relations by $k-1$ applications of the three logical operations. By the induction hypothesis, Q is arithmetical, hence it is represented in $\boldsymbol{\Omega}$ by some formula $\boldsymbol{\beta}$. Then P is represented in $\boldsymbol{\Omega}$ by the formula $\neg\boldsymbol{\beta}$, and is therefore arithmetical.

Case 3: $P = Q \rightarrow R$, where Q and R are each obtainable from elementary relations by fewer than k applications of the three logical operations. By the induction hypothesis, P and Q are arithmetical, hence represented in $\boldsymbol{\Omega}$ by formulas $\boldsymbol{\beta}$ and $\boldsymbol{\gamma}$ respectively. Then P is represented in $\boldsymbol{\Omega}$ by the formula $\boldsymbol{\beta} \rightarrow \boldsymbol{\gamma}$, and is therefore arithmetical.

Case 4: P is obtained by universal quantification from an $(n+1)$-ary relation Q:

$$Px \Leftrightarrow \forall x_{n+1} Q(x, x_{n+1}),$$

where Q is obtainable from elementary relations by $k-1$ applications of the three logical operations. By the induction hypothesis, Q is arithmetical, hence represented in $\boldsymbol{\Omega}$ by some $\boldsymbol{\beta} \in \boldsymbol{\Phi}_{n+1}$. Then it is easy to see that P is represented in $\boldsymbol{\Omega}$ by the formula $\forall \mathbf{v}_{n+1} \boldsymbol{\beta}$, and is therefore arithmetical. ■

5.12. Remarks

(i) Thm. 5.11 means that the class of arithmetical relations is the smallest class that contains all elementary relations and is closed under the logical operations.

(ii) That the proof of Thm. 5.11 was so easy is due in part to the notation we are using (cf. Warning 9.1.4).

The following corollary is extremely useful.

5.13. Corollary

If P is an n-ary r.e. relation, then it is arithmetical. Moreover, it is represented in $\mathbf{\Omega}$ *by a formula of the form*

$$\exists \mathbf{v}_{n+1} \exists \mathbf{v}_{n+2} \ldots \exists \mathbf{v}_{n+m}(\mathbf{r}=\mathbf{t}),$$

where $m \geqslant 0$.

PROOF

By the MRDP Thm. 9.5.4, P is diophantine. This means that P is obtained from an elementary relation by a finite number of (informal) existential quantifications. The second half of the proof of Thm. 5.11 shows that P is represented in $\mathbf{\Omega}$ by a formula having the required form. ■

5.14. Remark

Since the formula in Cor. 5.13 must be in $\mathbf{\Phi}_n$, all the variables occurring in $\mathbf{r}$ or $\mathbf{t}$ must be among $\mathbf{v}_1, \mathbf{v}_2, \ldots, \mathbf{v}_{n+m}$.

5.15. Corollary

Every recursive relation is arithmetical.

PROOF

A recursive relation is r.e. by Thm. 9.3.6, hence it is arithmetical by Cor. 5.13. ■

5.16. Remark

Since every elementary relation is recursive (see Rem. 9.5.5(i)), it follows from Rem. 5.12(i) and Cor. 5.15 that the class of arithmetical relations is the smallest class that contains all recursive relations and is closed under the logical operations.

5.17. Reminder

In what follows we use the terms *function* and *graph* in the same sense as in Ch. 9: an n-ary function is an n-ary operation on N; and its graph is the $(n+1)$-ary relation P such that, for all $\mathbf{x} \in N^n$ and all $y \in N$,

$$P(\mathbf{x}, y) \Leftrightarrow f\mathbf{x} = y.$$

5.18. Definition

An *arithmetical function* is a function whose graph is an arithmetical relation.

5.19. Theorem

Every recursive function is arithmetical.

PROOF

If f is a recursive function then by Thm. 9.3.10 its graph is r.e., hence by Cor. 5.13 it is arithmetical. ■

5.20. Problem

Let P be a k-ary arithmetical relation and let $f_1, f_2, \ldots, f_k$ be n-ary arithmetical functions. Let the n-ary relation Q be defined, for all $\mathbf{x} \in N^n$, by the equivalence

$$Q\mathbf{x} \Leftrightarrow P(f_1\mathbf{x}, f_2\mathbf{x}, \ldots, f_k\mathbf{x}).$$

Prove that Q is arithmetical. (Argue as in the proof of Thm. 9.4.6.)

§6. Coding

6.1. Preview

In a natural language we can talk of many things: of shoes and ships and sealing wax, of cabbages and kings – and of that very language itself. Can the same thing be done in $\mathcal{L}$, under its standard interpretation? Can $\mathcal{L}$ be used to 'talk' of its own expressions, of their properties, of relations among them and of operations upon them? At first glance this seems absurd: under its standard interpretation $\mathcal{L}$ 'talks' of numbers, numerical properties, relations and operations. However, we can make this idea work by using the device of *coding*: to each symbol and expression of $\mathscr{L}$ we assign a *code-number* (a.k.a. *Gödel number*) and then we can refer to expressions obliquely, via their code-numbers. Because $\mathscr{L}$, under its standard interpretation, 'talks' of numbers, it can be construed as referring obliquely to its own expressions, via their code-numbers.

The particular method of coding is of little importance; the only essential condition is that coding and decoding (encryption and decryption) must be *algorithmic* operations, of the kind that a computer can be programmed to do. Thus, it should be possible to program a computer so that, whenever an $\mathscr{L}$-expression is fed into it, the computer, after a finite number of computation steps, will output the code-number of the expression. Likewise, it should be possible to program a computer so that, whenever a number is fed into it, the computer, after a finite number of computation steps, will output a signal indicating whether that number is the code-number of an $\mathscr{L}$-expression; and, if so, also output that expression. (Here we have used the term *computer* in the sense explained in §2 of Ch. 9. Note that, strictly speaking, computer inputs and outputs are not numbers and $\mathscr{L}$-expressions as such, but suitable *representations* of them in a notation that the computer can handle.)

The coding we shall introduce here is different from that used in B&M (p. 327f). It will employ the binary ('base-2') representation of numbers.

6.2. Definition

(i) To distinguish between the ordinary decimal and the binary notation we shall use italic (slanted) digits '*0*' and '*1*' for the latter. Thus $\mathit{0} = 0$, $\mathit{1} = 1$, $\mathit{10} = 2$, $\mathit{11} = 3$, $\mathit{100} = 4$, etc.

(ii) If $k \geqslant 1$ and $a_1, a_2, \ldots, a_k$ are any numbers, with $a_1 > 0$, we define their *binary concatenation*

$$a_1 \mathbin{\char`\^} a_2 \mathbin{\char`\^} \ldots \mathbin{\char`\^} a_k$$

to be the number whose binary representation is obtained by concatenating the binary representations of $a_1, a_2, \ldots, a_k$ in this order. Thus, for example,

$$3\char`\^0\char`\^6 = \mathit{11}\char`\^\mathit{0}\char`\^\mathit{110} = \mathit{110110} = 32 + 16 + 4 + 2 = 54.$$

6.3. Definition

(i) To each primitive symbol $\mathbf{p}$ of $\mathcal{L}$ we assign a *code-number* $\#\mathbf{p}$, as follows:

$$\#\mathbf{0} = 2 = \mathit{10},$$

$$\#\mathbf{s} = 4 = 2^2 = \mathit{100},$$

$$\#+ = 8 = 2^3 = \mathit{1{,}000},$$

$$\#\times = 16 = 2^4 = \mathit{10{,}000},$$

$$\#= \; = 32 = 2^5 = \mathit{100{,}000},$$

$$\#\neg = 64 = 2^6 = \mathit{1{,}000{,}000},$$

$$\#\rightarrow = 128 = 2^7 = \mathit{10{,}000{,}000},$$

$$\#\forall = 256 = 2^8 = \mathit{100{,}000{,}000},$$

$$\#\mathbf{v}_i = 2^{8+i} \text{ for } i = 1, 2, \ldots .$$

(ii) If $k \geqslant 1$ and $\mathbf{p}_1, \mathbf{p}_2, \ldots, \mathbf{p}_k$ are primitive symbols of $\mathcal{L}$ then we assign to the $\mathcal{L}$-string $\mathbf{p}_1\mathbf{p}_2 \ldots \mathbf{p}_k$ the *code-number*

$$\#(\mathbf{p}_1\mathbf{p}_2 \ldots \mathbf{p}_k) = \#\mathbf{p}_1{}^\frown\#\mathbf{p}_2{}^\frown\ldots{}^\frown\#\mathbf{p}_k.$$

6.4. Remarks

(i) It is easy to see that a number is the code-number of a string iff its binary representation consists of one or more blocks, each of which consists of a single '*1*' followed by one or more '*0*'s. For example, 0, 3 (= *11*) and 5 (= *101*) are not code-numbers of any string.

(ii) Since $\mathcal{L}$-expressions – terms and formulas – are in particular $\mathcal{L}$-strings, Def. 6.3 assigns a code-number $\#\mathbf{t}$ to each term $\mathbf{t}$ and a code-number $\#\boldsymbol{\alpha}$ to each formula $\boldsymbol{\alpha}$. Note that in computing the code-number of an expression, the symbols of the latter must be taken in the order in which they occur in the original 'Polish' notation of $\mathcal{L}$. For example, the (false) equation $\mathbf{s}_0 = \mathbf{s}_1$ is the string $=\mathbf{0s0}$. Hence its code-number is

$$\#(=\mathbf{0s0}) \;=\; 32{}^\frown 2{}^\frown 4{}^\frown 2 \;=\; \mathit{100{,}000}{}^\frown\mathit{10}{}^\frown\mathit{100}{}^\frown\mathit{10}$$

$$= \mathit{1{,}000{,}001{,}010{,}010} = 4{,}096 + 64 + 16 + 2 = 4{,}178.$$

6.5. Convention

When a noun or nominal phrase referring to $\mathcal{L}$-expressions appears in small capitals, it should be read with the words 'code-number of' or 'code-number of a' prefixed to it. Thus, for example, 'TERM' is short for 'code-number of a term'.

Many relations and functions connected with the syntax of $\mathcal{L}$ can easily be seen to be recursive.

6.6. Examples

(i) Consider the property Tm defined by

$$\mathrm{Tm}(x) \Leftrightarrow_{\mathrm{df}} x \text{ is a TERM.}$$

It is clear that a computer can be programmed to check whether any number x fed into it is a TERM or not. (According to standard practice, the computer will first represent x in binary notation. The results of Prob. 8.2.1 can then be used to 'parse' this binary representation and check whether x is a TERM.) Thus Tm is a recursive property.

(ii) The property Fla, defined by

$$\mathrm{Fla}(x) \Leftrightarrow_{\mathrm{df}} x \text{ is a FORMULA},$$

is similarly seen to be recursive.

(iii) Consider the relation Frm, defined by

$$\mathrm{Frm}(x, y) \Leftrightarrow_{\mathrm{df}} x \text{ is a FORMULA belonging to } \mathbf{\Phi}_y.$$

In other words, $\mathrm{Frm}(x, y)$ holds iff $x = \#\boldsymbol{\alpha}$ for some formula $\boldsymbol{\alpha}$ such that all the free variables of $\boldsymbol{\alpha}$ are among $\mathbf{v}_1, \mathbf{v}_2, \ldots, \mathbf{v}_y$. Frm is clearly recursive.

The following example introduces a recursive function that will play an important role in the sequel.

6.7. Example

The *diagonal function* is the unary function d defined as follows

$$d(x) =_{\mathrm{df}} \begin{cases} \#[\boldsymbol{\alpha}(\mathbf{s}_x)] & \text{if } x \text{ is a FORMULA } \boldsymbol{\alpha}, \\ x & \text{if } x \text{ is not a FORMULA.} \end{cases}$$

How can $d(x)$ be calculated? First, we check whether x is a FORMULA. If it isn't, there is nothing further to do: $d(x)$ is x itself.

Now suppose x is a FORMULA. We have to take that formula $\boldsymbol{\alpha}$ of which x is the code-number and substitute $\mathbf{s}_x$ in it for $\mathbf{v}_1$ (cf. Def. 4.4); and $d(x)$ is then the code-number of the resulting formula, $\boldsymbol{\alpha}(\mathbf{s}_x)$. This calculation is quite easy to do if x is represented in binary notation. Each occurrence of $\mathbf{v}_1$ appears in this representation as a block of the form '*1000000000*'. We have to locate all blocks of this form that correspond to *free* occurrences of $\mathbf{v}_1$ in $\boldsymbol{\alpha}$, and replace each of them by the binary representation of $\mathbf{s}_x$, which consists of x successive blocks of the form '*100*' (corresponding to x successive occurrences of $\mathbf{s}$) followed by a single block '*10*' (corresponding to $\mathbf{0}$). When these replacements are made, we have got the binary representation of $d(x)$.

Clearly, a computer can be programmed to perform this procedure. Thus we have:

6.8. *Theorem*

The function d is recursive. For any formula $\boldsymbol{\alpha}$,

$$d(\#\boldsymbol{\alpha}) = \#[\boldsymbol{\alpha}(\mathbf{s}_{\#\boldsymbol{\alpha}})].$$

PROOF

For the recursiveness claim, see above. The equality follows directly from the definition of d. ■

§7. Tarski's Theorem

7.1. *Preview*

We have seen that various relations connected with the syntax of $\mathscr{L}$ are recursive. By Cor. 5.15, these relations are representable in $\boldsymbol{\Omega}$; thus they are expressible in $\mathscr{L}$ under its standard interpretation.

For example, we have seen that the property Tm of being a TERM is recursive; hence it is arithmetical. So (cf. Rems. 5.3 and 5.7) there is a formula $\boldsymbol{\alpha} \in \boldsymbol{\Phi}_1$ such that, for any number x,

$$\mathrm{Tm}(x) \Leftrightarrow \mathfrak{N} \models \boldsymbol{\alpha}[x] \Leftrightarrow \mathfrak{N} \models \boldsymbol{\alpha}(\mathbf{s}_x).$$

In this sense the formula $\boldsymbol{\alpha}$ *expresses* the property of being a TERM and the sentence $\boldsymbol{\alpha}(\mathbf{s}_x)$ 'says' that x is a TERM. Thus $\mathscr{L}$, under its standard interpretation $\mathfrak{N}$, is able to discourse of various aspects of its own syntax, albeit obliquely, by referring to its own expressions via their code-numbers.

Can the standard semantics of $\mathscr{L}$ likewise be discussed in $\mathscr{L}$? We shall show that it cannot.

7.2. Definition

For any set $\boldsymbol{\Sigma}$ of sentences, the property T_{Σ} is defined by

$$T_{\Sigma}(x) \Leftrightarrow_{\text{df}} x \text{ is a SENTENCE belonging to } \boldsymbol{\Sigma}.$$

7.3. Remarks

(i) Equivalently, T_{Σ} is the set $\#[\boldsymbol{\Sigma}]$ of all SENTENCES of $\boldsymbol{\Sigma}$.

(ii) In particular, T_{Ω} is the property of being a SENTENCE of $\boldsymbol{\Omega}$. In other words, $T_{\Omega}(x)$ holds iff x is a TRUE SENTENCE (see Def. 1.9(ii)).

7.4. Theorem (Tarski, 1933)

T_{Ω} is not arithmetical.

PROOF

By Thm. 6.8, the diagonal function d is recursive; hence by Thm. 5.19 it is arithmetical. Now, let P be the property obtained by composing T_{Ω} with d and then applying $\neg$; thus

$$(*) \qquad Px \Leftrightarrow_{\text{df}} \neg T_{\Omega}(d(x)).$$

If T_{Ω} were arithmetical, then by Prob 5.20 and Thm. 5.11 it would follow that P is arithmetical as well. This would mean that there is some formula $\boldsymbol{\alpha} \in \boldsymbol{\Phi}_1$ such that, for any number x,

$$(**) \qquad Px \Leftrightarrow \boldsymbol{\alpha}(\mathbf{s}_x) \in \boldsymbol{\Omega}.$$

Taking x to be $\#\boldsymbol{\alpha}$, we would therefore have:

$$\begin{aligned} \boldsymbol{\alpha}(\mathbf{s}_{\#\alpha}) \in \boldsymbol{\Omega} &\Leftrightarrow P(\#\boldsymbol{\alpha}) && \text{by } (**), \\ &\Leftrightarrow \neg T_{\Omega}(d(\#\boldsymbol{\alpha})) && \text{by } (*), \\ &\Leftrightarrow \neg T_{\Omega}(\#[\boldsymbol{\alpha}(\mathbf{s}_{\#\alpha})]) && \text{by Thm. 6.8}, \\ &\Leftrightarrow \boldsymbol{\alpha}(\mathbf{s}_{\#\alpha}) \notin \boldsymbol{\Omega} && \text{by Def. 7.2.} \end{aligned}$$

This contradiction proves that T_{Ω} cannot be arithmetical. ■

7.5. Remarks

(i) Let us paraphrase the proof just given. If the property P were arithmetical then it would be expressed (that is, represented in

$\mathbf{\Omega}$) by some formula $\boldsymbol{\alpha} \in \mathbf{\Phi}_1$. For any number x, the sentence $\boldsymbol{\alpha}(\mathbf{s}_x)$ 'says' that Px holds. By ($*$), this is the same as 'saying' that $d(x)$ is not a TRUE SENTENCE.

Now, taking x to be the FORMULA $\boldsymbol{\alpha}$ itself, we find that the sentence $\boldsymbol{\alpha}(\mathbf{s}_{\#\alpha})$ 'says' that $d(\#\boldsymbol{\alpha})$ is not a TRUE SENTENCE. By Thm. 6.8, this means that $\#[\boldsymbol{\alpha}(\mathbf{s}_{\#\alpha})]$ is not a TRUE SENTENCE; in other words, that the sentence $\boldsymbol{\alpha}(\mathbf{s}_{\#\alpha})$ itself is untrue.

Thus, $\boldsymbol{\alpha}(\mathbf{s}_{\#\alpha})$ would be 'saying' something like 'I am false'!

Clearly, this is closely related to the well-known Liar Paradox. Except that *here* there is no paradox: the argument in the proof shows that a formula representing P in $\mathbf{\Omega}$ cannot exist; hence P – and therefore also T_{Ω} – cannot be arithmetical.

(ii) Tarski's Theorem applies not only to the language $\mathcal{L}$ and its standard interpretation; indeed, it was originally proved in a far wider context. The argument used here can be adapted to show, roughly speaking, that any sufficiently powerful formal language-cum-interpretation – powerful enough to express certain key concepts regarding its own syntax – cannot adequately express the most basic notions of its own semantics. Hence it cannot adequately serve as its own metalanguage.

The rest of this section contains an outline of a somewhat stronger version of Tarski's Theorem.

7.6. *Definition*

Let f be an n-ary function and let $\boldsymbol{\alpha} \in \mathbf{\Phi}_{n+1}$. We say that $\boldsymbol{\alpha}$ *represents* f *numeralwise* in a theory $\mathbf{\Sigma}$ if, for any $\mathfrak{a} \in N^n$, the sentence

$$(***) \qquad \forall \mathbf{v}_{n+1}[\boldsymbol{\alpha}(\mathbf{s}_{\mathfrak{a}}) \leftrightarrow \mathbf{v}_{n+1} = \mathbf{s}_{f\mathfrak{a}}]$$

belongs to $\mathbf{\Sigma}$.

7.7. *Problem*

Let $\boldsymbol{\alpha}$ represent the n-ary function f numeralwise in the theory $\mathbf{\Sigma}$. For any formula $\boldsymbol{\beta}$ in $\mathbf{\Phi}_1$, define $\boldsymbol{\beta}'$ as the formula

$$\exists \mathbf{v}_{n+1}[\boldsymbol{\beta}(\mathbf{v}_{n+1}) \wedge \boldsymbol{\alpha}].$$

Prove that, for any $\mathfrak{a} \in N^n$, the sentence $\boldsymbol{\beta}(\mathbf{s}_{f\mathfrak{a}}) \leftrightarrow \boldsymbol{\beta}'(\mathbf{s}_{\mathfrak{a}})$ belongs to $\mathbf{\Sigma}$. (It is enough to show that this sentence is deducible from ($***$) in Fopcal.)

7.8. Definition

A formula $\gamma \in \mathbf{\Phi}_1$ is called a *truth definition inside* a theory $\mathbf{\Sigma}$ if, for each sentence φ, $\mathbf{\Sigma}$ contains the sentence

$$\gamma(\mathbf{s}_{\#\varphi}) \leftrightarrow \varphi.$$

7.9. Problem

(i) Prove that if the diagonal function d is representable numeralwise in a consistent theory $\mathbf{\Sigma}$, then there cannot exist a truth definition inside $\mathbf{\Sigma}$. (Given any $\gamma \in \mathbf{\Phi}_1$, use Prob. 7.7 to find a formula $\boldsymbol{\delta} \in \mathbf{\Phi}_1$ such that for every number a the sentence $\neg\gamma(\mathbf{s}_{d(a)}) \leftrightarrow \boldsymbol{\delta}(\mathbf{s}_a)$ is in $\mathbf{\Sigma}$; then take φ as $\boldsymbol{\delta}(\mathbf{s}_{\#\boldsymbol{\delta}})$.)

(ii) Prove that d is representable numeralwise in $\mathbf{\Omega}$; hence deduce that there is no truth definition inside $\mathbf{\Omega}$. (Since d is arithmetical, there is a formula $\boldsymbol{\alpha} \in \mathbf{\Phi}_2$ that represents the graph of d in $\mathbf{\Omega}$. Show that the same $\boldsymbol{\alpha}$ also represents d numeralwise in $\mathbf{\Omega}$.)

(iii) Using (ii), give a new proof of Thm. 7.4. (Show that if $T_{\mathbf{\Omega}}$ were represented in $\mathbf{\Omega}$ by a formula γ, then γ would be a truth definition inside $\mathbf{\Omega}$.)

(iv) Prove that if $\mathbf{\Sigma}$ is a sound theory (see Def. 2.14) there is no truth definition inside it.

§8. Axiomatizability

Recall (Def. 2.7) that a *set of postulates* (a.k.a. *extralogical axioms*) for a theory $\mathbf{\Sigma}$ is a set of sentences $\mathbf{\Gamma}$ such that $\mathbf{\Sigma} = \mathbf{Dc}\mathbf{\Gamma}$. Having a set of postulates is no big deal: every theory $\mathbf{\Sigma}$ has one, because (by Def. 2.5) $\mathbf{\Sigma} = \mathbf{Dc}\mathbf{\Sigma}$. In order to qualify as an *axiomatic* theory, $\mathbf{\Sigma}$ must be presented by means of a postulate set $\mathbf{\Gamma}$ specified by a *finite recipe*. This does not mean that $\mathbf{\Gamma}$ itself must be finite. (Of course, if $\mathbf{\Gamma}$ *is* finite then so much the better, for then its sentences can be specified directly by means of a finite laundry list.) Rather, it means that we are provided with an algorithm – a finite set of instructions – whereby the sentences of $\mathbf{\Gamma}$ can be generated mechanically, one after the other. By Church's Thesis, this is equivalent to saying that $T_{\mathbf{\Gamma}}$ must be given as an r.e. property.

8.1. Conventions

(i) When we say that a set $\mathbf{\Gamma}$ of sentences is *recursive* (or *r.e.*), we mean that $T_{\mathbf{\Gamma}}$ is a recursive (or r.e.) property.

(ii) When we say that $\mathbf{\Gamma}$ is *given as a recursive* (or *r.e.*) set, we mean that it is given in such a way as to enable us to program a computer to operate as a decide-$T_{\mathbf{\Gamma}}$ (or enumerate-$T_{\mathbf{\Gamma}}$) machine. Similarly, when we say that we *can find a recursive* (or *r.e.*) set of sentences $\mathbf{\Gamma}$, we mean that we can describe $\mathbf{\Gamma}$ in such a way as to indicate how a computer can be programmed to operate as a decide-$T_{\mathbf{\Gamma}}$ (or enumerate-$T_{\mathbf{\Gamma}}$) machine.

8.2. Definition

(i) A theory $\mathbf{\Sigma}$ is *axiomatic* if it is presented by means of a set of postulates $\mathbf{\Gamma}$, which is given as an r.e. set.
(ii) A theory $\mathbf{\Sigma}$ is *axiomatizable* if there exists an r.e. set $\mathbf{\Gamma}$ of postulates for $\mathbf{\Sigma}$.

8.3. Remark

Note that being *axiomatic* is an *intensional* attribute: it is not a property of a theory as such, in a Platonic sense, but describes the way in which a theory is *presented*. On the other hand, *axiomatizability* is an *extensional* attribute of a theory as such, irrespective of how it is presented.

8.4. Theorem

If $\mathbf{\Sigma}$ *is an axiomatizable theory then there exists a recursive set of postulates for* $\mathbf{\Sigma}$.

PROOF

By assumption, $\mathbf{\Sigma} = \mathbf{Dc}\mathbf{\Gamma}$, where $\mathbf{\Gamma}$ is an r.e. set of sentences.

Without loss of generality we may assume that $\mathbf{\Gamma}$ is infinite. (Otherwise, we can add to $\mathbf{\Gamma}$ an infinite r.e. set of Fopcal axioms, for example: $\mathbf{\Theta} = \{\mathbf{s}_n = \mathbf{s}_n : n \in N\}$. The set $\mathbf{\Gamma} \cup \mathbf{\Theta}$ is clearly an infinite r.e. set of postulates for our theory $\mathbf{\Sigma}$.)

By assumption, there exists an enumerate-$T_{\mathbf{\Gamma}}$ machine. Let

$$\#\gamma_0, \#\gamma_1, \ldots, \#\gamma_n, \ldots$$

be the order in which it enumerates the SENTENCES of $\mathbf{\Gamma}$. We define sentences $\boldsymbol{\delta}_n$ by induction on n as follows:

$$\boldsymbol{\delta}_0 = \gamma_0, \qquad \boldsymbol{\delta}_{n+1} = \gamma_{n+1} \wedge \boldsymbol{\delta}_n \text{ for all } n.$$

Thus, $\boldsymbol{\delta}_n = \boldsymbol{\gamma}_n \wedge \boldsymbol{\gamma}_{n-1} \wedge \ldots \wedge \boldsymbol{\gamma}_0$ for all n. We put $\boldsymbol{\Delta} = \{\boldsymbol{\delta}_n : n \in N\}$.

It is easy to see that $\boldsymbol{\Delta}$ is a set of postulates for $\boldsymbol{\Sigma}$. Indeed, it is evident that for each n we have $\boldsymbol{\Gamma} \vdash_0 \boldsymbol{\delta}_n$ as well as $\boldsymbol{\Delta} \vdash_0 \boldsymbol{\gamma}_n$. Hence $\mathbf{Dc}\boldsymbol{\Delta} = \mathbf{Dc}\boldsymbol{\Gamma} = \boldsymbol{\Sigma}$.

Clearly, using the enumerate-$T_{\boldsymbol{\Gamma}}$ machine we can construct a machine that enumerates the SENTENCES of $\boldsymbol{\Delta}$,

$$(*) \qquad \#\boldsymbol{\delta}_0, \#\boldsymbol{\delta}_1, \ldots, \#\boldsymbol{\delta}_n, \ldots$$

in this order. (The output of the enumerate-$T_{\boldsymbol{\Gamma}}$ machine can be converted by a simple further computation to yield this enumeration.)

Note, moreover, that in the enumeration $(*)$ the SENTENCES of $\boldsymbol{\Delta}$ are produced in increasing order: it is easy to see that

$$\#\boldsymbol{\delta}_{n+1} = \#(\boldsymbol{\gamma}_{n+1} \wedge \boldsymbol{\delta}_n) > \#\boldsymbol{\delta}_n \text{ for all } n.$$

This enables us to construct a decide-$T_{\boldsymbol{\Delta}}$ machine, as follows. Given any number x, monitor the enumeration $(*)$ until a number greater than x turns up – which is bound to happen, sooner or later, because the numbers in $(*)$ keep increasing. Then $T_{\boldsymbol{\Delta}}(x)$ holds iff by this time x itself has turned up in the enumeration $(*)$.

This procedure is clearly mechanizable; hence $\boldsymbol{\Delta}$ is a recursive set of postulates for $\boldsymbol{\Sigma}$. ∎

8.5. Remark

The proof of Thm. 8.4 shows that if $\boldsymbol{\Sigma}$ is not merely axiomatizable but an axiomatic theory, then we can actually find a recursive set of postulates for it.

To proceed, we shall need to assign a code-number to each non-empty finite sequence of formulas.

8.6. Definition

For any formulas $\boldsymbol{\varphi}_1, \boldsymbol{\varphi}_2, \ldots, \boldsymbol{\varphi}_n$, where $n \geqslant 1$, we put

$$\#(\boldsymbol{\varphi}_1, \boldsymbol{\varphi}_2, \ldots, \boldsymbol{\varphi}_n) = \#\boldsymbol{\varphi}_1{}^\frown 1^\frown \#\boldsymbol{\varphi}_2{}^\frown 1^\frown \ldots {}^\frown 1^\frown \#\boldsymbol{\varphi}_n.$$

8.7. Remark

Thus, the binary representation of $\#(\boldsymbol{\varphi}_1, \boldsymbol{\varphi}_2, \ldots, \boldsymbol{\varphi}_n)$ is obtained by stringing together the binary representations of the code-numbers $\#\boldsymbol{\varphi}_1$,

$\#\varphi_2, \ldots, \#\varphi_n$, in this order, but inserting a digit '*1*' between each one and the next. These additional '*1*'s serve as separators (like commas) showing where the binary representation of the code-number of one formula ends, and the next one begins. These separators are easily detected: they are always the first of two successive occurrences of '*1*'. (The second '*1*' belongs to the binary representation of the next formula.)

8.8. Definition

For any set of sentences $\mathbf{\Gamma}$ we define a binary relation Ded_{Γ} by:

> $\mathrm{Ded}_{\Gamma}(x, y) \Leftrightarrow_{\mathrm{df}} x$ is a SENTENCE and y is a SEQUENCE-OF-FORMULAS that constitutes a deduction of that sentence from $\mathbf{\Gamma}$.

8.9. Lemma

If $\mathbf{\Gamma}$ *is a recursive set of sentences then the relation* Ded_{Γ} *is recursive.*

PROOF

It is easy to see that the property of being an AXIOM of Fopcal in $\mathcal{L}$ is recursive: from the description of the axioms (Ax. 8.9.1–Ax. 8.9.8) it is clear that a computer can be programmed to decide whether any given number is an AXIOM. By assumption, the property T_{Γ} is recursive as well.

In order to determine whether $\mathrm{Ded}_{\Gamma}(x, y)$ holds for a given x and y, the following checks must be made.

(1) It must be verified that y is the code-number of a finite sequence of formulas.
(2) If it is, this sequence must next be scanned to verify that it is a deduction from $\mathbf{\Gamma}$; that is, that each formula in it is an axiom, or a member of $\mathbf{\Gamma}$, or obtainable by *modus ponens* from two formulas that occur earlier in the sequence.
(3) If this turns out to be so, then finally the last formula of the sequence must be checked to verify that it is a sentence and that its code-number is x.

Clearly, a computer can be programmed to perform the checks in (1) and (3). Since the property of being an AXIOM and the property T_{Γ} are recursive, it follows that the checks required in (2) can likewise be done by a suitably programmed computer. This shows that the relation $\mathrm{Ded}_{\boldsymbol{\Gamma}}(x, y)$ is recursive. ■

8.10. Theorem

A theory is axiomatizable iff it is an r.e. set of sentences.

PROOF

If $\boldsymbol{\Sigma}$ is axiomatizable then by Thm. 8.4 there is a recursive set of sequences $\boldsymbol{\Gamma}$ such that $\boldsymbol{\Sigma} = \mathbf{Dc}\boldsymbol{\Gamma}$; that is, $\boldsymbol{\Sigma}$ is the set of sentences deducible in Fopcal from $\boldsymbol{\Gamma}$. Thus, for all x,

$$T_{\boldsymbol{\Sigma}}(x) \Leftrightarrow \exists y\, \mathrm{Ded}_{\boldsymbol{\Gamma}}(x, y).$$

By Lemma 8.9, $\mathrm{Ded}_{\boldsymbol{\Gamma}}$ is recursive, hence r.e. (by Thm. 9.3.6). Therefore (by Thm. 9.3.8) $T_{\boldsymbol{\Sigma}}$ is an r.e. property.

Conversely, if the theory $\boldsymbol{\Sigma}$ is r.e., then $\boldsymbol{\Sigma}$ has an r.e. set of postulates: $\boldsymbol{\Sigma}$ itself, because $\boldsymbol{\Sigma} = \mathbf{Dc}\boldsymbol{\Sigma}$. ■

8.11. Remarks

(i) The proof of Thm. 8.10 (including the proofs of Thm. 8.4 and Lemma 8.9) shows that if $\boldsymbol{\Sigma}$ is not merely axiomatizable but an axiomatic theory, then a program can actually be produced for making a computer operate as an enumerate-$T_{\boldsymbol{\Sigma}}$ machine. Hence $\boldsymbol{\Sigma}$ can be given as an r.e. set in the sense of Conv. 8.1(i).

(ii) The theorem means that a theory is axiomatizable iff there exists a finite presentation of it, by means of a program for generating one by one all the SENTENCES of the theory.

8.12. Theorem

$\boldsymbol{\Omega}$ is not axiomatizable.

PROOF

By Tarski's Thm. 7.4, $T_{\boldsymbol{\Omega}}$ is not arithmetical; hence by Cor. 5.13 it is not an r.e. property. ■

8.13. Theorem

If P is weakly representable in an axiomatizable theory then P is an r.e. relation.

PROOF

Let P be an n-ary relation and let $\boldsymbol{\alpha}$ be a formula in $\boldsymbol{\Phi}_n$ that represents P weakly in an axiomatizable theory $\boldsymbol{\Sigma}$. By Def. 4.7(i) we have, for all $x \in N^n$,

$$Px \Leftrightarrow \boldsymbol{\alpha}(\mathbf{s}_x) \in \boldsymbol{\Sigma}.$$

This means that, for all $x \in N^n$,

$$Px \Leftrightarrow T_{\boldsymbol{\Sigma}}(\#[\boldsymbol{\alpha}(\mathbf{s}_x)]).$$

The n-ary function f defined by the identity $fx = \#[\boldsymbol{\alpha}(\mathbf{s}_x)]$ is clearly recursive. (To compute fx the n numerals $\mathbf{s}_x$ must be substituted for the variables $\mathbf{v}_1, \mathbf{v}_2, \ldots, \mathbf{v}_n$ in $\boldsymbol{\alpha}$; the code-number of the resulting sentence is fx. This computation can evidently be performed by a suitably programmed computer.)

By Thm. 8.10 $T_{\boldsymbol{\Sigma}}$ is r.e.; therefore by Thm. 9.4.6(iii) P is r.e. as well. ■

8.14. Problem

Prove that if P is strongly representable in a consistent axiomatizable theory, then P is a recursive relation. (First show that if $\boldsymbol{\alpha}$ represents P strongly in a theory, then $\neg\boldsymbol{\alpha}$ represents $\neg P$ strongly in that theory.)

§9. Baby arithmetic

9.1. Preview

In this section we introduce a sound axiomatic theory $\boldsymbol{\Pi}_0$, which we call 'baby arithmetic' because it formalizes only a very rudimentary corpus of arithmetic facts: it 'knows' the true addition table and multiplication table for numerals, and of course everything that can be deduced from them logically – but nothing more. Despite its weakness, it is sufficient for a very simple weak representation of all r.e. relations.

$\boldsymbol{\Pi}_0$ is based on the following four postulate schemes:

9.2. Postulate scheme 1

$$\mathbf{s}_m + \mathbf{s}_0 = \mathbf{s}_m.$$

9.3. Postulate scheme 2

$$\mathbf{s}_m+\mathbf{s}_{n+1}=\mathbf{s}(\mathbf{s}_m+\mathbf{s}_n).$$

9.4. Postulate scheme 3

$$\mathbf{s}_m\times\mathbf{s}_0=\mathbf{s}_0.$$

9.5. Postulate scheme 4

$$\mathbf{s}_m\times\mathbf{s}_{n+1}=\mathbf{s}_m\times\mathbf{s}_n+\mathbf{s}_m.$$

Here m and n are any numbers.

9.6. Remark

Evidently, all these postulates are true; hence $\mathbf{\Pi}_0$ is sound. Also, this theory is axiomatic, as the set of postulates 9.2–9.5 is evidently recursive.

From the postulates of $\mathbf{\Pi}_0$ we can deduce in Fopcal formal versions of the addition and multiplication tables.

9.7. Example

Let us show that $\mathbf{s}_1+\mathbf{s}_1=\mathbf{s}_2 \in \mathbf{\Pi}_0$. First, note that the equation

(1) $$\mathbf{s}_1+\mathbf{s}_1=\mathbf{s}(\mathbf{s}_1+\mathbf{s}_0)$$

is an instance of Post. 2, and so belongs to $\mathbf{\Pi}_0$. Also, the equation

(2) $$\mathbf{s}_1+\mathbf{s}_0=\mathbf{s}_1$$

is an instance of Post. 1, and hence belongs to $\mathbf{\Pi}_0$. Using Ax. 6 of Fopcal, we deduce from (2) the equation $\mathbf{s}(\mathbf{s}_1+\mathbf{s}_0)=\mathbf{s}\mathbf{s}_1$, which (in view of Def. 1.6) is

(3) $$\mathbf{s}(\mathbf{s}_1+\mathbf{s}_0)=\mathbf{s}_2.$$

Finally, using Ax. 5 and Ax. 7 of Fopcal, we deduce from (1) and (3) the equation

$$\mathbf{s}_1+\mathbf{s}_1=\mathbf{s}_2,$$

which must therefore belong to $\mathbf{\Pi}_0$, as claimed.

9.8. Problem

Prove that $\mathbf{\Pi}_0$ contains the sentence:

(i) $\mathbf{s}_m\mathbf{+}\mathbf{s}_n\mathbf{=}\mathbf{s}_{m+n}$ (the formal addition table),
(ii) $\mathbf{s}_m\mathbf{\times}\mathbf{s}_n\mathbf{=}\mathbf{s}_{mn}$ (the formal multiplication table),

for all $m, n \in N$. (Use weak induction on n.)

9.9. Lemma

If $\mathbf{t}$ *is a closed term and* $\mathbf{t}^{\mathfrak{N}} = n$, *then* $\mathbf{t}\mathbf{=}\mathbf{s}_n \in \mathbf{\Pi}_0$.

PROOF

We proceed by induction on deg $\mathbf{t}$, considering the five cases mentioned in Rem. 1.4(iii). In each case it is enough to show that the equation $\mathbf{t}\mathbf{=}\mathbf{s}_n$ is deducible in Fopcal from sentences known to belong to $\mathbf{\Pi}_0$.

Case 1: $\mathbf{t}$ is a variable. Inapplicable here, as $\mathbf{t}$ is assumed closed.

Case 2: $\mathbf{t}$ is $\mathbf{0}$. Then $n = 0$ and $\mathbf{s}_n = \mathbf{s}_0 = \mathbf{0}$ by Def. 1.6. So the equation $\mathbf{t}\mathbf{=}\mathbf{s}_0$ is $\mathbf{0}\mathbf{=}\mathbf{0}$, which is an instance of Ax. 5 of Fopcal, and hence belongs to $\mathbf{\Pi}_0$.

Case 3: $\mathbf{t}$ is $\mathbf{sr}$, where $\mathbf{r}$ is a closed term. Let $\mathbf{r}^{\mathfrak{N}} = m$. Then $n = m + 1$. By the induction hypothesis, the equation $\mathbf{r}\mathbf{=}\mathbf{s}_m$ is in $\mathbf{\Pi}_0$. From this equation we deduce (using Ax. 6 of Fopcal) the equation $\mathbf{sr}\mathbf{=}\mathbf{ss}_m$, which is in fact $\mathbf{t}\mathbf{=}\mathbf{s}_n$.

Case 4: $\mathbf{t}$ is $\mathbf{q}\mathbf{+}\mathbf{r}$, where $\mathbf{q}$ and $\mathbf{r}$ are closed terms. Let $\mathbf{q}^{\mathfrak{N}} = k$ and $\mathbf{r}^{\mathfrak{N}} = m$. Then $n = k + m$. By the induction hypothesis, the sentences $\mathbf{q}\mathbf{=}\mathbf{s}_k$ and $\mathbf{r}\mathbf{=}\mathbf{s}_m$ are in $\mathbf{\Pi}_0$. From these two sentences we deduce (again using Ax. 6 of Fopcal) $\mathbf{q}\mathbf{+}\mathbf{r}\mathbf{=}\mathbf{s}_k\mathbf{+}\mathbf{s}_m$, which is in fact

$$\mathbf{t}\mathbf{=}\mathbf{s}_k\mathbf{+}\mathbf{s}_m.$$

By Prob. 9.8(i), the equation

$$\mathbf{s}_k\mathbf{+}\mathbf{s}_m\mathbf{=}\mathbf{s}_n$$

also belongs to $\mathbf{\Pi}_0$. From these two equations we deduce (using Ax. 5 and Ax. 7 of Fopcal) the equation $\mathbf{t}\mathbf{=}\mathbf{s}_n$.

Case 5: $\mathbf{t}$ is $\mathbf{q}\times\mathbf{r}$, where $\mathbf{q}$ and $\mathbf{r}$ are closed terms. This is similar to Case 4. ■

9.10. Definition

A formula (or sentence) of the form

$$\exists \mathbf{x}_1 \exists \mathbf{x}_2 \ldots \exists \mathbf{x}_m (\mathbf{r}=\mathbf{t}),$$

where $m \geqslant 0$, is called a *simple existential* formula (or sentence).

9.11. Lemma

$\mathbf{\Pi}_0$ *contains all true simple existential sentences.*

PROOF

Let $\boldsymbol{\varphi}$ be a true simple existential sentence. We proceed by induction on the number m of quantifiers in $\boldsymbol{\varphi}$.

First, let $m = 0$. Then $\boldsymbol{\varphi}$ is an equation $\mathbf{r}=\mathbf{t}$, where $\mathbf{r}$ and $\mathbf{t}$ are closed terms. Since $\boldsymbol{\varphi}$ is true, it follows that $\mathbf{r}^{\mathfrak{N}} = \mathbf{t}^{\mathfrak{N}}$; that is, $\mathbf{r}$ and $\mathbf{t}$ have the same numerical value. Let n be this common numerical value. Then by Lemma 9.9 the equations $\mathbf{r}=\mathbf{s}_n$ and $\mathbf{t}=\mathbf{s}_n$ belong to $\mathbf{\Pi}_0$. Using the equality axioms of Fopcal, we can deduce from these two equations the equation $\mathbf{r}=\mathbf{t}$, which must therefore belong to $\mathbf{\Pi}_0$ as well.

For the induction step, let $\boldsymbol{\varphi}$ have $m + 1$ quantifiers. Then $\boldsymbol{\varphi}$ has the form $\exists \mathbf{x} \boldsymbol{\psi}$, where $\boldsymbol{\psi}$ is a simple existential formula with m quantifiers, and with no free variable other than $\mathbf{x}$.

Since $\boldsymbol{\varphi}$ is true, it is easy to see (cf. Lemma 5.6) that $\boldsymbol{\psi}(\mathbf{x}/\mathbf{s}_n)$ must be true for some number n. But $\boldsymbol{\psi}(\mathbf{x}/\mathbf{s}_n)$ is a simple existential sentence with m quantifiers; hence, by the induction hypothesis, it belongs to $\mathbf{\Pi}_0$. By EG (Rem. 8.10.2(iv)), $\boldsymbol{\psi}(\mathbf{x}/\mathbf{s}_n) \vdash \exists \mathbf{x} \boldsymbol{\psi}$. Thus $\boldsymbol{\varphi}$ must be in $\mathbf{\Pi}_0$. ■

9.12. Theorem

For any given n-ary r.e. relation P, we can find a formula of the form

$$\exists \mathbf{v}_{n+1} \exists \mathbf{v}_{n+2} \ldots \exists \mathbf{v}_{n+m} (\mathbf{r}=\mathbf{t}),$$

that belongs to $\mathbf{\Phi}_n$ *and represents P weakly in every sound theory that includes* $\mathbf{\Pi}_0$.

PROOF

By Cor. 5.13, we can find a formula $\boldsymbol{\alpha}$ of this form that represents P in $\boldsymbol{\Omega}$. Thus, for every $\mathfrak{x} \in N^n$,

$$P\mathfrak{x} \Leftrightarrow \boldsymbol{\alpha}(\mathbf{s}_{\mathfrak{x}}) \in \boldsymbol{\Omega}.$$

But $\boldsymbol{\alpha}(\mathbf{s}_{\mathfrak{x}})$ is a simple existential sentence. Hence, if $\boldsymbol{\Sigma}$ is a theory such that $\boldsymbol{\Pi}_0 \subseteq \boldsymbol{\Sigma} \subseteq \boldsymbol{\Omega}$, it follows from Lemma 9.11 that

$$\boldsymbol{\alpha}(\mathbf{s}_{\mathfrak{x}}) \in \boldsymbol{\Omega} \Leftrightarrow \boldsymbol{\alpha}(\mathbf{s}_{\mathfrak{x}}) \in \boldsymbol{\Sigma}.$$

Hence, for every $\mathfrak{x} \in N^n$,

$$P\mathfrak{x} \Leftrightarrow \boldsymbol{\alpha}(\mathbf{s}_{\mathfrak{x}}) \in \boldsymbol{\Sigma}. \qquad \blacksquare$$

9.13. Remarks

(i) By Thm. 8.13, *only* r.e. relations can be weakly represented in an axiomatizable theory. We have just shown that *every* r.e. relation is in fact weakly representable in $\boldsymbol{\Pi}_0$. Thus $\boldsymbol{\Pi}_0$ achieves as much as is possible for any axiomatizable theory as regards weak representation.

(ii) As we shall see (Thm. 11.13), there are even weaker axiomatic theories in which every r.e. relation is weakly representable. However, the postulates of $\boldsymbol{\Pi}_0$ have been devised so as to make this theory just strong enough for Lemma 9.11 to hold; hence r.e. relations are weakly represented in $\boldsymbol{\Pi}_0$ by formulas of a particularly simple form.

9.14. Problem

Let $\mathfrak{U}$ be an $\mathscr{L}$-structure whose domain U is a singleton $\{u\}$.

(i) Show that all the sentences of $\boldsymbol{\Pi}_0$ are satisfied in $\mathfrak{U}$.

(ii) Show that the sentence $\mathbf{s}_0 \neq \mathbf{s}_1$ is not in $\boldsymbol{\Pi}_0$.

(iii) Show that if the n-ary relation P is strongly representable in $\boldsymbol{\Pi}_0$, then P is a trivial relation: $P\mathfrak{x}$ holds either for *all* $\mathfrak{x} \in N^n$ or for *none*. In other words, P is N^n or $\varnothing$. (First show that if $\boldsymbol{\alpha} \in \boldsymbol{\Phi}_n$ and $\mathfrak{U} \models \boldsymbol{\alpha}(\mathbf{s}_{\mathfrak{a}})$ for some $\mathfrak{a} \in N^n$, then $\mathfrak{U} \models \boldsymbol{\alpha}(\mathbf{s}_{\mathfrak{x}})$ for every $\mathfrak{x} \in N^n$.)

We return to our discussion of Thm. 9.12. Let $\boldsymbol{\alpha}$ be the formula

$$\exists \mathbf{v}_{n+1} \exists \mathbf{v}_{n+2} \ldots \exists \mathbf{v}_{n+m} \boldsymbol{\varphi},$$

where $\boldsymbol{\varphi}$ is any formula belonging to $\boldsymbol{\Phi}_{n+m}$; thus $\boldsymbol{\alpha} \in \boldsymbol{\Phi}_n$. Let P be the n-ary relation represented by $\boldsymbol{\alpha}$ in $\boldsymbol{\Omega}$ and let $\mathfrak{a} \in N^n$. Thus we have $P\mathfrak{a} \Leftrightarrow \mathfrak{N} \models \boldsymbol{\alpha}[\mathfrak{a}]$ (see Rem. 5.7). Now, due to the particular form of $\boldsymbol{\alpha}$, it is easy to see that

$$\mathfrak{N} \models \boldsymbol{\alpha}[\mathfrak{a}] \Leftrightarrow \text{there are numbers } b_1, b_2, \ldots, b_m \text{ such that } \mathfrak{N} \models \boldsymbol{\varphi}\,[\mathfrak{a}, b_1, b_2, \ldots, b_m].$$

Therefore

$$(9.15) \quad P\mathfrak{a} \Leftrightarrow \text{there are numbers } b_1, b_2, \ldots, b_m \text{ such that } \mathfrak{N} \models \boldsymbol{\varphi}[\mathfrak{a}, b_1, b_2, \ldots, b_m].$$

This justifies the following

9.16. Definition

Let $\boldsymbol{\varphi}$ be a formula belonging to $\boldsymbol{\Phi}_{n+m}$ and let $\boldsymbol{\alpha}$ be the formula

$$\exists \mathbf{v}_{n+1} \exists \mathbf{v}_{n+2} \ldots \exists \mathbf{v}_{n+m} \boldsymbol{\varphi}.$$

Let P be the n-ary relation represented by $\boldsymbol{\alpha}$ in $\boldsymbol{\Omega}$. Let $\mathfrak{a} \in N^n$. Then by *an* $\boldsymbol{\alpha}$*-witness that* $P\mathfrak{a}$ we mean any m-tuple of numbers $\langle b_1, b_2, \ldots, b_m \rangle$ such that

$$\mathfrak{N} \models \boldsymbol{\varphi}[\mathfrak{a}, b_1, b_2, \ldots, b_m].$$

9.17. Remarks

(i) Thus (9.15) means that – under the assumptions made in Def. 9.16 – $P\mathfrak{a}$ holds iff there exists an $\boldsymbol{\alpha}$-witness that it does. Moreover, the sentence $\boldsymbol{\alpha}(\mathbf{s}_\mathfrak{a})$ may be regarded as 'saying' that there exists an $\boldsymbol{\alpha}$-witness that $P\mathfrak{a}$. Indeed, it is clear that $\boldsymbol{\alpha}(\mathbf{s}_\mathfrak{a})$ is true – that is, $\mathfrak{N} \models \boldsymbol{\alpha}(\mathbf{s}_\mathfrak{a})$ – iff such a witness exists.

(ii) In the special situation covered by Thm. 9.12, P is an r.e. relation, $\boldsymbol{\alpha}$ is a simple existential formula of a particularly neat form and $\boldsymbol{\varphi}$ is an equation $\mathbf{r}{=}\mathbf{t}$. In this case an $\boldsymbol{\alpha}$-witness that $P\mathfrak{a}$ is an m-tuple $\langle b_1, b_2, \ldots, b_m \rangle$ such that

$$(*) \qquad \mathfrak{N} \models (\mathbf{r}{=}\mathbf{t})[\mathfrak{a}, b_1, b_2, \ldots, b_m].$$

What does it take to show that such a witness exists? We may search systematically through the set N^m of all m-tuples of numbers. For each m-tuple $\langle b_1, b_2, \ldots, b_m \rangle$, we can test whether it is a witness of the kind we are looking for. This involves performing a finite number of additions and multiplications, to see whether (∗) holds; in other words, whether the equation $\mathbf{r}=\mathbf{t}$ is satisfied by a valuation (based on $\mathfrak{N}$) that assigns the values $\mathfrak{a}$, b_1, b_2, ..., b_m to the variables $\mathbf{v}_1, \mathbf{v}_2, \ldots, \mathbf{v}_{n+m}$. Of course, if $P\mathfrak{a}$ does not hold, then we can never find a witness that it does. But if $P\mathfrak{a}$ does hold, then a witness exists, and in order to recognize one we only need to be able to do the following things:

1. Add and multiply, to calculate the values of terms **r** and **t** under a given assignment of numerical values to their variables.
2. If both terms have the same value, recognize that this is so.

Now, these operations are so simple, that even the very modest power of the theory $\mathbf{\Pi}_0$ is sufficient for performing them *formally*, *within* this theory. In other words, if the sentence $\boldsymbol{\alpha}(\mathbf{s}_{\mathfrak{a}})$ – which 'says' formally that a witness of the required kind exists – is true, then it can be deduced in Fopcal from Post. 9.2–9.5.

§10. Junior arithmetic

10.1. Preview

By adding to the postulates of baby arithmetic three schemes dealing with inequalities, we obtain a somewhat more powerful axiomatic theory, $\mathbf{\Pi}_1$ (a.k.a. *junior arithmetic*), in which all recursive relations are strongly represented by relatively simply formulas. This will follow from a major result, the Main Lemma, which will also play an important role later on.

10.2. Definition

For any terms **r** and **t**, we put

$$\mathbf{r}\leqslant\mathbf{t} =_{\mathrm{df}} \exists\mathbf{z}(\mathbf{r}+\mathbf{z}=\mathbf{t}),$$

where **z** is the first variable in alphabetic order that occurs neither in **r** nor in **t**.

10.3. Remark

This is yet another mnemonic pun: by Ex. 5.8(ii), the formula $\mathbf{v}_1\leqslant\mathbf{v}_2$ represents in $\mathbf{\Omega}$ the relation $\leqslant$.

As postulates for $\mathbf{\Pi}_1$ we take Post. 1–4 (9.2–9.5), as well as the following three schemes:

10.4. Postulate scheme 5

$$\mathbf{s}_m \neq \mathbf{s}_n,$$

10.5. Postulate scheme 6

$$\forall \mathbf{v}_1(\mathbf{v}_1 \leqslant \mathbf{s}_n \leftrightarrow \mathbf{v}_1 = \mathbf{s}_0 \vee \mathbf{v}_1 = \mathbf{s}_1 \vee \ldots \vee \mathbf{v}_1 = \mathbf{s}_n),$$

10.6. Postulate scheme 7

$$\forall \mathbf{v}_1(\mathbf{s}_n \leqslant \mathbf{v}_1 \vee \mathbf{v}_1 \leqslant \mathbf{s}_n),$$

where n is any number, and (in Post. 5) m is any number such that $m \neq n$.

10.7. Remarks

(i) Evidently, $\mathbf{\Pi}_1$ is a sound axiomatic theory.

(ii) $\mathbf{\Pi}_1$ is a *proper* extension of $\mathbf{\Pi}_0$ because, for example, no instance of Post. 5 belongs to $\mathbf{\Pi}_0$ (cf. Prob. 9.14(ii)).

10.8. Problem

Show that the results of Prob. 3.9(i), (ii) and (iii) hold with '$\mathbf{\Omega}$' replaced by '$\mathbf{\Pi}_1$'.

10.9. Problem

(i) Let $^*\mathfrak{N}$ be the $\mathscr{L}$-structure such that:

1. $^*N = N \cup \{\infty\}$, where ∞ is an object that is not a natural number;
2. $^*0 = 0$;
3. $^*s$ is the extension of the ordinary successor function such that $^*s(\infty) = 0$;
4. $^*+$ is the extension of ordinary addition such that if $a = \infty$ or $b = \infty$ then $a \,^*\!+ b = \infty$;
5. $^*\times$ is the extension of ordinary multiplication such that if $a = \infty$ or $b = \infty$ then $a \,^*\!\times b = 0$.

Show that $^*\mathfrak{N}$ is a model for $\mathbf{\Pi}_1$.

(ii) Prove that the sentence $\forall \mathbf{v}_1(\mathbf{s}\mathbf{v}_1 \neq \mathbf{s}_0)$ is not in $\mathbf{\Pi}_1$.

10.10. Definition

For any variable $\mathbf{x}$, term $\mathbf{r}$ and formula $\boldsymbol{\alpha}$ we put

(i) $\exists \mathbf{x} \leqslant \mathbf{r} \boldsymbol{\alpha} =_{\text{df}} \exists \mathbf{x}(\mathbf{x} \leqslant \mathbf{r} \wedge \boldsymbol{\alpha})$,
(ii) $\forall \mathbf{x} \leqslant \mathbf{r} \boldsymbol{\alpha} =_{\text{df}} \forall \mathbf{x}(\mathbf{x} \leqslant \mathbf{r} \rightarrow \boldsymbol{\alpha})$.

10.11. Preliminaries

Let two n-ary r.e. relations P and P' be given. By Cor. 5.13, we obtain two formulas

$$\boldsymbol{\alpha} = \exists \mathbf{v}_{n+1} \exists \mathbf{v}_{n+2} \ldots \exists \mathbf{v}_{n+m}(\mathbf{r}=\mathbf{t}),$$

$$\boldsymbol{\alpha}' = \exists \mathbf{v}_{n+1} \exists \mathbf{v}_{n+2} \ldots \exists \mathbf{v}_{n+m'}(\mathbf{r}'=\mathbf{t}'),$$

that belong to $\boldsymbol{\Phi}_n$ and represent P and P' respectively in $\boldsymbol{\Omega}$. Without loss of generality we may assume that $m' = m$. Indeed, if $m' < m$, then by Prob. 8.5.12 we may insert a string of $m - m'$ additional ('vacuous') quantifiers $\exists \mathbf{v}_{n+m'+1} \ldots \exists \mathbf{v}_{n+m}$ in $\boldsymbol{\alpha}'$ and obtain a formula that is logically equivalent to $\boldsymbol{\alpha}'$ and, like it, represents P' in $\boldsymbol{\Omega}$. Similarly, if $m < m'$, we can insert additional quantifiers into $\boldsymbol{\alpha}$. Therefore we shall assume

$$\boldsymbol{\alpha} = \exists \mathbf{v}_{n+1} \exists \mathbf{v}_{n+2} \ldots \exists \mathbf{v}_{n+m}(\mathbf{r}=\mathbf{t}),$$

$$\boldsymbol{\alpha}' = \exists \mathbf{v}_{n+1} \exists \mathbf{v}_{n+2} \ldots \exists \mathbf{v}_{n+m}(\mathbf{r}'=\mathbf{t}').$$

From these two formulas we construct two new ones:

$$\boldsymbol{\beta} = \exists \mathbf{v}_{n+1} \leqslant \mathbf{y} \exists \mathbf{v}_{n+2} \leqslant \mathbf{y} \ldots \exists \mathbf{v}_{n+m} \leqslant \mathbf{y}(\mathbf{r}=\mathbf{t}),$$

$$\boldsymbol{\beta}' = \exists \mathbf{v}_{n+1} \leqslant \mathbf{y} \exists \mathbf{v}_{n+2} \leqslant \mathbf{y} \ldots \exists \mathbf{v}_{n+m} \leqslant \mathbf{y}(\mathbf{r}'=\mathbf{t}'),$$

where $\mathbf{y}$ is $\mathbf{v}_{n+m+1}$. Finally, we construct a fifth formula:

$$\boldsymbol{\gamma} = \exists \mathbf{y}(\boldsymbol{\beta} \wedge \neg \boldsymbol{\beta}').$$

Note that the free variables of $\boldsymbol{\beta}$ and $\boldsymbol{\beta}'$ are among $\mathbf{v}_1, \mathbf{v}_2, \ldots, \mathbf{v}_n$ and $\mathbf{y}$, and therefore $\boldsymbol{\gamma}$ is in $\boldsymbol{\Phi}_n$.

10.12. Main Lemma

Given any two n-ary r.e. relations P and P', let $\boldsymbol{\gamma}$ be the formula constructed above. Then for every $\mathfrak{a} \in N^n$ we have

$$P\mathfrak{a} \wedge \neg P'\mathfrak{a} \Rightarrow \boldsymbol{\gamma}(\mathbf{s}_{\mathfrak{a}}) \in \boldsymbol{\Pi}_1,$$

$$\neg P\mathfrak{a} \wedge P'\mathfrak{a} \Rightarrow \neg \boldsymbol{\gamma}(\mathbf{s}_{\mathfrak{a}}) \in \boldsymbol{\Pi}_1.$$

PROOF

For the simple but somewhat lengthy proof, see B&M, pp. 337–340. (The Main Lemma appears there as Lemma 7.9, but its proof requires two earlier results, Lemmas 7.7 and 7.8.) ■

10.13. Analysis

Let $\boldsymbol{\varphi}$ and $\boldsymbol{\varphi}'$ be the equations $\mathbf{r}{=}\mathbf{t}$ and $\mathbf{r}'{=}\mathbf{t}'$ that occur in the formulas $\boldsymbol{\alpha}$ and $\boldsymbol{\alpha}'$ respectively.

We take up the discussion begun in Rem. 9.17. Recall that $P\mathfrak{a}$ holds iff there exists an $\boldsymbol{\alpha}$-witness that it does. By Def. 9.16, such a witness is an m-tuple $\langle b_1, b_2, \ldots, b_m \rangle$ for which

$$\mathfrak{N} \models \boldsymbol{\varphi}[\mathfrak{a}, b_1, b_2, \ldots, b_m].$$

Moreover, $\boldsymbol{\alpha}(\mathbf{s}_\mathfrak{a})$ 'says' that such a witness exists.

Now let us find out what is 'said' by a sentence obtained from $\boldsymbol{\beta}$ by substituting numerals for its free variables. It is easy to see that

$$\begin{aligned}\boldsymbol{\beta}(\mathbf{s}_\mathfrak{a}, \mathbf{y}/\mathbf{s}_b) \in \boldsymbol{\Omega} &\Leftrightarrow \mathfrak{N} \models \boldsymbol{\beta}[\mathfrak{a}, \mathbf{y}/b] \\ &\Leftrightarrow \text{there are } b_1, b_2, \ \ldots, b_m \leqslant b \text{ such that} \\ &\qquad \mathfrak{N} \models \boldsymbol{\varphi}[\mathfrak{a}, b_1, b_2, \ldots, b_m].\end{aligned}$$

Thus $\boldsymbol{\beta}(\mathbf{s}_\mathfrak{a}, \mathbf{y}/\mathbf{s}_b)$ 'says': *There is an* $\boldsymbol{\alpha}$*-witness that* $P\mathfrak{a}$*, and this witness is bounded by the number* b. In other words: *Among the numbers* $\leqslant b$ *there can be found an* $\boldsymbol{\alpha}$*-witness that* $P\mathfrak{a}$. Exactly the same analysis applies to P', $\boldsymbol{\alpha}'$ and $\boldsymbol{\beta}'$.

What does the sentence $\boldsymbol{\gamma}(\mathbf{s}_\mathfrak{a})$ 'say'? Recalling that $\boldsymbol{\gamma} = \exists\mathbf{y}(\boldsymbol{\beta} \wedge \neg\boldsymbol{\beta}')$, we see that

$$\begin{aligned}\boldsymbol{\gamma}(\mathbf{s}_\mathfrak{a}) \in \boldsymbol{\Omega} &\Leftrightarrow \mathfrak{N} \models \boldsymbol{\gamma}[\mathfrak{a}] \\ &\Leftrightarrow \text{there is a number } b \text{ such that} \\ &\qquad \mathfrak{N} \models \boldsymbol{\beta}[\mathfrak{a}, \mathbf{y}/b] \text{ but } \mathfrak{N} \not\models \boldsymbol{\beta}'[\mathfrak{a}, \mathbf{y}/b].\end{aligned}$$

Thus $\boldsymbol{\gamma}(\mathbf{s}_\mathfrak{a})$ is true iff for some number b there is an $\boldsymbol{\alpha}$-witness, bounded by b, that $P\mathfrak{a}$, but there is no $\boldsymbol{\alpha}'$-witness bounded by b that $P'\mathfrak{a}$. Putting this a bit less accurately but more suggestively, $\boldsymbol{\gamma}(\mathbf{s}_\mathfrak{a})$ 'says':

An $\boldsymbol{\alpha}$*-witness that* $P\mathfrak{a}$ *is found before an* $\boldsymbol{\alpha}'$*-witness that* $P'\mathfrak{a}$.

Or, even more simply:

$P\mathfrak{a}$ *is* $\boldsymbol{\alpha}$*-witnessed before* $P'\mathfrak{a}$ *is* $\boldsymbol{\alpha}'$*-witnessed.*

The whole of N^n can be divided into four mutually exclusive regions, as follows (see Fig. 5):

$$\begin{aligned} \text{Region} \quad \text{I} &= P \wedge \neg P', \\ \text{Region} \quad \text{II} &= \neg P \wedge P', \\ \text{Region} \quad \text{III} &= P \wedge P', \\ \text{Region} \quad \text{IV} &= \neg P \wedge \neg P'. \end{aligned}$$

Let us consider the truth value of $\boldsymbol{\gamma}(\mathbf{s}_{\mathfrak{a}})$ in each of these regions (that is, for $\mathfrak{a}$ belonging to each region).

For $\mathfrak{a}$ in Region I, $P\mathfrak{a}$ holds, and hence is $\boldsymbol{\alpha}$-witnessed by some m-tuple $\langle b_1, b_2, \ldots, b_m \rangle$. If we choose b large enough (say as the largest among these b_i) then $P\mathfrak{a}$ has an $\boldsymbol{\alpha}$-witness bounded by b. But in this region $P'\mathfrak{a}$ does not hold, hence has no $\boldsymbol{\alpha}'$-witness, let alone a witness bounded by our b. Thus $P\mathfrak{a}$ is $\boldsymbol{\alpha}$-witnessed *before* $P'\mathfrak{a}$ is $\boldsymbol{\alpha}'$-witnessed, simply because the former witness exists and the latter does not. So $\boldsymbol{\gamma}(\mathbf{s}_{\mathfrak{a}})$ is true throughout Region I.

In Region II, the position is reversed. Here $P'\mathfrak{a}$ holds, and is therefore $\boldsymbol{\alpha}'$-witnessed; but $P\mathfrak{a}$ is not $\boldsymbol{\alpha}$-witnessed at all, let alone *before* $P'\mathfrak{a}$ is $\boldsymbol{\alpha}'$-witnessed. Hence $\boldsymbol{\gamma}(\mathbf{s}_{\mathfrak{a}})$ is false throughout Region II.

In Region III, both $P\mathfrak{a}$ and $P'\mathfrak{a}$ hold, and are therefore witnessed, but for some $\mathfrak{a}$ in this region $P\mathfrak{a}$ may be $\boldsymbol{\alpha}$-witnessed before $P'\mathfrak{a}$ is $\boldsymbol{\alpha}'$-witnessed, while for other $\mathfrak{a}$ in the same region this may not be the case. So there is no general uniform answer for this region: $\boldsymbol{\gamma}(\mathbf{s}_{\mathfrak{a}})$ may be true for some $\mathfrak{a}$ and false for others.

In Region IV, neither $P\mathfrak{a}$ nor $P'\mathfrak{a}$ holds, and hence neither is witnessed. So, $P\mathfrak{a}$ is not $\boldsymbol{\alpha}$-witnessed at all, let alone *before* $P'\mathfrak{a}$ is $\boldsymbol{\alpha}'$-witnessed. Hence the sentence $\boldsymbol{\gamma}(\mathbf{s}_{\mathfrak{a}})$ is false in this region.

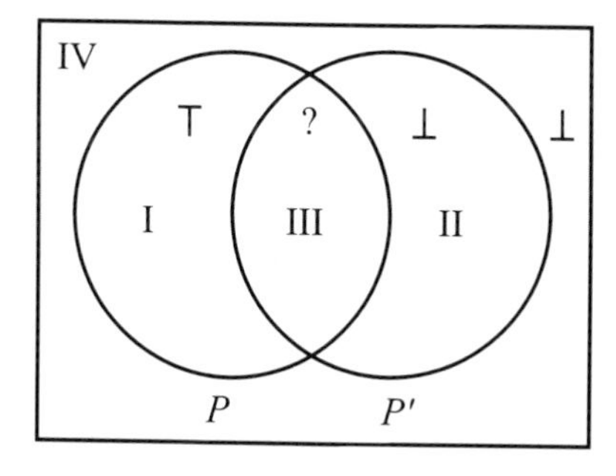

Fig. 5

Our Lemma says that for $\mathfrak{a}$ in Region I the sentence $\boldsymbol{\gamma}(\mathbf{s}_\mathfrak{a})$ is not only true, but even deducible from the postulates of $\mathbf{\Pi}_1$; and that for $\mathfrak{a}$ in Region II the sentence is not only false, but even refutable (that is, its negation is deducible) from these postulates.

The Lemma says nothing about the provability or refutability of $\boldsymbol{\gamma}(\mathbf{s}_\mathfrak{a})$ in the other two regions. As far as Region III is concerned, the reason is obvious: as we have seen, the sentence may not have a uniform truth value in this region, so we cannot expect any uniform result concerning its provability or refutability. But in Region IV the position is quite different, because our sentence is false throughout this region, just as in Region II. Why does the Lemma tell us nothing about this fourth region?

To understand the reason for this discrepancy, we must examine *what kind of evidence* is available for the truth or falsehood of $\boldsymbol{\gamma}(\mathbf{s}_\mathfrak{a})$ when $\mathfrak{a}$ is in Regions I, II, and IV.

In order to decide whether a given m-tuple $\langle b_1, b_2, \ldots, b_m \rangle$ of numbers is an $\boldsymbol{\alpha}$-witness that $P\mathfrak{a}$, we must be able to tell whether $\mathfrak{N} \models \boldsymbol{\varphi}[\mathfrak{a}, b_1, b_2, \ldots, b_m]$, where $\boldsymbol{\varphi}$ is the equation $\mathbf{r}=\mathbf{t}$.

As we saw in Rem. 9.17, if $\langle b_1, b_2, \ldots, b_m \rangle$ is indeed an $\boldsymbol{\alpha}$-witness that $P\mathfrak{a}$, then the operations required to recognize this fact can be performed formally *within* $\mathbf{\Pi}_0$, and *a fortiori* within $\mathbf{\Pi}_1$.

Now, if $\langle b_1, b_2, \ldots, b_m \rangle$ is *not* an $\boldsymbol{\alpha}$-witness that $P\mathfrak{a}$, then the operations required to recognize *this* fact involve not only adding and multiplying to compute the relevant values of $\mathbf{r}$ and $\mathbf{t}$, but also the ability to tell that these two values are *unequal*. Thanks to Post. 5, all this can be performed formally within $\mathbf{\Pi}_1$.

Thus, in $\mathbf{\Pi}_1$ it is possible to carry out formally all the operations required to tell whether or not any given m-tuple $\langle b_1, b_2, \ldots, b_m \rangle$ is an $\boldsymbol{\alpha}$-witness that $P\mathfrak{a}$.

In order to decide whether a given $\boldsymbol{\alpha}$-witness that $P\mathfrak{a}$ is bounded by a number b, we need to check whether each of the m components of the witness is $\leqslant b$.

Now, if $\mathfrak{a}$ is in Region I, then in order to verify that $\boldsymbol{\gamma}(\mathbf{s}_\mathfrak{a})$ is true we need only to check that a given m-tuple of numbers is an $\boldsymbol{\alpha}$-witness that $P\mathfrak{a}$, and is bounded by some given number b; and then to verify that each of the m-tuples bounded by b fails to be an $\boldsymbol{\alpha}'$-witness that $P'\mathfrak{a}$. Since there are only finitely many such m-tuples, all this requires a finite number of simple steps.

In order to obtain a *formal deduction* of $\boldsymbol{\gamma}(\mathbf{s}_\mathfrak{a})$, we need to *formalize* the process just described; and for this we need to have at our disposal a fairly modest set of postulates dealing formally with addition, multiplication and inequalities of both kinds (that is, $\neq$ and $\leqslant$). The postulates of $\mathbf{\Pi}_1$ are adequate for this.

In Region II the situation is broadly similar. If $\mathfrak{a}$ is in this region, then in order to verify that $\gamma(\mathbf{s}_\mathfrak{a})$ is false, we need to check that a given m-tuple is an $\boldsymbol{\alpha}'$-witness that $P'\mathfrak{a}$ and is bounded by a given number b; then we need to check, for each m-tuple bounded by b, that it *fails* to be an $\boldsymbol{\alpha}$-witness that $P\mathfrak{a}$. Finally, from these facts – namely, that $P'\mathfrak{a}$ has an $\boldsymbol{\alpha}$-witness bounded by b, but $P\mathfrak{a}$ has no such $\boldsymbol{\alpha}$-witness – we need to infer that $P\mathfrak{a}$ cannot be $\boldsymbol{\alpha}$-witnessed before $P'\mathfrak{a}$ is $\boldsymbol{\alpha}'$-witnessed. Again, all this amounts to a finite number of operations of additions and multiplications, together with some very elementary inferences about inequalities.

To obtain a *formal refutation* of $\gamma(\mathbf{s}_\mathfrak{a})$, we need to formalize this procedure. Again, the postulates of $\mathbf{\Pi}_1$ are adequate for this.

But in Region IV the situation is quite different. If $\mathfrak{a}$ is in this region, then in general there is no finite procedure of the kind described above (that is, consisting of additions, multiplications and simple inferences with inequalities) that would provide sufficient evidence that $\gamma(\mathbf{s}_\mathfrak{a})$ is false. Of course, the sentence is in fact false, but in general the only way to verify this would be to check that none of the infinitely many m-tuples of numbers is an $\boldsymbol{\alpha}$-witness that $P\mathfrak{a}$. This requires an infinite amount of calculation, and we cannot expect such an infinite procedure to be formalizable within an axiomatic theory such as $\mathbf{\Pi}_1$.

One final remark. There is nothing magical about the particular set of postulates of $\mathbf{\Pi}_1$. It is not these postulates that are of prime importance, but the Main Lemma. What we need is a sound axiomatic theory, preferably quite weak, for which the lemma can be proved. The theory $\mathbf{\Pi}_1$ was invented for the sake of the lemma. The postulates of the theory were selected by working back from the lemma, and discovering what postulates were needed to make the proof of the lemma work without too much difficulty. This is of course the kind of process described by Imre Lakatos in *Proofs and Refutations*.

10.14. Theorem

Given a recursive relation R, we can find a formula γ, of the form specified in Prel. 10.11, *that represents R strongly in any theory that includes $\mathbf{\Pi}_1$.*

PROOF

In the Main Lemma, take P and P' as R and $\neg R$, which are r.e. by Thm. 9.3.6. Then the lemma shows that $\boldsymbol{\gamma}$ represents R strongly in $\mathbf{\Pi}_1$, and hence also in any theory that includes $\mathbf{\Pi}_1$. ■

10.15. Problem

Let $\mathbf{\Sigma}$ be a theory that includes $\mathbf{\Pi}_1$. Show that every recursive function is representable numeralwise in $\mathbf{\Sigma}$. (If $\boldsymbol{\alpha}$ strongly represents the graph of the n-ary function f in $\mathbf{\Pi}_1$, prove that the formula

$$\forall \mathbf{y} \leqslant \mathbf{v}_{n+1}[\boldsymbol{\alpha}(\mathbf{v}_{n+1}/\mathbf{y}) \leftrightarrow \mathbf{y} = \mathbf{v}_{n+1}],$$

where $\mathbf{y}$ is $\mathbf{v}_{n+2}$, represents f numeralwise in $\mathbf{\Pi}_1$.)

Hence show that if $\mathbf{\Sigma}$ is consistent there cannot exist a truth definition inside it. (See Def. 7.8 and Prob. 7.9.)

10.16. Remark

The results of this section, particularly the Main Lemma, in a somewhat weaker form, are essentially due to Barkley Rosser.[1] The present stronger version is made possible by the MRDP Thm., which allows us to take $\boldsymbol{\alpha}$ and $\boldsymbol{\alpha}'$ as simple existential formulas.

§ 11. A finitely axiomatized theory

Whereas $\mathbf{\Pi}_0$ and $\mathbf{\Pi}_1$ were based on infinitely many postulates, our next theory, $\mathbf{\Pi}_2$, is based on the following nine.

11.1. Postulate I

$$\forall \mathbf{v}_1(\mathbf{s}\mathbf{v}_1 \neq \mathbf{s}_0).$$

11.2. Postulate II

$$\forall \mathbf{v}_1 \forall \mathbf{v}_2(\mathbf{s}\mathbf{v}_1 = \mathbf{s}\mathbf{v}_2 \rightarrow \mathbf{v}_1 = \mathbf{v}_2).$$

[1] His 1936 paper, 'Extensions of some theorems of Gödel and Church', is reprinted in M. Davis, *The Undecidable*.

11.3. Postulate III

$$\forall \mathbf{v}_1(\mathbf{v}_1 + \mathbf{s}0 = \mathbf{v}_1).$$

11.4. Postulate IV

$$\forall \mathbf{v}_1 \forall \mathbf{v}_2[\mathbf{v}_1 + \mathbf{s}\mathbf{v}_2 = \mathbf{s}(\mathbf{v}_1 + \mathbf{v}_2)].$$

11.5. Postulate V

$$\forall \mathbf{v}_1(\mathbf{v}_1 \times \mathbf{s}0 = \mathbf{s}0).$$

11.6. Postulate VI

$$\forall \mathbf{v}_1 \forall \mathbf{v}_2(\mathbf{v}_1 \times \mathbf{s}\mathbf{v}_2 = \mathbf{v}_1 \times \mathbf{v}_2 + \mathbf{v}_1).$$

11.7. Postulate VII

$$\forall \mathbf{v}_1(\mathbf{v}_1 \leqslant \mathbf{s}0 \rightarrow \mathbf{v}_1 = \mathbf{s}0).$$

11.8. Postulate VIII

$$\forall \mathbf{v}_1 \forall \mathbf{v}_2(\mathbf{v}_1 \leqslant \mathbf{s}\mathbf{v}_2 \rightarrow \mathbf{v}_1 \leqslant \mathbf{v}_2 \vee \mathbf{v}_1 = \mathbf{s}\mathbf{v}_2).$$

11.9. Postulate IX

$$\forall \mathbf{v}_1 \forall \mathbf{v}_2(\mathbf{v}_1 \leqslant \mathbf{v}_2 \vee \mathbf{v}_2 \leqslant \mathbf{v}_1).$$

11.10. Remarks

(i) The theory $\mathbf{\Pi}_2$ is clearly sound and axiomatic.

(ii) Instead of adopting these nine separate postulates, we could have taken their conjunction as a *single* postulate for $\mathbf{\Pi}_2$. Indeed, we shall make use of this option in the sequel. However, here we have preferred to present shorter separate postulates, for the sake of clarity.

(iii) $\mathbf{\Pi}_2$ is a modification of a finitely axiomatized theory proposed by Raphael Robinson in 1950.

11.11. Theorem

$\mathbf{\Pi}_1 \subseteq \mathbf{\Pi}_2$.

PROOF

It is quite easy to show that all the postulates of $\mathbf{\Pi}_1$ (Post. 1–7) can be deduced from Post. I–IX. (DIY, or see the details in B&M, pp. 341–342.) ■

11.12. Problem

(i) Let $^*\mathfrak{N}$ be the $\mathcal{L}$-structure such that:

1. $^*N = N \cup \{\infty\}$, where ∞ is an object that is not a natural number;
2. $^*0 = 0$;
3. $^*s$ is the extension of the ordinary successor function such that $^*s(\infty) = \infty$;
4. $^*+$ is the extension of ordinary addition such that if $a = \infty$ or $b = \infty$ then $a \, ^*{+} \, b = \infty$;
5. $^*\times$ is the extension of ordinary multiplication such that if $b \neq 0$ then $\infty \, ^*{\times} \, b = \infty$; $\infty \, ^*{\times} \, 0 = 0$; and $a \, ^*{\times} \, \infty = \infty$ for all a.

Show that $^*\mathfrak{N}$ is a model for $\mathbf{\Pi}_2$.

(ii) Prove that the sentence $\forall \mathbf{v}_1(\mathbf{s}\mathbf{v}_1 \neq \mathbf{v}_1)$ is not in $\mathbf{\Pi}_2$.

11.13. Theorem

(i) *Given an r.e. relation P, we can find a formula that represents P weakly in any sound theory.*

(ii) *Given a recursive relation R, we can find a formula that represents R weakly in any theory* $\mathbf{\Sigma}$ *such that* $\mathbf{\Sigma} \cup \mathbf{\Pi}_2$ *is consistent.*

PROOF

(i) Let P be a given n-ary r.e. relation. Take $\boldsymbol{\alpha}$ as the formula provided by Cor. 5.13 and Thm. 9.12. Let $\boldsymbol{\pi}$ be the conjunction of Posts. I–IX. We shall show that $\boldsymbol{\pi} \rightarrow \boldsymbol{\alpha}$ does the job.

$\mathbf{\Pi}_2$ is a sound theory, and by Thm. 11.11 it includes $\mathbf{\Pi}_1$, hence also $\mathbf{\Pi}_0$. Therefore by Thm. 9.12 $\boldsymbol{\alpha}$ represents P weakly in $\mathbf{\Pi}_2$.

Let $\mathfrak{a}$ be an n-tuple such that $P\mathfrak{a}$. Then $\boldsymbol{\alpha}(\mathbf{s}_\mathfrak{a}) \in \mathbf{\Pi}_2$. Since all the sentences of $\mathbf{\Pi}_2$ are deducible in Fopcal from $\boldsymbol{\pi}$, we have $\boldsymbol{\pi} \vdash \boldsymbol{\alpha}(\mathbf{s}_\mathfrak{a})$;

hence by DT $\vdash \boldsymbol{\pi} \rightarrow \boldsymbol{\alpha}(\mathbf{s}_\mathfrak{a})$. Thus the sentence $\boldsymbol{\pi} \rightarrow \boldsymbol{\alpha}(\mathbf{s}_\mathfrak{a})$ belongs to *every* theory, and in particular to every sound one.

Now let $\mathfrak{a}$ be such that $\neg P\mathfrak{a}$. Since $\boldsymbol{\alpha}$ represents P in $\boldsymbol{\Omega}$, we have $\boldsymbol{\alpha}(\mathbf{s}_\mathfrak{a}) \notin \boldsymbol{\Omega}$; in other words, $\boldsymbol{\alpha}(\mathbf{s}_\mathfrak{a})$ is false. But $\boldsymbol{\pi}$ is a true sentence, so $\boldsymbol{\pi} \rightarrow \boldsymbol{\alpha}(\mathbf{s}_\mathfrak{a})$ is false, and hence cannot belong to any sound theory. Thus we have shown that, for any sound theory $\boldsymbol{\Sigma}$ and any $\mathfrak{a} \in N^n$,

$$P\mathfrak{a} \Leftrightarrow \boldsymbol{\pi} \rightarrow \boldsymbol{\alpha}(\mathbf{s}_\mathfrak{a}) \in \boldsymbol{\Sigma}.$$

(ii) Let R be a given n-ary recursive relation. Take $\boldsymbol{\gamma}$ as the formula of Thm. 10.14. Then $\boldsymbol{\gamma}$ represents R strongly in $\boldsymbol{\Pi}_2$.

Let $\mathfrak{a}$ be an n-tuple such that $R\mathfrak{a}$. Then by an argument like the one used in the proof of (i) it follows that the sentence $\boldsymbol{\pi} \rightarrow \boldsymbol{\gamma}(\mathbf{s}_\mathfrak{a})$ belongs to every theory.

Now let $\mathfrak{a}$ be such that $\neg R\mathfrak{a}$. Then $\neg\boldsymbol{\gamma}(\mathbf{s}_\mathfrak{a}) \in \boldsymbol{\Pi}_2$, hence $\boldsymbol{\pi} \vdash \neg\boldsymbol{\gamma}(\mathbf{s}_\mathfrak{a})$. If $\boldsymbol{\Sigma}$ is a theory such that $\boldsymbol{\pi} \rightarrow \boldsymbol{\gamma}(\mathbf{s}_\mathfrak{a}) \in \boldsymbol{\Sigma}$, then from $\boldsymbol{\Sigma} \cup \{\boldsymbol{\pi}\}$ we can deduce both $\boldsymbol{\gamma}(\mathbf{s}_\mathfrak{a})$ and $\neg\boldsymbol{\gamma}(\mathbf{s}_\mathfrak{a})$, so $\boldsymbol{\Sigma} \cup \boldsymbol{\Pi}_2$ is inconsistent. In other words, if $\boldsymbol{\Sigma} \cup \boldsymbol{\Pi}_2$ is consistent then $\boldsymbol{\pi} \rightarrow \boldsymbol{\gamma}(\mathbf{s}_\mathfrak{a}) \notin \boldsymbol{\Sigma}$.

Thus we have shown that if $\boldsymbol{\Sigma}$ is a theory such that $\boldsymbol{\Sigma} \cup \boldsymbol{\Pi}_2$ is consistent then, for any $\mathfrak{a} \in N^n$,

$$R\mathfrak{a} \Leftrightarrow \boldsymbol{\pi} \rightarrow \boldsymbol{\gamma}(\mathbf{s}_\mathfrak{a}) \in \boldsymbol{\Sigma}. \qquad \blacksquare$$

§12. Undecidability

Let $\boldsymbol{\Sigma}$ be a set of sentences. The *decision problem for* $\boldsymbol{\Sigma}$ is the problem of finding an algorithm – a deterministic mechanical procedure – whereby, for any sentence $\boldsymbol{\varphi}$, it can be determined whether or not $\boldsymbol{\varphi} \in \boldsymbol{\Sigma}$. This is clearly equivalent to the problem of finding an algorithm whereby, for any number x, it can be determined whether or not $T_{\boldsymbol{\Sigma}}(x)$ holds (that is, whether or not x is a SENTENCE of $\boldsymbol{\Sigma}$). If such an algorithm is found, then this constitutes a *positive* solution to the decision problem for $\boldsymbol{\Sigma}$, and $\boldsymbol{\Sigma}$ is said to be *decidable*. If it is proved that such an algorithm cannot exist, this constitutes a *negative* solution to that decision problem, and $\boldsymbol{\Sigma}$ is said to be *undecidable*.

Note that if $\boldsymbol{\Sigma}$ is undecidable, it does *not* follow that there is some sentence for which it is impossible to decide whether or not it belongs to $\boldsymbol{\Sigma}$. Each such *individual* problem may well be solvable by some means or other. The undecidability of $\boldsymbol{\Sigma}$ only means that no algorithm will work for all sentences.

In order to make rigorous reasoning about decidability possible, this intuitive notion must be given a precise mathematical explication. Church's Thesis (a.k.a. the *Church–Turing Thesis*) states that such explication is provided by the notion of recursiveness. As mentioned in Rem. 9.3.11(ii), this thesis is supported by very weighty arguments, and has won virtually universal acceptance. Nevertheless, we shall keep our terminology free from commitment to Church's Thesis, by using the adverb 'recursively' where the thesis is needed to justify its omission.

12.1. Definition

If $\mathbf{\Sigma}$ is a set of sentences such that the property $T_{\mathbf{\Sigma}}$ is not recursive, we say that $\mathbf{\Sigma}$ is *recursively undecidable* and that the decision problem for $\mathbf{\Sigma}$ is *recursively unsolvable*.

From Tarski's Theorem 7.4 and Cor. 5.15 it follows at once that $\mathbf{\Omega}$ is recursively undecidable. This, as well as many other undecidability results, also follows from

12.2. Theorem

If $\mathbf{\Sigma}$ is a theory in which every recursive property is weakly representable, then $\mathbf{\Sigma}$ is recursively undecidable.

PROOF

Suppose $T_{\mathbf{\Sigma}}$ were recursive. Let the property P be defined by

(*) $$Px \Leftrightarrow_{\mathrm{df}} \neg T_{\mathbf{\Sigma}}(d(x)).$$

Since by Thm. 6.8 the function d is recursive, P would also be recursive by Thms. 9.4.6(ii) and 9.4.1. Therefore P would be weakly represented in $\mathbf{\Sigma}$ by some formula $\boldsymbol{\alpha} \in \mathbf{\Phi}_1$. Thus, for all $x \in N$,

(**) $$Px \Leftrightarrow \boldsymbol{\alpha}(\mathbf{s}_x) \in \mathbf{\Sigma}.$$

Taking x to be the number $\#\boldsymbol{\alpha}$, we get, exactly as in the proof of Thm. 7.4:

$$\begin{aligned} \boldsymbol{\alpha}(\mathbf{s}_{\#\alpha}) \in \mathbf{\Sigma} &\Leftrightarrow P(\#\boldsymbol{\alpha}) && \text{by (**)},\\ &\Leftrightarrow \neg T_{\mathbf{\Sigma}}(d(\#\boldsymbol{\alpha})) && \text{by (*)},\\ &\Leftrightarrow \neg T_{\mathbf{\Sigma}}(\#[\boldsymbol{\alpha}(\mathbf{s}_{\#\alpha})]) && \text{by Thm. 6.8},\\ &\Leftrightarrow \boldsymbol{\alpha}(\mathbf{s}_{\#\alpha}) \notin \mathbf{\Sigma} && \text{by Def. 7.2}. \end{aligned}$$

This contradiction proves that $T_{\mathbf{\Sigma}}$ cannot be recursive. ■

12.3. Corollary

Any sound theory is recursively undecidable.

PROOF

Immediate, by Thms. 9.3.6 and 11.13(i). ∎

12.4. Corollary

Any consistent theory in which every recursive property is strongly representable is recursively undecidable.

PROOF

Immediate, by Rem. 4.8(ii). ∎

12.5. Corollary

Any consistent theory that includes $\mathbf{\Pi}_1$ *is recursively undecidable.*

PROOF

Immediate, by Cor. 12.4 and Thm. 10.14. ∎

12.6. Corollary

If $\mathbf{\Sigma}$ *is a theory such that* $\mathbf{\Sigma} \cup \mathbf{\Pi}_2$ *is consistent, then* $\mathbf{\Sigma}$ *is recursively undecidable.*

PROOF

Immediate, by Thm. 11.13(ii). ∎

12.7. Corollary (Church's Theorem)

$\mathbf{\Lambda}$ *is recursively undecidable.*

PROOF

Immediate from Cor. 12.6, since $\mathbf{\Lambda} \cup \mathbf{\Pi}_2 = \mathbf{\Pi}_2$ is clearly consistent. ∎

12.8. Remarks

(i) The consistency of $\mathbf{\Pi}_2$ follows of course from its soundness; but it can also be proved by more elementary arguments, without invoking semantic notions.

(ii) If $\mathbf{\Sigma}$ is an *axiomatizable* theory that satisfies the condition of Thm. 12.2, then $T_{\mathbf{\Sigma}}$ is r.e. by Thm. 8.10, but not recursive. This applies, in particular, to $\mathbf{\Lambda}$, $\mathbf{\Pi}_0$, $\mathbf{\Pi}_1$ and $\mathbf{\Pi}_2$. These provide us with examples of r.e. properties that are not recursive.

12.9. Problem

Using Rem. 12.8(ii) and Prob. 8.14, obtain an alternative proof of Thm. 8.12, not using Tarski's Theorem.

12.10. Problem

Deduce Cor. 12.3 from Cor. 12.6.

12.11. Remarks

(i) Cor. 12.6 can be deduced from Cor. 12.5, as follows. Assume that $\mathbf{\Sigma}$ is a theory such that $\mathbf{\Sigma} \cup \mathbf{\Pi}_2$ is consistent.

In general, $\mathbf{\Sigma} \cup \mathbf{\Pi}_2$ is not a theory; but $\mathbf{\Delta} = \mathbf{Dc}(\mathbf{\Sigma} \cup \mathbf{\Pi}_2)$ is clearly a consistent theory that includes $\mathbf{\Pi}_2$, and hence also $\mathbf{\Pi}_1$. Therefore by Cor. 12.5 $\mathbf{\Delta}$ is recursively undecidable.

Let π be the conjunction of the nine postulates of $\mathbf{\Pi}_2$. Then, it is easy to show (DIY!) that, for any sentence φ,

$$\varphi \in \mathbf{\Delta} \Leftrightarrow \pi \rightarrow \varphi \in \mathbf{\Sigma}.$$

Recall that $\#(\pi \rightarrow \varphi) = 128 \hat{} \#\pi \hat{} \#\varphi$. Therefore, for all x,

$$T_{\mathbf{\Delta}}(x) \Leftrightarrow T_{\mathbf{\Sigma}}(fx), \qquad \text{where } fx = 128 \hat{} \#\pi \hat{} x.$$

Clearly, f is a recursive function. If $T_{\mathbf{\Sigma}}$ were recursive then $T_{\mathbf{\Delta}}$ would likewise be recursive, which is impossible because $\mathbf{\Delta}$ is recursively undecidable. Therefore $T_{\mathbf{\Sigma}}$ cannot be recursive, so $\mathbf{\Sigma}$ is recursively undecidable.

(ii) This illustrates the method of *reduction*. If $\mathbf{\Sigma}_1$ and $\mathbf{\Sigma}_2$ are theories such that for all x

$$T_{\mathbf{\Sigma}_1}(x) \Leftrightarrow T_{\mathbf{\Sigma}_2}(fx),$$

where f is a recursive function, then f is said to be a *reduction of* Σ_1 to Σ_2. If Σ_1 is known to be recursively undecidable, then it follows that $\mathbf{\Sigma}_2$ must also be recursively undecidable.

Starting from the results we have proved here, the method of reduction is used to obtain many other undecidability results, not only for theories in the present language $\mathscr{L}$, but in other languages as well. It turns out that almost every interesting mathematical theory is recursively undecidable. Which is just as well, for otherwise mathematicians could be made redundant and replaced by computers.

§ 13. First-order Peano arithmetic

The theory $\mathbf{\Pi}$, generally known as *first-order Peano arithmetic (FOPA)*, is based on the set of postulates comprising the first six postulates of $\mathbf{\Pi}_2$ and the following scheme:

13.1. Postulate scheme of induction

$$\forall \mathbf{v}_2 \forall \mathbf{v}_3 \ldots \forall \mathbf{v}_n [\boldsymbol{\alpha}(\mathbf{s}_0) \rightarrow \forall \mathbf{v}_1 \{\boldsymbol{\alpha} \rightarrow \boldsymbol{\alpha}(\mathbf{s}\mathbf{v}_1)\} \rightarrow \forall \mathbf{v}_1 \boldsymbol{\alpha}],$$

for every number $n \geqslant 1$ and any formula $\boldsymbol{\alpha} \in \mathbf{\Phi}_n$.

13.2. Remark

It is clear that $\mathbf{\Pi}$ is axiomatic. We shall soon see that it is also sound.

To explain the meaning of these new postulates, we need the following two definitions, the first of which extends the notation introduced in Def. 5.4 to arbitrary $\mathscr{L}$-structures.

13.3. Definition

(i) Let $\boldsymbol{\alpha} \in \mathbf{\Phi}_n$, let $^*\mathfrak{N}$ be an $\mathscr{L}$-structure and let $\mathfrak{a} = \langle a_1, a_2, \ldots, a_n \rangle$ be an n-tuple of individuals in the domain $^*N$. If $\boldsymbol{\alpha}$ is satisfied by some – and hence every – valuation σ based on $^*\mathfrak{N}$ such that $\mathbf{v}_i{}^\sigma = a_i$ for $i = 1, 2, \ldots, n$, we write:

$$`{}^*\mathfrak{N} \models \boldsymbol{\alpha}[\mathfrak{a}]\text{'}.$$

(ii) For any $\mathscr{L}$-structure $^*\mathfrak{N}$, any formula $\boldsymbol{\alpha} \in \mathbf{\Phi}_n$ (with $n \geqslant 1$) and any $a_2, a_3, \ldots, a_n \in {}^*N$, we put

$$M({}^*\mathfrak{N}, \boldsymbol{\alpha}; a_2, a_3, \ldots, a_n) =_{\text{df}} \{a_1 \in {}^*N : {}^*\mathfrak{N} \models \boldsymbol{\alpha}[\mathfrak{a}]\}.$$

(iii) The set $M({}^*\mathfrak{N}, \boldsymbol{\alpha}; a_2, a_3, \ldots, a_n)$ is said to be *defined in* $^*\mathfrak{N}$ *by* $\boldsymbol{\alpha}$, *with parameter values* $a_2, a_3, \ldots, a_n$. Sets of this form are said to be *parametrically definable in* $^*\mathfrak{N}$.

13.4. Definition

If $^*\mathfrak{N}$ is an $\mathcal{L}$-structure and X is any subset of $^*N$, we say that X is *inductive in* $^*\mathfrak{N}$ if it satisfies the condition:

If $^*0 \in X$, *and for every* $x \in X$ *also* $^*s(x) \in X$, *then* $X = {}^*N$.

13.5. Remarks

(i) A straightforward application of the BSD shows that

$$^*\mathfrak{N} \models \forall \mathbf{v}_2 \forall \mathbf{v}_3 \ldots \forall \mathbf{v}_n [\boldsymbol{\alpha}(\mathbf{s}_0) \rightarrow \forall \mathbf{v}_1 \{\boldsymbol{\alpha} \rightarrow \boldsymbol{\alpha}(\mathbf{s}\mathbf{v}_1)\} \rightarrow \forall \mathbf{v}_1 \boldsymbol{\alpha}]$$

is equivalent to the condition that for all $a_2, a_3, \ldots, a_n \in {}^*N$ the set $M(^*\mathfrak{N}, \boldsymbol{\alpha}; a_2, a_3, \ldots, a_n)$ is inductive in $^*\mathfrak{N}$.

Thus, all instances of the induction postulates 13.1 hold in $^*\mathfrak{N}$ iff all sets that are parametrically definable in $^*\mathfrak{N}$ are inductive in $^*\mathfrak{N}$.

(ii) The Principle of Induction says that *every* subset of N is inductive in $\mathfrak{N}$. It follows that all instances of 13.1 are true (that is, they hold in $\mathfrak{N}$) and hence $\mathbf{\Pi}$ is sound.

(iii) However, the present first-order induction scheme 13.1 falls far, far short of expressing (under the standard interpretation) the full power of the Principle of Induction. The latter states that *all* subsets of N are inductive (in $\mathfrak{N}$). It is a *second-order* principle, and was stated as such in Peano's 1889 axiomatization of arithmetic (cf. Rem. 6.1.8). Note that by Cantor's Thm. 3.6.8 there are *uncountably many* subsets of N.

On the other hand, our first-order induction postulates only manage to state (under the standard interpretation) the inductiveness of subsets of N that are parametrically definable in $\mathfrak{N}$ – that is, sets of the form $M(\mathfrak{N}, \boldsymbol{\alpha}; a_2, a_3, \ldots, a_n)$. However, it is easy to see (by an argument similar to that used in proving Thm. 6.3.9) that there are only denumerably many such subsets of N.

FOPA is in this sense merely a pale first-order shadow of the theory outlined by Peano.

(iv) Nevertheless, $\mathbf{\Pi}$ is an extremely strong theory. Although by Thm. 8.12 we know that $\mathbf{\Pi}$ must be a *proper* subtheory of $\mathbf{\Omega}$, and there must therefore exist true sentences that are not in $\mathbf{\Pi}$, it requires very great ingenuity to discover such sentences.

The first examples of true sentences that do not belong to $\mathbf{\Pi}$ were given by Gödel in 1931. (We shall present his results in the next two sections.) However, his sentences state interesting facts only when read obliquely, as referring to $\mathscr{L}$-expressions via their code-numbers; and these facts are then of purely logical (rather than general mathematical) interest.

It was only in 1977 that J. Paris and L. Harrington invented a method for producing true sentences that do not belong to $\mathbf{\Pi}$ and, when read *directly* rather than obliquely, express reasonably interesting mathematical facts, of the kind that can be of interest to an honest mathematician, not just to a logician.

13.6. Theorem

$\mathbf{\Pi}_2 \subseteq \mathbf{\Pi}$.

PROOF

It is enough to show that the last three postulates of $\mathbf{\Pi}_2$ (Post. VII–IX) belong to $\mathbf{\Pi}$. This is not difficult. (DIY or see B&M, p. 343f.) ■

13.7. Problem

Prove that $\forall \mathbf{v}_1(\mathbf{s}\mathbf{v}_1 \neq \mathbf{v}_1) \in \mathbf{\Pi}$. Hence by Prob. 11.12(ii) $\mathbf{\Pi}$ is a *proper* extension of $\mathbf{\Pi}_2$.

13.8. Remarks

(i) Let $^*\mathfrak{N}$ be a model of $\mathbf{\Pi}$. Then $^*\mathfrak{N}$ is, in particular, also a model of $\mathbf{\Pi}_1$; hence by Prob. 10.8 there is a unique embedding f of $\mathfrak{N}$ in $^*\mathfrak{N}$. Without loss of generality, we can assume that $^*\mathfrak{N}$ is actually an extension of $\mathfrak{N}$. This amounts to assuming that $N \subseteq {}^*N$ and that $fn = n$ for every number n. Thus by Def. 3.4 we have:

$$^*0 = 0, \quad ^*s(m) = m + 1, \quad m \mathbin{^*\!+} n = m + n, \quad m \mathbin{^*\!\times} n = mn,$$

for all numbers m and n.

(ii) For some structural information about nonstandard models of $\mathbf{\Pi}$, see B&M, p. 345 (Prob. 9.14 there). The same information applies, in particular, to nonstandard models of $\mathbf{\Omega}$.

13.9. Problem

Let $^*\mathfrak{N}$ be a nonstandard model of $\mathbf{\Pi}$. Without loss of generality, assume that $^*\mathfrak{N}$ is an extension of $\mathfrak{N}$.

(i) Show that N is not parametrically definable in $^*\mathfrak{N}$. (See Def. 13.3(iii).)

(ii) Hence prove, more generally, that no infinite subset of N is parametrically definable in $^*\mathfrak{N}$.

§ 14. The First Incompleteness Theorem

14.1. Preview

In an epoch-making paper published in 1931, Gödel presented two main results, known as the First and Second Incompleteness Theorems.[1]

Actually, the First Incompleteness Theorem came in two versions. One version, which applies to sound theories – and therefore depends on *semantic* notions – is explained in the introduction to the paper. Thanks to the MRDP Thm. 9.5.4, proved in 1970, it is now possible to obtain a somewhat stronger form of this semantic version: we shall prove it as Thm. 14.2 below.

In the main body of the paper, Gödel proves another version of the First Theorem, which does not depend on semantic notions. It applies to theories that are *ω-consistent*. (A theory $\mathbf{\Sigma}$ is *ω-inconsistent* if for some formula $\boldsymbol{\alpha} \in \mathbf{\Phi}_1$, it contains the sentences $\boldsymbol{\alpha}(\mathbf{s}_n)$ for all n as well as the sentence $\neg \forall \mathbf{v}_1 \boldsymbol{\alpha}$. The inconsistent theory is clearly ω-inconsistent, but the converse is not true.) In 1936 Rosser showed that this version of the First Theorem can be extended to theories that are just consistent, but not necessarily ω-consistent. His proof employed a result which is the prototype of our Main Lemma 10.12.

Using the MRDP Thm., the Gödel–Rosser Theorem can also be strengthened. This stronger form is proved below as Thm. 14.6.

The Second Incompleteness Theorem is stated by Gödel, but its proof is only briefly outlined. In the next section we shall give a mere outline of the proof.

By Thm. 8.12, $\mathbf{\Omega}$ is not axiomatizable. It follows at once that every sound axiomatizable theory $\mathbf{\Sigma}$ must be a *proper* sub-theory of $\mathbf{\Omega}$, and hence incomplete. Thus there must exist a true sentence that does not belong to $\mathbf{\Sigma}$. The following theorem shows that, given a sound axiomatic theory $\mathbf{\Sigma}$, we can find such a sentence, of a particularly simple form.

[1] A translation of his paper, 'On formally undecidable propositions of *Principia mathematica* and related systems I', is printed in van Heijenoort, *From Ferge to Gödel*.

14.2. Theorem (Semantic version of First Incompleteness Theorem)
Given a sound axiomatic theory $\mathbf{\Sigma}$, *we can find a true sentence* $\boldsymbol{\varphi}$ *of the form* $\forall \mathbf{x}_1 \forall \mathbf{x}_2 \ldots \forall \mathbf{x}_m (\mathbf{p} \neq \mathbf{q})$ *that does not belong to* $\mathbf{\Sigma}$.

PROOF

By Thm. 8.10 (cf. also Rem. 8.11(i)), we obtain $T_{\mathbf{\Sigma}}$ as an r.e. property. We put

$$Px \Leftrightarrow_{\text{df}} T_{\mathbf{\Sigma}}(d(x)).$$

Then by Thm. 9.4.6(iii) P is r.e. as well. Hence, by Cor. 5.13, we can find a formula $\boldsymbol{\alpha} \in \mathbf{\Phi}_1$ of the form $\exists \mathbf{v}_2 \exists \mathbf{v}_3 \ldots \exists \mathbf{v}_{m+1} (\mathbf{r}=\mathbf{t})$ that represents P in $\mathbf{\Omega}$. Let

$$\boldsymbol{\beta} =_{\text{df}} \forall \mathbf{v}_2 \forall \mathbf{v}_3 \ldots \forall \mathbf{v}_{m+1} (\mathbf{r} \neq \mathbf{t}).$$

Clearly, $\boldsymbol{\beta}$ is logically equivalent to $\neg \boldsymbol{\alpha}$, and represents $\neg P$ in $\mathbf{\Omega}$. Thus, for any number x,

$$\boldsymbol{\beta}(\mathbf{s}_x) \in \mathbf{\Omega} \Leftrightarrow \neg T_{\mathbf{\Sigma}}(d(x)).$$

Taking $x = \#\boldsymbol{\beta}$, we have:

$$\begin{aligned} \boldsymbol{\beta}(\mathbf{s}_{\#\boldsymbol{\beta}}) \in \mathbf{\Omega} &\Leftrightarrow \neg T_{\mathbf{\Sigma}}(d(\#\boldsymbol{\beta})) \\ &\Leftrightarrow \neg T_{\mathbf{\Sigma}}(\#[\boldsymbol{\beta}(\mathbf{s}_{\#\boldsymbol{\beta}})]) && \text{by Thm. 6.8,} \\ &\Leftrightarrow \boldsymbol{\beta}(\mathbf{s}_{\#\boldsymbol{\beta}}) \notin \mathbf{\Sigma} && \text{by Def. 7.2.} \end{aligned}$$

Let $\boldsymbol{\varphi}$ be $\boldsymbol{\beta}(\mathbf{s}_{\#\boldsymbol{\beta}})$. Then $\boldsymbol{\varphi}$ is indeed of the form $\forall \mathbf{x}_1 \forall \mathbf{x}_2 \ldots \forall \mathbf{x}_m (\mathbf{p} \neq \mathbf{q})$. (Here $\mathbf{x}_1, \mathbf{x}_2, \ldots, \mathbf{x}_m$ are $\mathbf{v}_2, \mathbf{v}_3, \ldots, \mathbf{v}_{m+1}$ respectively; and the terms $\mathbf{p}$ and $\mathbf{q}$ are obtained from $\mathbf{r}$ and $\mathbf{t}$ respectively by substituting the numeral $\mathbf{s}_{\#\boldsymbol{\beta}}$ for the variable $\mathbf{v}_1$.) Also, we have just shown that

$$\boldsymbol{\varphi} \in \mathbf{\Omega} \Leftrightarrow \boldsymbol{\varphi} \notin \mathbf{\Sigma}.$$

This means that either

(*) $$\boldsymbol{\varphi} \in \mathbf{\Omega} \text{ and } \boldsymbol{\varphi} \notin \mathbf{\Sigma},$$

or

(**) $$\boldsymbol{\varphi} \notin \mathbf{\Omega} \text{ and } \boldsymbol{\varphi} \in \mathbf{\Sigma}.$$

However, (**) is impossible by the soundness of $\mathbf{\Sigma}$; so (*) must be the case. ■

14.3. Remarks

(i) If $\mathbf{\Sigma}$, instead of being axiomatic, is assumed to be merely axiomatizable, then the proof shows that there *exists* a sentence $\boldsymbol{\varphi}$ with the properties stated in the theorem, without telling us how to obtain it.

(ii) In the proof of Thm. 14.2 we established not only that $\boldsymbol{\varphi} \notin \mathbf{\Sigma}$ but also that $\boldsymbol{\varphi} \in \mathbf{\Omega}$; hence $\neg\boldsymbol{\varphi} \notin \mathbf{\Omega}$. Since $\mathbf{\Sigma}$ is assumed to be sound, it follows that $\neg\boldsymbol{\varphi} \notin \mathbf{\Sigma}$ as well. Thus neither $\boldsymbol{\varphi}$ nor its negation is in $\mathbf{\Sigma}$, showing $\mathbf{\Sigma}$ to be incomplete. For this reason Thm. 14.2 is an incompleteness theorem.

(iii) Gödel says of $\boldsymbol{\varphi}$ that it is *[formally] undecidable in* $\mathbf{\Sigma}$. We prefer to say that $\boldsymbol{\varphi}$ is *undecided by* $\mathbf{\Sigma}$, so as to avoid confusion with the term *undecidable* explained in § 12.

14.4. Analysis

We know that $T_{\mathbf{\Sigma}}$ is the property of being a SENTENCE of $\mathbf{\Sigma}$. Moreover, tracing through the proof of Thm. 8.10, we see that – for an axiomatic theory $\mathbf{\Sigma}$ – $T_{\mathbf{\Sigma}}$ was obtained as an r.e. property by noting that, for any x,

$$T_{\mathbf{\Sigma}}(x) \Leftrightarrow x \text{ is a SENTENCE deducible from the postulates of } \mathbf{\Sigma}.$$

(The postulates referred to here are an r.e. set of postulates in terms of which $\mathbf{\Sigma}$ is presented.) Since $\boldsymbol{\beta}$ represents $\neg P$ in $\mathbf{\Omega}$, the sentence $\boldsymbol{\beta}(\mathbf{s}_x)$ can be taken to 'say' (under the standard interpretation): *$d(x)$ is not a* SENTENCE *deducible from the postulates of* $\mathbf{\Sigma}$.

In particular, when we take x to be $\#\boldsymbol{\beta}$, the sentence $\boldsymbol{\beta}(\mathbf{s}_x)$ is our $\boldsymbol{\varphi}$ and $d(x)$ is $\#\boldsymbol{\varphi}$. Thus $\boldsymbol{\varphi}$ 'says': *$\#\boldsymbol{\varphi}$ is not a* SENTENCE *deducible from the postulates of* $\mathbf{\Sigma}$. Or, briefly, $\boldsymbol{\varphi}$ 'says':

I am not deducible from the postulates of $\mathbf{\Sigma}$.

Compare this with the proof of Tarski's Theorem, analysed in Rem. 7.5(i). There we saw that if $T_{\mathbf{\Omega}}$ were arithmetical, there would exist a sentence that 'says' *I am untrue*. This would reproduce the Liar Paradox in $\mathcal{L}$. But in fact there was no paradox, since such a sentence cannot exist; and this only showed that $T_{\mathbf{\Omega}}$ is *not* arithmetical.

The Gödel sentence $\boldsymbol{\varphi}$ in the proof of Thm. 14.2. certainly *does* exist: we have in fact shown how to obtain it. Nor does it assert its own falsity; rather, it asserts its own *undeducibility* from the postulates of $\mathbf{\Sigma}$. Since $\mathbf{\Sigma}$ is sound, the postulates of $\mathbf{\Sigma}$ are all true. It follows that $\boldsymbol{\varphi}$

cannot lie; for if it lied, it *would* be deducible from these true postulates, and hence it would be true! Thus φ is true and *just because of this* it is undeducible from the postulates of $\mathbf{\Sigma}$. Or, if you like, it is true because it is undeducible from these postulates.

Here too there is no paradox: the Liar Paradox is merely *skirted*.

So far, we have subjected φ to the oblique version of the standard interpretation, the reading that takes φ to refer to expressions of $\mathcal{L}$ via their code-numbers. It transpires that the $\mathcal{L}$-expression to which it refers is φ itself. Read in this way, from a logical point of view, φ is a very interesting sentence.

Now let us read φ directly. Deformalizing φ (cf. Ex. 5.8) we see that under the standard interpretation it expresses a fact of the form

$$\forall x_1 \forall x_2 \ldots \forall x_m (f\mathfrak{x} \neq g\mathfrak{x}),$$

where f and g are n-ary polynomials in the sense of Def. 9.5.2(ii). An equation $f\mathfrak{x} = g\mathfrak{x}$, where f and g are two such polynomials, is called *diophantine*, after Diophantus, the third-century(?) author of a book on arithmetic. By a *solution* of the equation we mean an n-tuple $\mathfrak{a}$ of natural numbers such that $f\mathfrak{a} = g\mathfrak{a}$.

So φ asserts the unsolvability of the diophantine equation $f\mathfrak{x} = g\mathfrak{x}$, and the proof of Thm. 14.2 produces, for any given sound axiomatic theory $\mathbf{\Sigma}$, a particular diophantine equation that is really unsolvable, but whose unsolvability cannot be deduced from the postulates of $\mathbf{\Sigma}$.

However, from a mathematical (rather than purely logical) point of view, there is in general no reason why the equation $f\mathfrak{x} = g\mathfrak{x}$, or the fact that it is unsolvable, should be of any particular interest.

From now on we shall consider the issue of completeness with regard to axiomatizable theories that are consistent, but need not be sound.

14.5. Theorem

Every axiomatizable complete theory is recursively decidable.

PROOF

Let $\mathbf{\Sigma}$ be an axiomatizable complete theory. Then by Thm. 8.10 $T_{\mathbf{\Sigma}}$ is an r.e. property.

Also, if x is any number then, by the completeness of $\mathbf{\Sigma}$: $\neg T_{\mathbf{\Sigma}}(x)$ iff x is not a SENTENCE, or x is a SENTENCE whose negation belongs to $\mathbf{\Sigma}$. Thus

$$\neg T_{\mathbf{\Sigma}}(x) \Leftrightarrow \neg \operatorname{Frm}(x, 0) \vee T_{\mathbf{\Sigma}}(64 \frown x).$$

Here Frm is the recursive relation defined in Ex. 6.6(iii). Note that $\mathrm{Frm}(x, 0)$ holds iff x is a SENTENCE. Note also that by Def. 6.3 $T_{\Sigma}(64^\frown x)$ holds iff x is a SENTENCE whose negation belongs to $\mathbf{\Sigma}$.

Clearly, $64^\frown x$ is a recursive function of x. Also, by Thm. 9.3.6 $\neg$ Frm is r.e. since Frm is recursive. Hence by Thms. 9.4.3 and 9.4.6(iii) it follows that $\neg T_{\Sigma}$ is an r.e. property.

Therefore by Thm. 9.3.6 T_{Σ} is recursive. ■

By Cor. 12.5 it now follows that every consistent axiomatizable theory $\mathbf{\Sigma}$ that includes $\mathbf{\Pi}_1$ must be incomplete; so there must exist a sentence $\boldsymbol{\varphi}$ such that neither $\boldsymbol{\varphi} \in \mathbf{\Sigma}$ nor $\neg\boldsymbol{\varphi} \in \mathbf{\Sigma}$. The following theorem shows that, given a consistent axiomatic extension of $\mathbf{\Pi}_1$, we can find such a sentence whose form is relatively simple.

14.6. Theorem (Strengthened version of Gödel–Rosser First Incompleteness Theorem)

Given any axiomatic theory $\mathbf{\Sigma}$ that includes $\mathbf{\Pi}_1$, we can find a formula $\boldsymbol{\gamma} \in \mathbf{\Phi}_1$, of the form described in Prel. 10.11 *with $n = 1$, such that if either of the sentences $\boldsymbol{\gamma}(\mathbf{s}_{\#\gamma})$, $\neg\boldsymbol{\gamma}(\mathbf{s}_{\#\gamma})$ belongs to $\mathbf{\Sigma}$ then so does the other, and hence $\mathbf{\Sigma}$ is inconsistent.*

PROOF

As in the proof of Thm. 14.2, we obtain T_{Σ} as an r.e. property. We now put, for any number x,

$$Px \Leftrightarrow_{\mathrm{df}} T_{\Sigma}(64^\frown d(x)), \qquad P'x \Leftrightarrow_{\mathrm{df}} T_{\Sigma}(d(x)).$$

Clearly, P and P' are r.e. properties. So we can construct the formulas $\boldsymbol{\alpha}, \boldsymbol{\alpha}', \boldsymbol{\beta}, \boldsymbol{\beta}'$ and $\boldsymbol{\gamma}$ as described in Prel. 10.11, with $n = 1$.

Note that, by Def. 6.3 and Thm. 6.8, it follows from the definitions of P and P' that

$$P(\#\gamma) \Leftrightarrow \neg\boldsymbol{\gamma}(\mathbf{s}_{\#\gamma}) \in \mathbf{\Sigma}, \qquad P'(\#\gamma) \Leftrightarrow \boldsymbol{\gamma}(\mathbf{s}_{\#\gamma}) \in \mathbf{\Sigma}.$$

Now assume $\boldsymbol{\gamma}(\mathbf{s}_{\#\gamma}) \in \mathbf{\Sigma}$. Then $P'(\#\gamma)$. If it were the case that $\neg\boldsymbol{\gamma}(\mathbf{s}_{\#\gamma}) \notin \mathbf{\Sigma}$ then $\neg P(\#\gamma)$ would also hold; therefore we would have $\neg P(\#\gamma) \wedge P'(\#\gamma)$.

So by the Main Lemma 10.12 we would have $\neg\boldsymbol{\gamma}(\mathbf{s}_{\#\gamma}) \in \mathbf{\Pi}_1 \subseteq \mathbf{\Sigma}$. Thus $\neg\boldsymbol{\gamma}(\mathbf{s}_{\#\gamma}) \in \mathbf{\Sigma}$ after all, and hence $\mathbf{\Sigma}$ is inconsistent in this case.

Similarly, suppose that $\neg\boldsymbol{\gamma}(\mathbf{s}_{\#\gamma}) \in \mathbf{\Sigma}$. Then $P(\#\gamma)$ holds. If it were the case that $\boldsymbol{\gamma}(\mathbf{s}_{\#\gamma}) \notin \mathbf{\Sigma}$, then $\neg P'(\#\gamma)$ would also hold, and we would have $P(\#\gamma) \wedge \neg P'(\#\gamma)$.

So by the Main Lemma we would have $\gamma(\mathbf{s}_{\#\gamma}) \in \mathbf{\Pi}_1 \subseteq \mathbf{\Sigma}$. Thus $\gamma(\mathbf{s}_{\#\gamma}) \in \mathbf{\Sigma}$ after all, and $\mathbf{\Sigma}$ is inconsistent in this case as well. ■

14.7. Remark

If $\mathbf{\Sigma}$ is not assumed to be axiomatic but merely axiomatizable, then the proof shows that there *exists* a formula γ with the stated properties, without telling us how to obtain it.

14.8. Analysis

Consider the properties P and P' defined in the proof of Thm. 14.6. By definition, $P'x$ holds iff $d(x)$ is a SENTENCE belonging to $\mathbf{\Sigma}$, and Px holds iff $d(x)$ is a SENTENCE whose *negation* is in $\mathbf{\Sigma}$.

Thus, if $\mathbf{\Sigma}$ is consistent Px and $P'x$ are incompatible. Referring back to the definition of the four regions in Analysis 10.13, this means that, for a consistent $\mathbf{\Sigma}$, Region III is empty. (The two discs in Fig. 5 do not overlap.)

On the other hand, if $\mathbf{\Sigma}$ is the inconsistent theory, then Px and $P'x$ hold for exactly the same numbers x – namely, for any x such that $d(x)$ is a SENTENCE. Thus in this case Regions I and II are empty. (The two discs in Fig. 5 coincide.)

Also, from Analysis 10.13 we find that (under the standard interpretation) the Gödel–Rosser sentence $\gamma(\mathbf{s}_{\#\gamma})$ 'says':

> *An* $\boldsymbol{\alpha}$*-witness that* $P(\#\gamma)$ *is found before an* $\boldsymbol{\alpha}'$*-witness that* $P'(\#\gamma)$.

However, as we observed in the proof of Thm. 14.6, $P(\#\gamma)$ means, by definition, that the sentence $\neg\gamma(\mathbf{s}_{\#\gamma})$ is deducible from the given postulates of $\mathbf{\Sigma}$; or, in other words, that $\gamma(\mathbf{s}_{\#\gamma})$ itself is *refutable* from these postulates. Also, $P'(\#\gamma)$ means that $\gamma(\mathbf{s}_{\#\gamma})$ is deducible from the postulates of $\mathbf{\Sigma}$.

Thus $\gamma(\mathbf{s}_{\#\gamma})$ 'says':

(∗) *An* $\boldsymbol{\alpha}$*-witness that I am refutable from the postulates of* $\mathbf{\Sigma}$ *is found before an* $\boldsymbol{\alpha}'$*-witness that I am deducible from these postulates*.

The proof of Thm. 14.6 shows that $\#\gamma$ cannot belong to either of the Regions I and II. Let us see why this is so.

Suppose $\#\gamma$ were in Region I. Then, as we saw in Analysis 10.13, $\gamma(\mathbf{s}_{\#\gamma})$ must be true. Therefore (∗) is a true statement. This implies that $\neg\gamma(\mathbf{s}_{\#\gamma})$ is in $\mathbf{\Sigma}$. On the other hand, the Main Lemma tells us that if $\#\gamma$ were in Region I then $\gamma(\mathbf{s}_{\#\gamma})$ would be in $\mathbf{\Pi}_1$ and hence in $\mathbf{\Sigma}$, making $\mathbf{\Sigma}$ inconsistent – in which case Region I is empty! So $\#\gamma$ cannot be in Region I.

Now suppose $\#\gamma$ were in Region II. Then the Main Lemma tells us that $\gamma(\mathbf{s}_{\#\gamma})$ is refutable from the postulates of $\mathbf{\Pi}_1$, hence also from those of $\mathbf{\Sigma}$. Therefore there is an $\boldsymbol{\alpha}$-witness that $\gamma(\mathbf{s}_{\#\gamma})$ is refutable from the latter postulates. But since $\#\gamma$ is in Region II, we know from Analysis 10.13 that $\gamma(\mathbf{s}_{\#\gamma})$ is false, so (*) is a false statement. This implies that although an $\boldsymbol{\alpha}$-witness for the refutability of $\gamma(\mathbf{s}_{\#\gamma})$ in $\mathbf{\Sigma}$ can indeed be found, this does not happen *before* an $\boldsymbol{\alpha}'$-witness for the provability of $\gamma(\mathbf{s}_{\#\gamma})$ in $\mathbf{\Sigma}$ is *also* found. This means that $\gamma(\mathbf{s}_{\#\gamma})$ is both refutable and provable from the postulates of $\mathbf{\Sigma}$, again making $\mathbf{\Sigma}$ inconsistent, in which case Region II is empty. So $\#\gamma$ cannot be there either.

So $\#\gamma$ must be in Region III or in Region IV. The former happens if $\mathbf{\Sigma}$ is the inconsistent theory. In this case $\gamma(\mathbf{s}_{\#\gamma})$ may be true or false, depending on the precise form of $\boldsymbol{\alpha}$ and $\boldsymbol{\alpha}'$, and in particular on the (inconsistent) set of postulates by means of which $\mathbf{\Sigma}$ is given.

If $\mathbf{\Sigma}$ is a consistent theory, then Region III is empty, so $\#\gamma$ belongs to Region IV. From Analysis 10.13 we know that in this case $\gamma(\mathbf{s}_{\#\gamma})$ is a false sentence. This can also be seen from the proof of Thm. 14.6, which shows that if $\mathbf{\Sigma}$ is consistent then $\gamma(\mathbf{s}_{\#\gamma})$ is neither provable nor refutable from the postulates of $\mathbf{\Sigma}$. Therefore (*) is an untrue statement, and $\gamma(\mathbf{s}_{\#\gamma})$ is a false sentence.

§ 15. The Second Incompleteness Theorem

We take Thm. 14.6 as our point of departure. So let $\mathbf{\Sigma}$ be an axiomatic theory that includes $\mathbf{\Pi}_1$. We let P, P', $\boldsymbol{\alpha}$, $\boldsymbol{\alpha}'$, $\boldsymbol{\beta}$, $\boldsymbol{\beta}'$ and γ be as specified in the proof of that theorem.

Part of what the theorem establishes is that

(1) *If* $\mathbf{\Sigma}$ *is consistent then* $\neg\gamma(\mathbf{s}_{\#\gamma}) \notin \mathbf{\Sigma}$.

We now look for a *formalization* of (1); in other words, we wish to find an $\mathcal{L}$-sentence that, under the standard interpretation, 'states' (1). This is in fact quite easy.

First, the words '*if* ... *then*' are obviously formalized by the implication symbol $\rightarrow$.

Next, let us look at the clause '$\neg\gamma(\mathbf{s}_{\#\gamma}) \notin \mathbf{\Sigma}$'. It states that sentence

$\neg\boldsymbol{\gamma}(\mathbf{s}_{\#\gamma})$, whose code-number is $64^\frown d(\#\gamma)$, is not in $\boldsymbol{\Sigma}$. Referring to the definition of P in the proof of Thm. 14.6, we see that this amounts to saying that $\neg P(\#\gamma)$. But P is represented in $\boldsymbol{\Omega}$ by the formula $\boldsymbol{\alpha}$. Thus the statement that $\neg P(\#\gamma)$ is expressed formally by the sentence $\neg\boldsymbol{\alpha}(\mathbf{s}_{\#\gamma})$, which 'says': $P(\#\gamma)$ *does not hold*. As we have just seen, this means that $\neg\boldsymbol{\gamma}(\mathbf{s}_{\#\gamma}) \notin \boldsymbol{\Sigma}$.

Now let us look at the clause '$\boldsymbol{\Sigma}$ is consistent'. This is equivalent to saying that the sentence $\mathbf{0}\neq\mathbf{0}$ – the negation of the simplest logical axiom – is not in $\boldsymbol{\Sigma}$. An easy calculation, using Def. 6.3, shows that $\#(\mathbf{0}=\mathbf{0}) = 32^\frown 2^\frown 2 = 522$. Since $\mathbf{0}=\mathbf{0}$ is a sentence, substituting any term for $\mathbf{v}_1$ in it leaves it unchanged, so by Thm. 6.8 we get $d(522) = \#(\mathbf{0}=\mathbf{0}) = 522$. Therefore $\#(\mathbf{0}\neq\mathbf{0}) = 64^\frown d(522)$. So, by the definition of P, to say that $\mathbf{0}\neq\mathbf{0} \notin \boldsymbol{\Sigma}$ amounts to saying that $\neg P(522)$. This statement is expressed formally by the sentence $\neg\boldsymbol{\alpha}(\mathbf{s}_{522})$, which 'says': $P(522)$ *does not hold*. As this amounts to saying that $\boldsymbol{\Sigma}$ is consistent, we put

$$\textbf{Consis}_{\Sigma} =_{\text{df}} \neg\boldsymbol{\alpha}(\mathbf{s}_{522}).$$

We have now got an $\mathscr{L}$-sentence that expresses (1) formally; it is

(2) $$\textbf{Consis}_{\Sigma} \to \neg\boldsymbol{\alpha}(\mathbf{s}_{\#\gamma}).$$

Moreover, since (1) is a true statement – we have *proved* it! – it follows that (2) is a true sentence; in other words, it belongs to $\boldsymbol{\Omega}$.

In fact, (2) belongs not only to $\boldsymbol{\Omega}$ but even to FOPA. This can be proved by examining the whole chain of (informal) reasoning that was used to establish (1), and showing that it can be formalized: reproduced step by step as a formal deduction in Fopcal from the postulates of FOPA.

This process is rather tedious, as the chain of reasoning that established (1) was very long: it includes the proofs of Thm. 14.6 itself as well as of the theorems on which it depended. But each step is quite easy. What makes the whole thing possible is the great strength of the postulates of FOPA. We shall not present the proof here, but ask you to accept the fact that

(3) $$\textbf{Consis}_{\Sigma} \to \neg\boldsymbol{\alpha}(\mathbf{s}_{\#\gamma}) \in \boldsymbol{\Pi}.$$

Referring to Prel. 10.11 (with $n = 1$), it is easy to see that for any number k we have both $\boldsymbol{\gamma}(\mathbf{s}_k) \vdash \exists\mathbf{y}\boldsymbol{\beta}(\mathbf{s}_k)$ and $\exists\mathbf{y}\boldsymbol{\beta}(\mathbf{s}_k) \vdash \boldsymbol{\alpha}(\mathbf{s}_k)$. Hence $\vdash\neg\boldsymbol{\alpha}(\mathbf{s}_k)\to\neg\boldsymbol{\gamma}(\mathbf{s}_k)$. Using this fact for $k = \#\gamma$, it follows from (3) that

(4) $$\textbf{Consis}_{\Sigma} \to \neg\boldsymbol{\gamma}(\mathbf{s}_{\#\gamma}) \in \boldsymbol{\Pi}.$$

So far, we have assumed $\mathbf{\Sigma}$ to be an axiomatic theory that includes $\mathbf{\Pi}_1$. Now let $\mathbf{\Sigma}$ be an axiomatic theory that includes $\mathbf{\Pi}$; then it certainly includes $\mathbf{\Pi}_1$, so (4) holds. Moreover, since $\mathbf{\Pi} \subseteq \mathbf{\Sigma}$, we have

$$\text{(5)} \qquad \mathsf{Consis}_{\Sigma} \to \neg\gamma(\mathbf{s}_{\#\gamma}) \in \mathbf{\Sigma}.$$

15.1. Theorem (Second Incompleteness Theorem)

Let $\mathbf{\Sigma}$ be an axiomatic theory that includes FOPA. If $\mathbf{\Sigma}$ is consistent, then the sentence Consis_{Σ}, *which expresses this fact formally, is not in* $\mathbf{\Sigma}$.

PROOF

If $\mathsf{Consis}_{\Sigma} \in \mathbf{\Sigma}$ then by (5) also $\neg\gamma(\mathbf{s}_{\#\gamma}) \in \mathbf{\Sigma}$. But then by Thm. 14.6 it follows that $\mathbf{\Sigma}$ is inconsistent. ■

15.2. Remarks

(i) The Second Incompleteness Theorem can be extended to all sufficiently strong formal theories, in $\mathcal{L}$ and other languages. All that is required is that the theory in question is axiomatic, and includes an appropriate 'translation' of $\mathbf{\Pi}$. For example, this result applies to all the usual formalizations of set theory, such as ZF.

(ii) The result means that the consistency of any sufficiently strong consistent axiomatic theory cannot be proved by means of arguments that are wholly formalizable within that theory.

(iii) This poses a grave difficulty for the formalist view of mathematics. For a brief discussion of this, see B&M, p. 358f.

(iv) In particular, if ZF is consistent, a proof of this fact cannot be carried out within ZF itself. For this reason, it is extremely unlikely that an intuitively convincing consistency proof for ZF can ever be found.

Gödel's two Incompleteness Theorems have had a profound and far-reaching effect on the subsequent development of logic and philosophy, particularly the philosophy of mathematics.

Appendix: Skolem's Paradox

§ 1. Set-theoretic reductionism

Zermelo's 1908 paper,[1] in which he proposed his axioms for set theory, begins with the words:

> 'Set theory is that branch of mathematics whose task is to investigate mathematically the fundamental notions "number", "order", and "function", taking them in their pristine, simple form, and to develop thereby the logical foundation of all arithmetic and analysis; thus it constitutes an indispensable component of the science of mathematics.'

This comes close to saying—but does not quite say—that set theory is the sole foundation of the whole of mathematics. But soon such radical claims were voiced. In 1910 Hermann Weyl[2] put forward the view that the whole of mathematics ought to be reduced to axiomatic set theory. Each notion in the other branches of mathematics must be defined explicitly in terms of previously defined notions. This regress stops with set theory; ultimately all mathematical notions are to be defined in set-theoretic terms.

> 'So set theory appears to us today, in logical respects, as the proper foundation of mathematical science, and we will have to make a halt with set theory if we wish to formulate principles of definition which are not only sufficient for elementary geometry, but also for *the whole of mathematics*.'

The basic set-theoretic notions (*set* and *membership*) cannot be defined explicitly, for this would lead to infinite regress. They – alone of all mathematical notions – have to be characterized implicitly by means

[1] Cited in § 2 of Ch. 1.

[2] The paper, 'Über die Definitionen der mathematischen Grundbegriffe' is reprinted in his *Gesammelte Abhandlungen* (1968). In this paper Weyl outlines a characterization of the notion *definite property*, which he was to make more precise eight years later in *Das Kontinuum* (cited in § 2 of Ch. 1). The lines quoted here were translated by Michael Hallett.

of an axiom system. Thus axiomatic set theory (more or less along the lines proposed by Zermelo) becomes the ultimate framework for the whole of mathematics.

Although Weyl was to change his mind, the reductionist view he had expressed in 1910 was rapidly becoming very widespread among mathematicians.

It was this reductionism that Skolem set out to criticize in 1922. His short paper[1] – text of an address delivered at a congress of Scandinavian mathematicians – contains a lucid presentation of an astonishing wealth of logical and set-theoretic ideas and insights.[2] But in Skolem's own view the most important result in his paper is what came to be known as *Skolem's Paradox*. It is the first of the fundamental limitative results in logic. In a Concluding Remark he comments on it:

> 'I had already communicated it orally to F. Bernstein in Göttingen in the winter of 1915–16. There are two reasons why I have not published anything about it until now: first, I have in the meantime been occupied with other problems; second, I believed that it was so clear that axiomatization in terms of sets was not a satisfactory ultimate foundation of mathematics that mathematicians would, for the most part, not be very much concerned with it. But in recent times I have seen to my surprise that so many mathematicians think that these axioms of set theory provide the ideal foundation for mathematics; therefore it seemed to me that the time has come to publish a critique.'

§ 2. Hugh's world

In what follows we shall deal with ZF set theory; and for the sake of simplicity we shall exclude individuals, so that all objects are assumed to be sets. But a similar treatment, with very few minor modifications, can be applied to the other axiomatizations of set theory, with or without individuals.

As mentioned in § 2 of Ch. 1, in order to make axiomatic set theory conform to the highest standard of rigour and to bar the linguistic as well as the logical antinomies, the theory must be *formalized*.

We shall assume that ZF is formalized in a first-order langauge $\mathscr{L}$ with equality, whose only extralogical symbol is a binary predicate

[1] Cited in § 2 of Ch. 1.

[2] Including the conjectures that it would 'no doubt be very difficult' to prove the consistency of Zermelo's axioms; and that the Continuum Hypothesis is 'quite probably' undecided by them. These conjectures have indeed been vindicated: the former in 1931 by Gödel's Second Incompleteness Theorem (see § 15 of Ch. 10); and the latter in 1963 by P. J. Cohen's result (cf. Rem. 6.2.14).

symbol **ε**. In the intended interpretation of $\mathscr{L}$, the variables range over all sets and **ε** is interpreted as denoting the relation $\in$ of membership between sets. We shall write, for example, '$\mathbf{x}\ \boldsymbol{\varepsilon}\ \mathbf{y}$' rather than '$\boldsymbol{\varepsilon}\mathbf{xy}$'.

Let **ZF** be the formalized version of ZF. The postulates and theorems of ZF are expressed in **ZF** by $\mathscr{L}$-sentences. For example, the Principle of Extensionality (for sets) is expressed by

$$\text{(PX)} \qquad \forall\mathbf{x}\forall\mathbf{y}\{\forall\mathbf{z}[\mathbf{z}\ \boldsymbol{\varepsilon}\ \mathbf{x}\leftrightarrow\mathbf{z}\ \boldsymbol{\varepsilon}\ \mathbf{y}]\rightarrow\mathbf{x}=\mathbf{y}\},$$

where **x**, **y** and **z** are distinct variables. (In **ZF** there is no need for classes; instead, one can use properties, expressed by $\mathscr{L}$-formulas.)

From the formal postulates of **ZF**, formal versions of the theorems of set theory can be deduced in Fopcal.

In particular, from the postulates of **ZF** we can deduce a formal version of the theorem that there exists an uncountable set. This theorem follows logically from the existence of a denumerable set – for example, ω (Thm. 4.3.4 and Def. 4.5.13) – and Cantor's Thm. 3.6.8.

Let us assume that **ZF** is consistent. If it isn't – which in any case is highly unlikely – then the very idea of reducing to it the whole of mathematics is quite pointless.

Since the language $\mathscr{L}$ is denumerable, it follows from Thm. 8.13.9 that **ZF** has a model $\mathfrak{U}$ (an $\mathscr{L}$-structure, or $\mathscr{L}$-interpretation, under which all the sentences of **ZF** are true) whose universe U is countable (cf. Def. 4.5.13).[1]

It is easy to show that U cannot be finite. This can be done even without invoking the Axiom of Infinity. Instead, it is enough to point out that the formal version of Prob. 3.3.3 must hold in $\mathfrak{U}$. So we may assume that U is denumerable.

Note that we are not saying that *every* model of **ZF** has a denumerable universe; only that among the models of this theory (assuming it is consistent) there is a model $\mathfrak{U}$ whose universe is denumerable.

What does the model $\mathfrak{U}$ consist of? First, there is the universe U, which serves as the range of values for the variables of $\mathscr{L}$. In other words, the members of U (that is, the individuals of the structure $\mathfrak{U}$) are what the structure $\mathfrak{U}$ interprets as 'sets'. We shall say that the members of U are $\mathfrak{U}$-*sets*.

Second, there is the binary relation $\boldsymbol{\varepsilon}^{\mathfrak{U}}$. For brevity, let us put

[1] In 1922 Fopcal had not been finalized (this was done in 1928 by David Hilbert and Wilhelm Ackermann). When Skolem assumes **ZF** to be 'consistent', he means that it is satisfiable. He then invokes the Löwenheim–Skolem Theorem (which he proves directly, using relatively elementary means) to obtain a denumerable model for **ZF**.

$E = \boldsymbol{\varepsilon}^{\mathfrak{U}}$. E is a binary relation on U, that is, a binary relation among $\mathfrak{U}$-sets; it serves as the interpretation of $\boldsymbol{\varepsilon}$ in the structure $\mathfrak{U}$. We shall say that E is the relation of $\mathfrak{U}$-*membership*. We shall write, for example, 'aEb' when we wish to say that the $\mathfrak{U}$-set a bears the relation E to the $\mathfrak{U}$-set b.

The $\mathfrak{U}$-sets are not necessarily sets in the usual intuitive sense, and the relation E is not necessarily a relation of membership in the usual intuitive sense. Rather, $\mathfrak{U}$-sets are sets *in the sense of the model* $\mathfrak{U}$, and the relation E of $\mathfrak{U}$-membership is the relation of membership *in the sense of* $\mathfrak{U}$. Nevertheless, since $\mathfrak{U}$ is a model of **ZF**, all the postulates of **ZF** are true in $\mathfrak{U}$; in other words, they hold for $\mathfrak{U}$-sets and $\mathfrak{U}$-membership just as they presumably hold for 'true' sets and 'true' membership. The same applies of course to all the theorems of **ZF**, that is, to all $\mathscr{L}$-sentences deducible from the postulates.

Let us imagine an internal observer, called *Hugh*, who 'lives' in the structure $\mathfrak{U}$. Hugh can observe the $\mathfrak{U}$-sets; they are the objects of his world. He can also observe whether or not aEb holds for any such objects a and b. Let us also imagine that we can communicate with Hugh and transmit to him $\mathscr{L}$-formulas, and in particular the postulates of **ZF**. He can then check and confirm that, as far as his observations go, these postulates – and indeed all $\mathscr{L}$-sentences deduced from them using Fopcal – are true under the interpretation $\mathfrak{U}$, in which the variables are regarded as ranging over U and the predicate symbol $\boldsymbol{\varepsilon}$ is interpreted as denoting the relation E.

Hugh has heard that **ZF** is 'axiomatic set theory'. He therefore comes to the conclusion that the theory is really about the objects of his world and the relation E. He comes to believe that the 'sets' and the 'membership relation' about which the theory speaks are these objects and the relation E (which for us are merely $\mathfrak{U}$-sets and $\mathfrak{U}$-membership). We try to tell him that the theory is intended to be about *real* sets and the *real* membership relation $\in$. But he has no reason to believe us. For one thing, he has no notion of what we call 'real' sets and 'real' membership – they are not real *to him*. Moreover, since his observations confirm that the postulates of **ZF** are true under *his* interpretation, why should he believe us that the theory is 'really' about some other reality?

Note that the whole idea of an axiomatic theory is that nothing must be assumed concerning the objects and relations about which the theory speaks, except what is stipulated by the postulates of the theory. An axiomatic theory cannot say more than what can be logically deduced from its postulates. The postulates, and they alone, must determine whether or not a given interpretation of the extra-logical symbols of the theory is legitimate: an interpretation is legitimate iff it satisfies the postulates.

Hugh – whose outlook is confined to his small provincial world – cannot understand our talk of 'real' sets and 'real' membership. But we – broad-minded people living in the big world – can understand his talk of 'sets' and 'membership'. We only have to remember that by 'set' he means what we think of as a $\mathfrak{U}$-set, and by 'membership' he means the relation E.

Actually, we can even translate his talk of [what are in reality] $\mathfrak{U}$-sets and the relation E to talk about *genuine* sets and membership. This is done as follows. For each $\mathfrak{U}$-set a, let us define:

$$\hat{a} = \{x : xEa\}. \tag{1}$$

We call $\hat{a}$ the *E-extension* of a. Clearly, $\hat{a}$ is a genuine set, in fact a subset of U; and we have, for all x

$$x \in \hat{a} \Leftrightarrow xEa. \tag{2}$$

Moreover, the correspondence between $\mathfrak{U}$-sets and their respective E-extensions is one-to-one. This follows from the fact that $\mathfrak{U}$, being a model of **ZF**, must satisfy the postulate **PX**. If a and b are two $\mathfrak{U}$-sets such that the sets $\hat{a}$ and $\hat{b}$ are equal, then it follows from (2) that a and b have exactly the same $\mathfrak{U}$-members. But the postulate **PX**, as interpreted in $\mathfrak{U}$, says that any two $\mathfrak{U}$-sets that have exactly the same $\mathfrak{U}$-members are equal. Hence a and b are equal.

Any statement about $\mathfrak{U}$-sets and the relation E can be rephrased in terms of E-extensions (which are real sets) and real membership.

§3. The paradox and its resolution

We have already observed that all the theorems of **ZF** must be true in $\mathfrak{U}$. Among these theorems there is, as we have noted, a sentence that says 'there exists an uncountable set'. In fact, Hugh – who is a competent logician and has been able to deduce this theorem – can point at a particular $\mathfrak{U}$-set c that instantiates the theorem: he can show that c has 'uncountably many members'. Naturally, we know that what Hugh regards as 'members' of c are really just $\mathfrak{U}$-members of c; in other words, they are $\mathfrak{U}$-sets that bear the relation E to c. But how can this be? The *whole* universe U of $\mathfrak{U}$ contains only denumerably many objects; therefore for any a there can only be countably many objects bearing the relation E to a. So how can there be *un*countably many objects bearing the relation E to c?

This seeming contradiction is *Skolem's Paradox*.

In fact, the contradiction is only apparent. The resolution of the paradox depends on the fact that many important set-theoretical notions, such as countability, are *relative*. Thus, a $\mathfrak{U}$-set c may be uncountable *in the sense of the structure* $\mathfrak{U}$, although when viewed from the outside c has only countably many $\mathfrak{U}$-members.

Let us explain how this comes about. First, let us recall what it means for a set to be countable. By Prob. 4.5.14, a set C is countable iff there exists an injective function from C to the set ω of finite ordinals (which in set theory play the role of natural numbers). Recall that such a function is itself a set. To say that f is an injective function from C to ω means that f is a set of ordered pairs of the form $\langle x, \xi \rangle$ with $x \in C$ and $\xi \in \omega$, such that for each $x \in C$ there is *exactly* one $\xi \in \omega$ for which $\langle x, \xi \rangle \in f$, and for each $\xi \in \omega$ there is *at most* one $x \in C$ for which $\langle x, \xi \rangle \in f$.

So, to say that C is countable means that *there exists* a set f having the properties just mentioned. But we must realize that *existence of such-and-such a set* may mean quite different things, depending on whether we interpret this phrase inside the structure $\mathfrak{U}$ or in the outside 'real' world.

We have seen above that to each $\mathfrak{U}$-set a there corresponds the real set $\hat{a}$, which is a subset of U. Now, it is easy to see that the converse is not generally true: if A is an arbitrary subset of U, there may not exist any $\mathfrak{U}$-set a such that $\hat{a} = A$. Indeed, the mapping that maps each $\mathfrak{U}$-set a to its E-extension $\hat{a}$ is an injection from the set U to its own power set; so by Cantor's Theorem it cannot be surjective.[1]

Let A be a subset of U, that is, a set of $\mathfrak{U}$-sets. Then A is an object in *our* world, the world of external observers. But if A is not $\hat{a}$ for any $\mathfrak{U}$-set a, then there is no object in the world $\mathfrak{U}$ of the internal observer Hugh that corresponds to A. The set A is then purely *external*, it corresponds to nothing in Hugh's ontology.

[1] Note the ironic double role played by Cantor's Theorem. On the one hand, the fact that Cantor's Theorem holds *inside* $\mathfrak{U}$ (that is, under the interpretation $\mathfrak{U}$) gave rise to the paradox in the first place, because it was used to give us an uncountable set (in the sense of $\mathfrak{U}$). Now we are using the fact that Cantor's Theorem holds 'in the real world' in order to resolve the paradox.

Let us see how these observations help to resolve the paradox. In his universe, Hugh finds an object $\omega^{\mathfrak{U}}$ that is 'the set of finite ordinals' in his sense ($\omega^{\mathfrak{U}}$ satisfies, in the interpretation $\mathfrak{U}$, the formal set-theoretic definition of *the set of finite ordinals*). Of course, $\omega^{\mathfrak{U}}$ may not 'really' be the set of finite ordinals; but it is quite easy to see that its E-extension is in fact denumerable. Now, Hugh has found another object ($\mathfrak{U}$-set) c, which serves as the $\mathfrak{U}$-*power-set of* $\omega^{\mathfrak{U}}$, and he can prove that c *is uncountable*. We, on the other hand, can prove that c has only countably many $\mathfrak{U}$-members. Who is right?

In fact, both he and we are right. *He* is right because there does not exist any $\mathfrak{U}$-set φ that constitutes an injection from c to $\omega^{\mathfrak{U}}$ *in the sense of the interpretation* $\mathfrak{U}$. *We*, on the other hand, are right because the set $\hat{c}$ (the E-extension of c) is countable in the sense of our external world. In fact, we can prove that there exists an injection f from $\hat{c}$ to the E-extension of $\omega^{\mathfrak{U}}$. However, this f is purely external; it exists in the *outside* world, but it cannot be the E-extension of any $\mathfrak{U}$-set. Indeed, if f were not purely external then it would be quite easy to show that c is countable in the sense of $\mathfrak{U}$.

So the paradox is resolved – but not very happily. It is disappointing to find that axiomatic set theory, if consistent, has such perverse models, in which an object that is really quite modest in size can seem huge.

As Skolem himself pointed out, countability is by no means the only important set-theoretic notion that is relative in this sense. For example, the notion of finiteness is also relative: we can have a model $\mathfrak{U}$ (even a denumerable one) in which a $\mathfrak{U}$-set a may be finite in the internal sense of $\mathfrak{U}$, while in fact a has infinitely many $\mathfrak{U}$-members.

Indeed, by an argument like that used in the proof of Skolem's Thm. 10.3.8 we can show that **ZF** has a model $\mathfrak{U}$ (with denumerable universe) such that the object $\omega^{\mathfrak{U}}$, the $\mathfrak{U}$-set-of-finite-ordinals, is *nonstandard*. This means that – in addition to $\mathfrak{U}$-members of the form $\boldsymbol{n}^{\mathfrak{U}}$ for each natural number n (that is, $\mathfrak{U}$-cardinals corresponding to the natural numbers) – $\omega^{\mathfrak{U}}$ also has $\mathfrak{U}$-members that do not correspond to any natural number. If α is such a *nonstandard* $\mathfrak{U}$-member of $\omega^{\mathfrak{U}}$ then α is a $\mathfrak{U}$-finite-ordinal: it satisfies in $\mathfrak{U}$ the formal definition of the notion *finite ordinal* (the formalization of the first part of Def. 4.3.1). In particular, α is $\mathfrak{U}$-finite. But, as seen from outside $\mathfrak{U}$, α actually has infinitely many $\mathfrak{U}$-members, and so $\hat{\alpha}$ is really (really?) an infinite set! (Cf. Warning 6.1.9.)

This has an important bearing on the issue raised in Rem. 10.3.10 in connection with Skolem's Theorem. The theorem says that the structure $\mathfrak{N}$ of natural numbers cannot be characterized uniquely (up to isomorphism) in the *first-order* language of arithmetic.

Now, Dedekind showed that the system of natural numbers can be characterized uniquely in set-theoretic terms (cf. Rem. 4.3.8(i)). Following him, Peano also formulated his axiomatization of that system using variables ranging over all sets of natural numbers (cf. Rem. 10.13.5(iii)). These, then, are characterizations of the system of natural numbers within an ambient set theory. And they seem to work, in the sense that in a sufficiently strong set theory it can be shown that Peano's axioms have (up to isomorphism) a unique model (cf. Rem. 6.1.8).

However, these set-theoretic characterizations are all *relative*: they merely pass the buck to set theory. And now we see that set theory itself has strange (nonstandard) models. Hugh may be very pleased to find that in his world there is (essentially) just one 'system of natural numbers' satisfying Peano's second-order postulates. But we, from our external vantage point, can see that this $\mathfrak{U}$-system-of-natural-numbers is in fact (in fact?) *nonstandard*, containing infinite *un*natural numbers, which merely seem finite to Hugh.

It turns out that axiomatic set theory is unable to characterize some of the most basic notions of mathematics, including intuitive set-theoretic notions – except in a merely verbal sense. If mathematics – and in particular the arithmetic of natural numbers – is more than mere verbal discourse, then its reduction to axiomatic set theory somehow fails to do it full justice.

Author index

Reference given to page numbers

General index

References are given to the places where a term is defined, re-defined or explained. A reference of the form x.y is to Section y of Chapter x. A reference of the form x.y.z is to item z in Section y of Chapter x.